Fat Absorption

Volume II

Editor

Arnis Kuksis

Professor
Banting and Best Department of Medical Research
University of Toronto
Toronto, Canada

CRC Press
Taylor & Francis Group
Boca Raton London New York

CRC Press is an imprint of the
Taylor & Francis Group, an **informa** business

First published 1987 by CRC Press
Taylor & Francis Group
6000 Broken Sound Parkway NW, Suite 300
Boca Raton, FL 33487-2742

Reissued 2018 by CRC Press

Library of Congress Cataloging-in-Publication Data
Main entry under title:
Fat absorption.

Includes bibliographies and index.
1. Fat—Metabolism. 2. Intestinal absorption.
I. Kuksis, Arnis. [DNLM: 1. Lipids—metabolism.
QU 85 F252]
QP752.F3F38   1986   612' .397   86-13633
ISBN 0-8493-6412-4 (v.1)
ISBN-0-8493-6413-2 (v.2)

A Library of Congress record exists under LC control number: 86013633

ISBN 13: 978-1-315-89288-7 (hbk)
ISBN 13: 978-1-351-07198-7 (ebk)

Visit the Taylor & Francis Web site at http://www.taylorandfrancis.com and the
CRC Press Web site at http://www.crcpress.com

PREFACE

Fatty acid esters of glycerol make up the bulk of dietary fat, which provides up to 40% of the daily caloric intake of man. In the form of the neutral glycerol esters, dietary fatty acids constitute the major storage form of energy in the body. As polar glycerolipids, dietary fatty acids serve in a variety of structural functions and participate in complex biochemical and physiological processes. Both energy balance and physiological function of the organism are affected by the relative efficiency of digestion and absorption of dietary fat and fat soluble vitamins and xenobiotics. The dietary fats and other lipids carried in them consist of a variety of chemical classes which comprise numerous molecular species, including homologous series of geometric and positional isomers of unsaturated fatty acids and stereoisomers of acylglycerols. These species are digested and absorbed at markedly different rates, and their assimilation may take place via different mechanisms and divergent metabolic routes. Furthermore, different dietary fats exhibit characteristic physiologic and metabolic effects. Since some of these effects are believed to be beneficial and others detrimental to the well being of man, elaborate efforts are currently made to manipulate and manage the composition, digestion, and absorption of dietary fat for both prophylactic and therapeutic purposes.[1,2] Obviously, this task requires detailed knowledge and understanding of the biochemical and metabolic transformations involved in the assimilation of dietary fats and other lipids.

As a result of this interest and research, a large amount of information has accumulated over the last few decades on the molecular transformations of the glycerolipids during their assimilation by the intestinal mucosa. Some of this information has been presented in various recent reviews on specific aspects of fat digestion, absorption, and chylomicron secretion. Thus, Borgström[3] has dealt with the properties of the lipolytic enzymes and their interaction with bile salts and lipids. Patton,[4] and Carey, Small, and Bliss[5] have reviewed the physicochemical behavior of model lipids in model systems. They have correlated this information with the physiological behavior of dietary fat throughout the alimentary canal and have challenged the micellar hypothesis of fat absorption (see, however, Borgström[6]). Thomson[7] and Thomson and Dietschy[8] have discussed the principles governing the movement of lipids across unstirred water layers. The basic mechanisms involved in the intestinal resynthesis of the dietary fat have been considered by Brindley,[9] Johnston,[10] and Mansbach and Parthasarathy,[11] while Glickman[12] and Redgrave,[13] as well as Bisgaier and Glickman,[14] have summarized the principle developments in our understanding of the composition and biosynthesis of intestinal lipoproteins. The morphological and ultrastructural aspects of the cellular assembly and transport of chylomicrons by the mucosal cell have been reviewed by Sabesin and Frase[15] and Friedman and Cardell.[16] In most instances the emphasis has been placed on the physicochemical nature of the assimilation process. The digestion and absorption of different dietary fats,[17,18] fat-soluble vitamins,[19,20] xenobiotics,[21,22] and other less well-known components of dietary fats have been assessed only partially. Previous reviews are also deficient in the limited discussion of the enzymology of the endogenous lipolysis and resynthesis of dietary fat, the extent of dilution of the dietary fat by endogenous fat, the role of different fats in the formation and secretion of chylomicrons, and the advantages of utilizing modern analytical methodology and physiologically more appropriate experimental systems.

In this book, we have expanded upon the more obscure areas by providing an authoritative and comprehensive source of information on the biochemical and metabolic aspects of digestion and absorption of different dietary fats and other lipids, with minimal discussion of the physical chemistry of the process, which has been covered in great detail in previous reviews. Volume I considers the digestion and absorption of fatty acids and acylglycerols. Included here are reviews on modern methodology of glycerolipid analyses, composition of enterocyte membrane lipids, and the purification and substrate specificity of intestinal lipases and acyltransferases. Volume II deals with the digestion and absorption of sterols, fat-soluble vitamins, and xenobiotics. The effect of dietary fats and fat-soluble vitamins and xenobiotics on chylomicron and lipoprotein secretion by the intestine is also reviewed. The volume concludes with a review of the regulation of dietary fat assimilation by gastrointestinal peptides. Both volumes are composed of chapters that can be read and understood independently although a limited cross-reference is also found. The books show that the relative rates and mechanisms of lipolysis and resynthesis of the different dietary fats and other lipids constitute critical steps in the overall assimilation process. Since dietary fat also affects the chemical composition and turnover of the plasma and microsomal membranes of the villus cell, the process of fat absorption can no longer be considered a metabolic shunt. The dietary fat also influences the size and apopeptide composition of the chylomicrons and other intestinal lipoproteins, which determine the rate of clearance of fat from the intestinal mucosa and plasma. It is therefore clear that a full understanding of the process of fat assimilation must be sought beyond the classical physicochemical and biochemical transformations. The reviews are intended for both researchers and practitioners in the biomedical field who require detailed knowledge of the biomedical and metabolic transformations involved in the intestinal digestion and resynthesis of dietary fats and other lipids.

Arnis Kuksis

REFERENCES

1. Levy, R. I., Rifkind, B. M., Dennis, B. H., and Ernst, N., Eds., Nutrition, lipids and coronary heart disease. A global view, in *Nutrition in Health and Disease*, Vol. 1, Raven Press, New York, 1983.
2. Goldberg, A. C. and Schonfeld, G., Effects on lipoprotein metabolism, *Annu. Rev. Nutr.*, 5, 195, 1985.
3. Borgström, B., Pancreatic lipase, in *Lipases*, Borgström, B. and Brockman, H. L., Eds., Elsevier, Amsterdam, 1984.
4. Patton, J. S., Gastrointestinal lipid digestion, in *Physiology of the Gastrointestinal Tract*, Johnson, L. R., Ed., Raven Press, New York, 1981, chap. 45.
5. Carey, M. C., Small, D. M., and Bliss, C. M., Lipid digestion and absorption, *Annu. Rev. Physiol.*, 45, 651, 1983.
6. Borgström, B., The micellar hypothesis of fat absorption: must it be revisited?, *Scand. J. Gastroenterol.*, 20, 389, 1985.
7. Thomson, A. B. R., Intestinal absorption of lipids: influence of the unstirred water layer and bile acid micelle, in *Disturbances in Lipid and Lipoprotein Metabolism*, Dietschy, J. M., Gotto, A. M., Jr., and Ontko, J. A., Eds., American Physiological Society, Washington, D.C., 1978, chap. 2.
8. Thomson, A. B. R. and Dietschy, J. M., Intestinal lipid absorption: major extracellular and intracellular events, in *Physiology of the Gastrointestinal Tract*, Johnson, L. R., Ed., Raven Press, New York, 1981, chap. 46.
9. Brindley, D. N., Absorption and transport of lipids in the small intestine, in *Intestinal Permeation*, Vol. 4, Kramer, M., and Lauterbach, F., Eds., Workshop Conference Hoechst, Excerpta Medica, Amsterdam, The Netherlands, 1978, 350.
10. Johnston, J. M., Esterification reactions in the intestinal mucosa and lipid absorption, in *Disturbances in Lipid and Lipoprotein Metabolism*, Dietschy, J. M., Gotto, A. M., Jr., and Ontko, J. A., Eds., American Physiological Society, Washington, D.C., 1978, 57.
11. Mansbach, C. M., II and Parthasarathy, S. A., A re-examination of the fate of glyceride-glycerol in neutral lipid absorption and transport, *J. Lipid Res.*, 23, 1009, 1982.
12. Glickman, R. M., Fat absorption and malabsorption, *Clin. Gastroenterol.*, 12, 323, 1983.
13. Redgrave, T. G., Formation and metabolism of chylomicrons, in *Gastrointestinal Physiology IV*, Vol. 28, *(Int. Rev. Physiol. Ser.)*, Young, D. B., Ed., University Park Press, Baltimore, Md., 1983, 103.
14. Bisgaier, C. L. and Glickman, R. M., Intestinal synthesis, secretion and transport of lipoproteins, *Annu. Rev. Physiol.*, 45, 625, 1983.
15. Sabesin, S. M. and Frase, S., Electron microscopic studies of the assembly, intracellular transport, and secretion of Chylomicrons by rat intestine, *J. Lipid Res.*, 18, 496, 1977.
16. Friedman, H. I. and Cardell, R. R., Jr., Alterations in the endoplasmic reticulum and Golgi complex of intestinal epithelial cells during fat absorption and after termination of this process: a morphological and morphometric study, *Anat. Rec.*, 188, 77, 1977.
17. Emken, E. A., Nutrition and biochemistry of *trans* and positional fatty acid isomers in hydrogenated oils, *Annu. Rev. Nutr.*, 4, 339, 1984.
18. Clandinin, T., Field, C. J., Hargreaves, K., Morson, L., and Zsigmond, E., Role of diet in subcellular structure and function, *Can. J. Physiol. Pharmacol.*, 63, 546, 1985.
19. Hollander, D., Intestinal absorption of vitamins A, E, D and K, *J. Lab. Clin. Med.*, 97, 449, 1981.
20. Weber, F., Absorption mechanisms for fat-soluble vitamins and the effect of other food constituents, in *Nutrition in Health and Disease and International Development: Symposia from the XIIth International Congress of Nutrition*, Alan R. Liss, New York, 1981, 119.
21. Pfeiffer, C. J., Gastroenterologic response to environmental agents — absorption and interactions, in *Handbook of Physiology, Section 9*, Lee, D. H. K., Ed., American Physiological Society, Washington, D.C., 1977, 349.
22. Kuksis, A., Intestinal digestion and absorption of fat-soluble environmental agents, in *Intestinal Toxicology*, Schiller, C. M., Ed., Raven Press, New York, 1984, 69.

THE EDITOR

Arnis Kuksis holds a Ph.D. in Biochemistry from Queen's University, and a Career Investigatorship from the Medical Research Council of Canada. He is a Professor of Biochemistry in the Banting and Best Department of Medical Research, University of Toronto, Toronto, Canada, where he also directs the Medical Research Council of Canada Regional Chromatography and Mass Spectrometry Laboratory.

In addition to biochemistry, Dr. Kuksis' training and experience include lipid chemistry and analysis, which he has extensively applied to studies of structure and function of lipoprotein and membrane lipids. Much of this work has been focused on the regulatory roles of cholesterol and acyltransferases in absorption of dietary fat in health and disease. Dr. Kuksis has over 200 original publications in these research areas, and has contributed numerous review articles on the composition and chromatography of lipids from natural sources.

ACKNOWLEDGMENTS

I would like to express my appreciation to Dr. Nina Morley for her excellent editorial assistance. In addition, my thanks extend to Dr. Peter Child and Mr. Steven Pind for their help in reviewing some of the manuscripts.

CONTRIBUTORS

Bjorn Akesson, M.D., Ph.D.
Assistant Professor
Department of Clinical Chemistry
University of Lund
Lund, Sweden

Frank P. Bell, Ph.D.
Research Scientist
Atherosclerosis-Thrombosis Unit
The Upjohn Company
Kalamazoo, Michigan

Peter Child, Ph.D.
Assistant Professor
Department of Medical Biophysics
University of Toronto
Toronto, Ontario, Canada

Joseph D. Fondacaro, Ph.D.
Assistant Director
Department of Pharmacology
Smith Kline & French Laboratories
Philadelphia, Pennsylvania

Alan G. D. Hoffman, M.D., Ph.D.,
  F.R.C.P.C.
Chief Medical Resident
Toronto General Hospital
Toronto, Ontario, Canada

Arnis Kuksis, Ph.D.
Professor
Banting and Best Department of
  Medical Research
University of Toronto
Toronto, Ontario, Canada

Peter Michelsen, Ph.D.
Assistant Professor
Division of Organic Chemistry
Chemical Center
University of Lund
Lund, Sweden

FAT ABSORPTION

Volume I

Volume II

TABLE OF CONTENTS

Chapter 1

THE ABSORPTION OF CHOLESTEROL AND PLANT STEROLS BY THE INTESTINE

Peter Child

TABLE OF CONTENTS

I. INTRODUCTION

Cholesterol, a 27-carbon alcohol present in all mammalian tissues, plays an essential role in both health and disease. It is a major structural component of mammalian membranes and is the direct precursor of both bile acids and steroid hormones.

Despite its essential and ubiquitous nature, excessive accumulations of cholesterol in the blood and vascular tissue are implicated in the genesis of atherosclerosis. Accumulation of plasma cholesterol arising from dietary excesses gives rise to the largest group of hypercholesterolemics and includes most patients with atherosclerosis. Because there is ample evidence to suggest that the levels of plasma cholesterol are determined, in part, by the intestinal absorption of dietary sterol, an understanding of the manner in which cholesterol is absorbed is of great clinical importance.

One of the first direct studies of cholesterol absorption was carried out in 1930 by Schonheimer and co-workers.[1] In this study, they fed cholesterol to mice and found that the liver sterol content increased following the meal. When sitosterol, or its 22,23-unsaturated analog, stigmasterol was given, however, the liver sterol did not rise as it had done with cholesterol. This experiment established two major facets of the metabolism of sterols: (1) dietary sterol is absorbed and is able to influence hepatic sterol content, and (2) dietary sitosterol is handled differently than cholesterol.

In the 50 years that have passed since the publication of that report, our knowledge of sterol absorption has expanded tremendously. The general sequence of events has been defined and the poor absorbability of sitosterol has been thoroughly documented. However, while it is safe to say that we know the overall sequence of events in the absorptive process, there is no part of which we have a complete mechanistic understanding. Nevertheless, much of the published work contains information contributing to such an understanding. It is the function of this chapter to review the evidence contained in such works in an attempt to define the rate-limiting steps in the absorptive process, the factors influencing absorption, and the molecular interactions involved. The review will concentrate on literature published since 1981, at which time two major monographs[23] on the subject appeared. Where possible, however, references prior to this period will be included to provide appropriate historical support and balance for the arguments.

II. OVERVIEW OF THE CHARACTERISTICS OF STEROL ABSORPTION

A. Chemical Structures

The most common dietary sterols are cholesterol and the plant sterols, campesterol and sitosterol. They are identical to each other with the exception of the substituents at C_{24} of the side chain. Their structure, shown in Figure 1, consists of four fused rings labeled A, B, C, and D and having the numbering system indicated.[4] The C/D ring junction is characterized by the *trans* orientation of the proton at position 14 relative to the methyl group at position 19. This orientation is clearly evident in the lower figure. The side chain, attached to ring D, is oriented *cis* to the same methyl group. The sterols also possess a double bond between carbons 5 and 6, and a hydroxyl group at position 3 (*β*-orientation).

Sitosterol differs from cholesterol in the presence of an ethyl group at C_{24} of the side chain. Sitosterol derived from soybeans, a common dietary source, is exclusively the 24-*α*-ethyl isomer of cholesterol.[5] This sterol, often referred to as *β*-sitosterol, will be called simply sitosterol in this report.

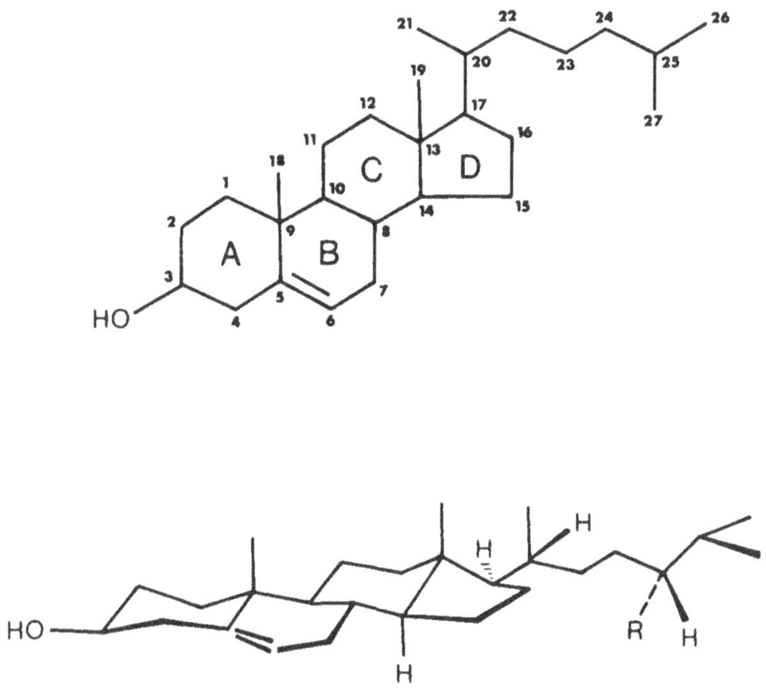

FIGURE 1.   The structure and numbering system of sterols. (Upper figure) Cholesterol structure shown with the conventional numbering system; (lower figure) approximate solution conformation of the sterol. R is a proton for cholesterol, a methyl group
for campesterol, and an ethyl group in sitosterol. The R group is in the α-orientation,
as are the protons at positions 14 and 17. The proton at position 20 is in the β-orienta-
tion. The hydroxyl group is equatorial.

The 24-methyl compound derived from soybeans has been shown by pNMR[5] to be a
roughly equimolar mixture of the α-(campesterol) and β-(dihydrobrassicasterol) epi-
mers. Throughout this chapter the 24-methyl soy sterols will be referred to as campesterol for simplicity. Space-filling models of cholesterol and sitosterol are shown in
Figure 2. It is evident from the upper structure that cholesterol has a generally cone-
shaped form, the cross-sectional area of the side chain being smaller than that of the
ring system. Sitosterol, on the other hand, is a more cylindrical shape by virtue of the
extra ethyl group in the side chain.

B. Overview of Sterol Absorption In Vivo

The pathways of sterol transfer from intestinal lumen to the lymph are generally
similar to those of glycerol lipids, although the rates of transport differ considerably.
Several good reviews of these pathways have been put forward recently[2,6,16] and therefore, only a brief overview of the process will be given in this survey.

The absorption of the bulk of lipid material in mammals occurs in the small intestine. The small intestine may be operationally divided into three regions: the duodenum
(proximal segment), the jejunum (from the duodenum to the halfway point), and the
ileum (distal half). The absorptive surfaces of these regions of the gastrointestinal tract
display a multitude of villar projections, which are covered with a single layer of villus
epithelial cells. These absorptive cells are shed into the intestinal lumen when they
become damaged and are replaced by cells derived from the crypt regions found at the
base of each villus. Thus, a complete turnover of the villus cell population occurs every
48 hr. The villus cells themselves are of a cylindrical shape and are capped with a brush

FIGURE 2.   Space-filling models of cholesterol and sitosterol. (Upper figure) Cholesterol; (lower figure) sitosterol. In both structures the side chains lie to the right-hand side and the hydroxyl group lies to the far left. The two methyl groups characteristic of the β-face of the four-membered ring system are clearly visible in the upper left and upper center of each model.

border membrane. When the cells are in their functional orientation within the mucosal wall only the brush borders are exposed to the luminal contents and provide the actual absorptive surface of the villi. As with the villi themselves, the brush border invaginations increase the absorptive surface area. More comprehensive reviews of intestinal morphology may be found elsewhere.[7,14,17]

Sterols, along with the glycerolipids, are absorbed primarily by the proximal 1/3 of the small intestine. This has been demonstrated in man following intubation experiments[18] and has also been shown in the rat,[19,20] rabbit, and guinea pig.[21] By feeding a radioactive dose of cholesterol and sitosterol to a rat, Sylven and Nordstrom[20] further localized the absorptive event to the tip of the intestinal villi. Pieces of intestinal tissue were frozen and cross sectioned in a cryostat after a meal of labeled sterol. The label was localized exclusively in the upper portions of the villi.

The greater absorption in the jejunal region of the small intestine may be expected from its position in the intestinal tract, but Thompson[22] has also reported structural differences between the jejunal and ileal epithelium. Using segments of isolated rabbit intestinal tissue, he was able to show a reduced cholesterol uptake in ileal sections compared to those derived from the jejunum. The basis for the reduction was suggested to arise from the reduced absorptive surface area of the ileal sections. Chow and Hollander[23] reached a similar conclusion from their experiments using evert gut sacs from rat. Absorption of sterols from the colon has been shown to be negligible, by a direct instillation of micellar solutions into the colonic region in the rat.[24]

Sterol esters must be converted to their free hydroxyl form before they can be absorbed.[25,26] Little or no absorption of cholesterol was observed when it was esterified to branched chain fatty acids, which were not hydrolyzed in the intestinal lumen.[27] Similarly, the absorption of cholesterol esterified to common fatty acids correlated with the rate of their hydrolysis in the lumen of the intestine of the rat.[28] The nonhydrolyzable derivatives, cholesterol trimethyl acetate and cholesterol-3-D-glucoside, are not absorbed.[28] Borgstrom[29] has shown that hydrolysis-resistant cholesterol ethers may be absorbed and transferred to lymph, but the rate of uptake decreases with increasing length to the alkyl chain.

Sterols, unlike glycerolipids, require bile salts for their absorption. This was reported as early as 1924[30] and has subsequently been confirmed in several laboratories.[19,31-35] From the evidence presented to date, it appears that sodium taurocholate is the bile salt of choice for the solubilization and absorption of sterols. Swell and co-workers[32] found that the absorption of cholesterol was completely suppressed in the rat when taurocholate was replaced with sodium dehydrocholate (does not form micelles) in a fed emulsion. Similarly, taurocholate was found to be superior to taurodeoxycholate, taurochenodeoxycholate, and the ionic detergent, dodecyl sulfate, in promoting the jejunal absorption of vitamin D_3 in the rat.[19]

It is currently believed, although not without reservation, that sterol absorption in the intestine occurs from an aqueous micellar phase in which the sterol is solubilized in an aggregate of bile salt, phospholipid, and the hydrolysis products of glycerolipids.[2,36-41] Micelles containing sterols, fatty acids, monoacyl-glycerols, phospholipids, and bile salts have been suggested[42,43] to be either spherical or cylindrical, depending on the ratio of phospholipid to bile salt. In either case, the bile salts are thought to lie with their ionized side chains exposed to the aqueous solvent, possibly with their ring systems intercalated between neutral lipid and phospholipid molecules as dimers within the micelle.[43] Sterols are thought to dissociate from the micelles in which they are solubilized to be taken up by the jejunal mucosa as free monomers.[36,41] According to this concept, the rate-limiting step in the uptake of sterol is the diffusion of the micelle across a layer of unstirred water surrounding the villi, and the function of the micelle is to provide a constant supply of monomers to the brush border surface. Absorption of the entire micelle in a pinocytotic process has been ruled out in a series of studies.[14,44]

Little is known of the molecular events occurring at the surface of the brush border membrane during the process of absorption. It has been suggested that the transfer of sterols across the membrane of the brush border is a passive process.[2,36] One model for the events occurring at this stage proposed an exchange diffusion[45] in which an entering molecule of cholesterol displaces a molecule from the opposite side of the bilayer and forces it into the cytoplasm. This concept is supported by reports[45,46] of an absence of build-up of cholesterol in the brush border membrane during sterol absorption.

Once inside the villus cell, the dietary sterols are rapidly distributed throughout the intracellular membranes.[45,47-49] It has been demonstrated that within 2 hr of the ingestion of a sterol meal, 75% of the absorbed sterol is found in the microsomal fraction of the rat intestinal mucosa.[45] This fraction was suggested to be derived from the membranes of the endoplasmic reticulum. The sterols are expected to be transmitted to the smooth endoplasmic reticulum for esterification and assembly into lipoproteins in a manner analogous to that occurring during the absorption of glycerolipids.[14] The handling of these dietary components differs in rate, however, with the reesterification of fatty acids to triacylglycerols occurring several times faster than the esterification of cholesterol. After feeding cholesterol and oleic acid or trioleoylglycerol to rats, Sjostrand and Borgstrom[49] found the bulk of the absorbed lipid in smooth membrane droplets, termed apical vesicles, in the region of the absorptive villus cells below the microvillar membrane. The high concentration of free fatty acids was taken as evidence that these structures were the site of reesterification. Esterified cholesterol appeared much more slowly in the apical vesicles, subsequent to an initial appearance in a smooth membrane fraction isolated from the cells.

A large portion of the absorbed sterol is esterified with fatty acids of dietary origin.[14] The esterase responsible was initially suggested[50] to be the cholesterol ester hydrolase (E.C. 3.1.1.13) derived from the pancreas in the rat. More recently, an esterase dependent on the formation of fatty acyl CoAs (E.C. 2.3.1.26) has been found in the microsomal fraction derived from the intestine of rat[51] and man.[52] It has subsequently been localized in the rough endoplasmic reticulum.[53] This enzyme is believed[51,52] to have sufficient activity to account for all of the cholesterol esters found in the intestinal lymph of man and rat.

Whether the esterification of sterol occurs before, during, or after the assembly of lipoproteins is not known with certainty. It has been suggested, however, that enrichment of the cholesterol content of the endoplasmic reticulum membrane during absorption stimulates sterol esterification.[53]

The process of chylomicron formation apparently represents a major rate-limiting step in the transport of sterol. It has been determined in the rat,[54] monkey,[55] and man[56,57] that a constant proportion of the fed sterol is taken up by the mucosal wall (about 50% of the fed dose) regardless of the dose. Although the actual amount absorbed increases with the dose fed, the amount delivered to the lymph appears to remain constant,[55,58-61] indicating that the incorporation of sterol into chylomicrons or their secretion from the cell is a rate-limiting step.

The relative sequence of events occurring within the intestinal villus cells of the rat during the absorption of cholesterol can be seen in Figure 3, which represents a compilation of data derived from several reports[26,63] which have studied the movement of radioactive cholesterol from the intestinal lumen into the lymph. As expected, the initial appearance of radioactivity occurs in the brush border membrane itself, followed quickly by transport into the cellular cytoplasm.[26] In these experiments, the brush border membranes were isolated by standard techniques following a cholesterol meal, and the radioactivity was assayed in this fraction and in a membrane-free supernatant derived from the intestinal mucosa.

Radioactivity was found in the cholesterol ester fraction of the cells a short time after its appearance in the supernatant. It is of interest to note that the actual mass of cholesterol ester in the absorptive villus cells during absorption apparently remains constant,[26,62] although from the radioactivity data, the turnover is clearly increased.

The lymph delivery of the absorbed sterol begins about 2 hr after the feeding and the peak delivery period can be seen in Figure 3 to occur several hours after the initial entry into the villus cells.

In the lymphatic delivery experiment performed by Sylven and Borgstrom,[63] cholesterol was fed in trioleoylglycerol at similar, but not identical, dietary loads to those used by David and co-workers.[26] To compensate for these differences, the lymphatic delivery has been shown here at two dietary levels. The 2-hr lag period appeared to be present at both levels. This is in marked contrast to the rate of appearance of the dietary triacylglycerol, which was detectable in lymph within 1 hr.[63]

Secretion of the chylomicrons occurs by a process thought to involve a microtubular network within the villus cell. The secretion of chylomicrons has been shown[64,65] to be partially inhibited by colchicine and other destabilizers of the microtubular system.

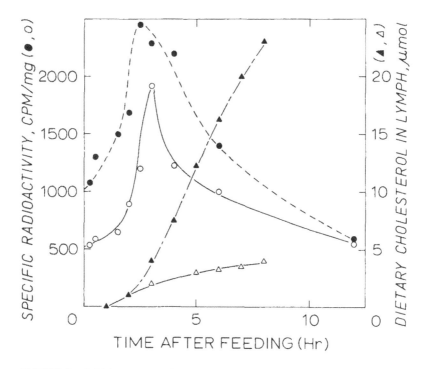

FIGURE 3.   Initial uptake and lymph delivery of dietary cholesterol in the rat. The specific radioactivity of ^{14}C-cholesterol in the brush border fraction (CPM per milligram sterol) and in the cholesterol ester fraction (CPM per milligram sterol ester in the intestinal mucosa) was measured and reported by David et al.[26] after feeding the labeled cholesterol (3 mg) in groundnut oil. The lymphatic delivery was reported by Sylven and Borgstrom.[63] The radioactive cholesterol was fed in triolein carrier. Symbols: ●, ^{14}C-CPM per milligram cholesterol in purified brush border; O, ^{14}C-CPM per milligram cholesterol ester; ▲, dietary cholesterol in lymph following feeding of a 100-μmol dose in triolein; △, dietary cholesterol in lymph after a 12.5-μmol dose. All data used with permission of the authors.

Reaven and Reaven[66] demonstrated that in the presence of colchicine, the number of microtubules in the cell decreased concomitantly with an increase in the cellular content of lipid. This process will be discussed more thoroughly in a subsequent chapter.

Following secretion from the villus cells, lipoproteins pass through the basement membrane and into the lamina propria. Transport down the core of the villi is via central lacteals, which merge to form the mesenteric lymph system. From this point, lymph derived from the intestine is mixed with that of hepatic origin in the cysternae chylae, and is transported to the bloodstream via the thoracic duct.

This outline of the general route of sterol absorption is thought to be similar in all mammals for all sterol types, except those containing polar groups, such as the bile acids and hydroxylated sterols, which may be transported from the villus cells by the portal blood system.[67,68]

C. Dietary Factors Influencing Cholesterol Absorption

A great variety of ingestible compounds is known to influence the levels of plasma cholesterol. Many of these operate by reducing either initial uptake or the subsequent lymphatic delivery of dietary sterol. For further background information the reader is directed to discussions by Kritchevsky and Pollack[3] and by Thomson and Dietschy.[2]

Table 1
INFLUENCE OF DIETARY FAT ON INTESTINAL CHOLESTEROL
ABSORPTION IN THE RAT

Dietary addition	Method of measurement of sterol absorption	Influence on cholesterol absorption	Comments	Ref.
Triolein or tripalmitin	Isotope equilibration method	None compared to lard-fed control		69
Trierucin or tristearin	Isotope equilibration method	Reduced about 50% compared to lard-fed control	More endogenous cholesterol in lumen of trierucin-fed rats	69
Tristearin		Reduced compared to lard-fed control		70
Tripalmitin, tristearin, triolein, or trilinolein	Lymph delivery	Decreased with saturated fats	Cholesterol absorption correlated with triglyceride absorption	71
Triolein stabilized by vegetable lecithin	Mass cholesterol in mesenteric lymph	Lymph output reduced	Initial uptake of cholesterol not affected	72
1% cholesterol in safflower oil	Cholesterol content of mucosa		Increased content of mucosal cholesterol — also increased during total fast	74

1. Triacylglycerols

Recent reports examining the influence of dietary fat are summarized in Table 1. Consistent evidence has been obtained, using the rat as experimental model, that the intestinal absorption of cholesterol is slower in the presence of fully saturated fats than with unsaturated. In three studies[69-71] tristearin reduced cholesterol absorption compared to lard-fed controls. Trierucin had an influence on cholesterol absorption similar to that of tristearin in one study,[69] but the behavior of tripalmitin was more like that of triolein. Palmitoyl- and stearoyl-containing triacylglycerols reduced cholesterol absorption to the same extent in the studies of Feldman et al.[71]

Infusions of triolein, stabilized with either vegetable lecithin or dipalmitoyl phosphatidylcholine (PC), reduced the lymphatic output of cholesterol[72] compared to controls given no lipid. Based on earlier work[73] there was thought to be no influence of the fat on the initial entry of cholesterol into the mucosal wall (not directly measured), but the triolein infusions led to a large depletion of the mucosal cholesterol content.[72] Sylven and Borgstrom[63] have previously reported an increase in the lymphatic output of endogenous cholesterol in rats following a bolus dose of triolein.

2. Bile Salts and Other Detergents

Recent reports dealing with the influence of bile acids and other detergents are tabulated in Table 2. There is consistent evidence in these reports to indicate that the addition of cholic acid (3α, 7α, 12α-trihydroxy) or chenodeoxycholic acid (3α, 7α-dihydroxy) to standard diets has no significant effect on the initial absorption of cholesterol or its subsequent lymphatic delivery[75-79] in rat, mouse, or humans with gallstones. In the mouse studies,[78,79] chenodeoxycholate slightly reduced cholesterol absorption, but in both cases the differences were small. There is evidence to suggest that deoxycholate (3α, 12α-dihydroxy) may reduce the initial intestinal uptake of cholesterol in humans.[77] This may be related to the ability of this particular bile acid to solubilize large amounts of cholesterol compared to cholic or chenodeoxycholic acids.[80] Urso-

Table 2
INFLUENCE OF BILE SALTS AND OTHER DETERGENTS ON
CHOLESTEROL ABSORPTION

Animal species	Method of measurement of sterol absorption	Dietary addition	Influence on cholesterol absorption	Comments	Ref.
Rat		Chenodeoxycholate	None	Liver morphology	75
		Ursodeoxycholate (high levels for 14 days)	None compared to standard diet	monitored	
Humans with gallstones	Plasma isotope ratio method	Chenodeoxycholate (13 mg/kg/day)	None		76
		Ursodeoxycholate (13 mg/kg/day)	None compared to bile acid-free diet		
	Intestinal perfusion	Cholate	None	No effect on	77
		Chenodeoxycholate	None	serum choles-	
		Deoxycholate	Absorption reduced by 50%	terol	
Mouse		Chow + chenodeoxycholate (0.2%, w/w)	Slightly reduced compared to cholate-supplemented diet		78
Mouse	Dual isotope ratio, sterol balance	Cholate Chenodeoxycholate Ursodeoxycholate	Compared to standard chow diet absorption decreased	Greatest effect with ursodeoxycholate	79
Rat	Lymph delivery of infused micellar solution	Pluronic L-81 (3—4 weeks' feeding)	Reduced	Effect reversible	81

deoxycholate (3α, 7β-dihydroxy) reportedly reduced cholesterol uptake by 50% compared to cholic acid in the mouse,[79] but was found to have no effect in humans with gallstones[76] or rats (without gallstones).[75]

A study on the influence of long-term feeding of the detergent, pluronic L-81, deserves special comment. Following a 3- to 4-week prefeeding period, the uptake of cholesterol from a perfused emulsion was found to be markedly reduced.[81] The authors found no evidence for an impairment of the initial uptake, but reported a greatly increased mucosal cholesterol content. Of special interest was the apparent reversibility of the effects of the detergent. The authors suggested that pluronic L-81 acted to inhibit the assembly or secretion of lipoproteins in the intestinal mucosa.

3. Fiber and Lipid Binding Agents

The results of these studies are summarized in Table 3. Cholestyramine has been consistently found to reduce the uptake and lymph delivery of cholesterol in the rat. These effects were evident at low (1% of diet) and high (5% of diet) levels in chronic feeding experiments, but also in single dose feedings.[82] In the latter, a 47% reduction in lymphatic delivery of dietary cholesterol was obtained after coadministration of a 50-mg dose of cholestyramine. These effects occur without apparently influencing the mucosal cholesterol content.[74] Colestipol-hydrochloride similarly reduces cholesterol absorption (lymph delivery) in the rat.[83] In these experiments, the coadministration of cholic acid removed the effect of colestipol, confirming the mode of action of this drug as a binder of bile acids in the intestinal lumen.

The addition of pectin, alfalfa, or cellulose to the diet also reduces cholesterol absorption in the rat.[84] Pectin was found to be slightly more effective than cellulose (5%

Table 3
INFLUENCE OF FIBER AND LIPID BINDING AGENTS ON CHOLESTEROL
ABSORPTION

Animal species	Method of measurement of sterol absorption	Dietary addition	Influence on cholesterol absorption	Comments	Ref.
Rat		Cholestyramine	Reduced		74,82,84
	Lymph delivery	Chitosan	Reduced		82
	Lymph delivery	Colestipol-HCl	Reduced		83
	Lymph delivery	Pectin, alfalfa, or cellulose (15%, w/w)	Reduced	Only wheat bran and cellulose decreased transit times	84
		Wheatbran or yeast glycan (15%, w/w)	Marginal effect	No effect on serum cholesterol	84
	Intestinal content analysis	Pectin (5%, w/w) or cellulose (5%, w/w)	Reduced		85
Monkey	Fecal excretion	Alfalfa saponins	Decreased		86
Rat		Soy protein	Reduced compared to casein-fed rats	Serum cholesterol reduced	87
		Ginseng saponins	Increased	Serum cholesterol increased	88
Guinea pig		Cabbage	Increased	Serum cholesterol increased	89
Rat	Perfused intestine	Guar gum (20 g/kg)	No effect of prior feeding — reduction when fed concurrently		90

w/w of diet) in reducing the initial uptake of cholesterol by the intestine.[85] The absorption-reducing effect of alfalfa saponins has also been confirmed in the monkey.[86]

Wheat bran and yeast glycan had only a marginal influence over lymphatic cholesterol delivery in the rat after chronic feeding at high levels (15% of diet).[84] It was of particular interest in this study that wheat bran and cellulose both significantly decreased intestinal transit times, but pectin did not. Both pectin and cellulose were, however, effective in reducing cholesterol absorption. The absorption of cholesterol in this experiment was clearly not simply a function of the time of residence in the gut. None of the treatments reduced serum cholesterol.

Soy protein, when fed chronically to rats, was found to be more effective than casein in reducing the intestinal absorption of cholesterol.[87] The amino acid hydrolysates of both proteins were ineffective.

Two treatments have been reported which increase intestinal absorption of cholesterol. Ginseng saponins fed to rats[88] and cabbage fed to guinea pigs[89] both gave rise to increased serum cholesterol levels subsequent to their influence on absorption.

4. Miscellaneous Dietary Factors

Reports studying a variety of lipid and nonlipid soluble dietary additions are tabulated in Table 4. Of particular interest is the reported reduction of cholesterol absorption in humans by the chronic administration of calcium gluconate.[91] The absorption of ^{13}C-labeled cholesterol doses, measured as ^{13}C recovered in feces, was reduced (two subjects assessed) by the administration of 1 g of the salt before each meal during a 1-week period. In contrast to the effects of a mixture of sitosterol and campesterol which was also assessed in this study, calcium gluconate decreased the excretion of deuterium-labeled plasma cholesterol, an effect suggested by the authors to be a result of reduced biliary secretion.

Table 4
INFLUENCE OF MISCELLANEOUS DIETARY FACTORS ON CHOLESTEROL ABSORPTION

Animal species	Method of measurement of sterol absorption	Dietary addition	Influence on cholesterol absorption	Comments	Ref.
Rat		L-Cystine (5%, w/w)	None		70
		4-*O*-Methyl ascochlorin	None	Serum cholesterol reduced	93
Hamster		3-OH-3-methyl glutaric acid (100 mg/kg, 10 weeks)	Suggested reduction	Serum cholesterol reduced	94
Monkey		Gossypol acetic acid (10 mg/kg/day, 6 months)	Suggested reduction	Serum cholesterol reduced	92
Chick		Vitamin A or tocopherol (high levels)	No effect	Serum cholesterol reduced by vitamin A	95
Human	Fecal excretion	Calcium gluconate (3 g/day)	Reduced	Only two subjects tested	91
Rat		Olefin/maleic acid co-polymers	Reduced		96
Rabbit	Dual isotope plasma ratio method	*N*(1-Oxo-9-octadecenyl)-DL-tryptophan (Z) ethyl ester (15—150) mg/kg)	Reduced in cholesterol-fed animals, not in chow-fed		97
Human	Sterol balance	Sucrose polyester (8—35 g/day)	Reduced		98

Also noteworthy is the drug, gossypol acetic acid, which has been found to reduce plasma cholesterol levels in monkeys, possibly through an influence on cholesterol absorption.[92]

5. Plant Sterols and Other Sterol Derivatives

Of the nearly 50 reports that have appeared in the past 4 years concerning agents acting to reduce the absorption of cholesterol, 20% of them deal with the efficacy of the plant sterols and other related analogs. These studies are summarized in Table 5.

The chronic administration of sitosterol has been found to reduce cholesterol absorption in rat,[99-101] rabbit,[102,103] mice,[78] and man,[91] generally confirming the results found before the 1981 period (see Pollack and Kritchevsky[3]). Some exceptions to this trend appear in experiments using the rat. It should be remembered that several earlier studies (see Sylven and Borgstrom[63] and included references) were not able to demonstrate a depression of cholesterol absorption in response to sitosterol feeding.

Ikeda and Sugano[99] found in their recent studies that sitosterol reduced cholesterol absorption when administered chronically as part of the chow diet, but not when coadministered with cholesterol as a subsaturated micellar solution in ligated *in situ* loops of rat intestine. In earlier studies carried out under similar conditions[101] sitosterol reduced the uptake of cholesterol, but 10- to 100-fold excesses of the plant sterol were required to bring about the effect.

Stigmasterol and sitosterol were equally effective in reducing the absorption of cholesterol fed as a test emulsion to rats.[100] Fucosterol had no such effect, despite the fact that the rate of absorption of this sterol was closely similar to that of stigmasterol and sitosterol (3 to 4% of the fed dose).[100] Sitostanol was reported[103] to be more effective than sitosterol in reducing cholesterol absorption in the rabbit. This observation confirmed earlier reports of these authors using the rat as an experimental model. Shellfish

Table 5

INFLUENCE OF PLANT STEROLS AND OTHER STEROL DERIVATIVES ON
CHOLESTEROL ABSORPTION

Animal species	Method of measurement of sterol absorption	Dietary addition	Influence on cholesterol absorption	Comments	Ref.
Rat	Dual isotope plasma ratio and jejunal wall cholesterol	Cholesterol + sitosterol	Reduced	Initial intestinal uptake and appearance in plasma both reduced	101
	Tissue cholesterol	Cholesterol + sitosterol (0.5%, w/w for 1 week)	Reduced	Liver cholesterol reduced compared to cholesterol-fed control	99
	Lymph delivery	Sitosterol (50 mg) Stigmasterol (50 mg) } Reduced by 54%		All 3 test sterols poorly absorbed	100
		Fucosterol (50 mg)	No effect		
	Thoracic lymph delivery	Shell fish sterols	Reduced 25—40%	Test sterols poorly absorbed	104
Rabbit		Chow + cholesterol + sitosterol	Reduced	Sitostanol more effective than sitosterol in reducing cholesterol absorption	103
		Chow + cholesterol + sitostanol	Reduced		
	Intestinal microsomal cholesterol	Coconut oil + cholesterol + sitosterol	Reduced slightly		102
		Coconut oil + sitosterol	Reduced by 48%		
Human	Fecal cholesterol	Phytosterols (9 g/day)	Reduced		91
Mouse		Chow + cholic acid + sitosterol	Reduced by 30%	Decreased biliary cholesterol secretion on addition of sitosterol to diet	78

sterols have also been found to effectively reduce cholesterol absorption when administered to rats.[104] This sterol mixture contained a large percentage of 28-carbon compounds including brassicasterol (an unsaturated C_{28} epimer of campesterol) and 24-methylene cholesterol.

D. Relative Rates of Absorption of Cholesterol Analogs

We will here consider those reports in which the intestinal absorption of two or more sterol analogs have been examined under similar experimental conditions. The absorption of one sterol can then serve as "internal standard" for the absorption of the other. The ratio is usually more reliable than the absolute values and provides important information on the features of the sterol molecule that determine absorbability. Before undertaking such a compilation of "relative rate" data, it is important to bear in mind, however, that in some studies the sterol pairs were coadministered and in others their absorptions were assessed separately. Furthermore, the ratio of the two sterols in the diet in the coadministration studies was not, in all cases, 1:1. Therefore, mutual interference during the absorption of the sterols or an influence of the dietary ratio may render the results not comparable in the strictest sense.

The most commonly studied sterol pair is cholesterol and its 24-ethyl analog, sitosterol. From Table 6, it is evident that cholesterol is absorbed by the rat on the average nine times faster than sitosterol, when their appearance is measured in lymph or in plasma. The selectivity measured in the jejunal wall itself is about one half of that seen

Table 6
RELATIVE RATES OF ABSORPTION OF CHOLESTEROL ANALOGS

Sterols	Ratio[a]	N[b]	Comments	Ref.
Cholesterol/sitosterol	9.4 ± 4.5	10	Measured in rat lymph, blood	63, 105—109
	4.2 ± 3.0	6	Measured in rat intestinal wall	54, 99, 107, 109—111
Campesterol/sitosterol	2.6 ± 1.0	7	Several species	21, 112—116
Cholesterol/cholestanol	1.75 ± 0.18	5	Rat	117—119
Sitosterol/sitostanol	5.1	2	Rat	120
Sitosterol/stigmasterol	2.6	1	Rat	116
	0.59	1	Chick	115
Cholesterol/vitamin D	1.15 ± 0.18	4	Rat	19, 58, 121

[a]   The values were calculated by dividing the percentage absorption of the sterol in the numerator by that in the denominator. The ratios from all reports were averaged and are expressed as mean ± standard deviation.

[b]   N represents the number of reported ratios used in the calculation.

in lymph or blood. The value given in Table 6 for selectivity expressed during uptake into the intestinal wall represents a compilation of data obtained by analyses of the mucosa, feces corrected for nonabsorptive losses of sterol, and net mucosal uptake measured by perfusion of the small intestine. These methods generally monitor the initial uptake of sterol into the intestinal wall. The difference in the mean ratio found in the intestinal mucosa and that measured in lymph may be caused by the nonselective adsorption of dietary sterol to the gut wall. It may also reflect a selective secretion of cholesterol by the villus cells into the lymph.[107,113] The lack of significant differences between ratios of sterols found in the intestinal wall after ½-hr exposure[109,111] when lymph delivery is negligible, and after 4 to 6 hr[107,122] when lymph delivery is at its peak, illustrates that the actual mucosal ratio is not artificially lowered during the transport of sterol out of the mucosal wall. Assuming negligible adherent material, the ratio of dietary cholesterol/sitosterol in the intestinal wall may reflect the magnitude of the selectivity expressed during the initial stages of absorption.

The results obtained with cholesterol and sitosterol in the rat generally reflect those in human. A value of 14.3 for the absorptive ratio of cholesterol/sitosterol was measured by a dual isotope plasma ratio method[123] and a ratio of 3.5 by a fecal excretion analysis.[124]

The data given in Table 6 for the sterol pair, campesterol/sitosterol, represent that pooled for seven different species, including rodents, dogs, and fowl. Despite the wide variety of subjects and methods of analysis used, the variation in the ratio found in the intestinal wall is remarkably small.

In a series of studies carried out using the rat, it has been determined that Δ^5 sterols are absorbed faster than their corresponding 5α-H saturated derivatives. This effect was consistent in both the cholesterol and sitosterol series (Table 6).

The absorption of the ring B-cleaved derivative of cholesterol, vitamin D_3, appears to be closely similar to that of cholesterol itself in the rat (Table 6). Thus, the nonrigid nature of ring B does not appear to significantly influence the absorptive properties of the sterol in this case.

The effect of side chain alkylation upon absorption is apparently retained in the vitamin D series as with the parent compounds having intact ring systems. Vitamin D_3 is absorbed by the intestinal wall in the rat two times faster than its 24-methyl analog, vitamin D_2, when instilled in an *in situ* segment.[19] It should be pointed out, however, that the direct precursor of vitamin D_2, ergosterol, is taken up by the intestine up to ten times slower than cholesterol.[125,126] This level of discrimination was clearly not evident between vitamins D_2 and D_3. Evidence that cleavage of ring B may reduce the absorptive influence of side chain alkylation has also been presented based on results using purified brush border membranes in vitro.[127]

A clear determination of the influence of side chain unsaturation is not possible at this time. Studies of sitosterol and its 22,23-unsaturated counterpart, stigmasterol, indicate that sitosterol is absorbed more readily in the rat,[116] but not in the chicken.[115]

The absorption of the 5,7-diene and 7-ene analogs of cholesterol (7-dehydrocholesterol and lathosterol) have been studied in the rabbit, rat, and guinea pig. A determination of absorption in the rabbit by fecal recovery (uncorrected for degradative losses) demonstrated[128] an 81 to 90% absorption of cholesterol. This followed feeding of the pure sterol as 1% of a continuous chow diet. The absorption of lathosterol was 80%. The measurement of the sterols was in this case based on the UV absorbance of 7-dehydrocholesterol and the distinction between fast-acting sterols (7-dehydrocholesterol and lathosterol) and slow-acting sterols (cholesterol) in the Lieberman-Burchard color reaction. In these experiments cholestanol was absorbed to a lesser extent than the unsaturated sterols. The rapid absorption of lathosterol and 7-dehydrocholesterol by the rabbit intestine has been studied by Lemmon et al.[129] and Biggs et al.[130] Biggs and co-workers,[130] using [³H]lathosterol and [³H]cholesterol, have shown that the assimilation of cholesterol occurs more rapidly than with lathosterol. The latter workers reported that in rabbit, the percentage of the fed dose (20 mg) present as digitonin-precipitable sterols in plasma (1 day after feeding) was about 7 for [³H]lathosterol and 25 for [³H]cholesterol. A more rapid absorption of cholesterol than 7-dehydrocholesterol or lathosterol was confirmed when measured in plasma by Cook et al.[128] The subcellular distribution of 7-dehydrocholesterol has been shown to be identical to that of cholesterol following absorption in the guinea pig.[30,131]

Although absorptive studies using 7-dehydrocholesterol and lathosterol in whole animals are complicated by their metabolic conversion into cholesterol,[128-131] the evidence suggests the following order of absorption: cholesterol ≃ 7-dehydrocholesterol > lathosterol. The differences between them appear to be small, as might be expected from their structural similarity.

Using this information and the ratios given in Table 6, the rates of intestinal absorption of the sterol analogs may be said to decrease in the following order: cholesterol ≃ vitamin D_3 ≃ 7-dehydrocholesterol > cholestanol > campesterol > sitosterol ≃ stigmasterol ≃ ergosterol > sitostanol. At this time it is not possible to distinguish among sitosterol, stigmasterol, and ergosterol. From this order, it is apparent that the absorption in the intact animal is extremely sensitive to sterol side chain alkylation and to reduction of the Δ^5 double bond. Further oxidation of ring B or complete cleavage, as in the D vitamins, or unsaturation in the side chain appears to have little effect on the absorption of members of the cholestane series.

In general, support of these conclusions are the results described by Vahouny, Connor, and co-workers,[104,132] analyzing the absorption of an entirely different series of these sterols in humans.[132] Thus, in this series also, side chain unsaturation did not greatly reduce absorption, but side chain alkylation was a dominant factor.

III. IN-DETAIL EXAMINATION OF THE INDIVIDUAL STEPS

In the following section, literature dealing with lumenal, membrane, cytosolic, and enzymatic aspects of sterol absorption will be considered in finer detail than was possible in the general overview given in the previous section. In these discussions, the influence of each of the major transformations on the absorption of cholesterol itself, on the ability of the intestine to selectively absorb cholesterol over other sterol analogs, and on the hypocholesterolemic effect of plant sterols will be considered.

A. Event Occurring in the Intestinal Lumen

1. Luminal Events Influencing the Absorption of Cholesterol

The events leading to the solubilization of lipid in the small intestine and their delivery to the microvillar membrane in an absorbable form are exceedingly complex and not well understood. It is currently accepted that one of the principal vehicles for the delivery of cholesterol to the water-brush border interface is the micelle composed of bile salt, phospholipid, sterol, and the hydrolysis products of glycerolipids: fatty acids and sn-2-monoacylglycerols.[2,36,37] These arise from the aggregation of dietary cholesterol, fatty acid, and monoglyceride with bile salts and phospholipid delivered through the bile duct into the duodenum. The principal bile salts in man and rat are taurocholate and taurochenodeoxycholate salts which are present in 4 to 10 mM concentrations in the duodenum and jejunum of human subjects following a meal.[2,37] Phospholipid concentrations are typically 0.5 to 1 mg/mℓ.[37]

It is clear that the rate of uptake of cholesterol and other lipids by the brush border membrane will be the product of the permeability coefficient of the membrane for the solute and their concentration at the aqueous-membrane interface. Micellized cholesterol is one form in which the solute may be delivered to within close proximity of the membrane surface.

Dietschy and Westergaard[36,41] have hypothesized that the ability of micelles to deliver absorbable solutes may be governed by their ability to diffuse across a layer of unstirred water that surrounds the intestinal membrane. This layer of water, suggested to be 130 to 330 μm thick, is therefore expected under some conditions to provide a barrier to absorption in addition to that provided by the brush border membrane itself. Thus, the initial rate of absorption of cholesterol may be influenced by both the effective resistance to micellar diffusion of the unstirred water layer and the passive permeability characteristics of the membrane. Under conditions where the resistance of the unstirred water is low, the rate of initial uptake would be essentially determined by the permeability characteristics of the membrane.

There is ample evidence to suggest that the concept is a reasonable approximation of the state of events occurring during the uptake of lipids by isolated intestinal preparations in vitro.[2] The underlying theoretical basis and the experimental support is extensively described in reviews by Dietschy[36] and by Thomson and Dietschy,[2] and only a brief overview will be given here. In accordance with the model, it has been shown experimentally in vitro that for short chain fatty acids which have high aqueous solubilities, variation of the unstirred layer thickness (by altering the stirring rate of the bulk solution) has little effect on the rate of uptake. For long chain fatty acids (C$_{16}$ to C$_{18}$) and cholesterol which have low aqueous solubilities, the resistance of this layer has been shown to be rate limiting to absorption in a wide variety of intestinal preparations.[2] Under these conditions, where the translocation of the substrates into the membrane is expected to be rapid as a result of the high membrane permeability coefficient for these compounds, the micelle serves as a reservoir to replenish the supply of substrate molecules at the membrane interface at a rate which is governed by the rate of micellar diffusion through the unstirred layer.

The techniques applied by Dietschy and co-workers have also provided strong evidence that in vitro, the absorption of cholesterol occurs from a monomer phase which exists in equilibrium with the cholesterol in micellar form. Equations describing sterol uptake indicated that in an absorptive process occurring through an aqueous cholesterol monomer population, the rate of absorption should be proportional to the mass of lipid in the aqueous phase. An absorptive process requiring direct contact between micelle and the microvillar membrane on the other hand, should, by similar arguments, be dependent on the mass of cholesterol in the micellar phase. The experimental results showed that the initial rate of cholesterol uptake by evert gut sacs decreased with conditions acting to reduce the aqueous monomer concentration (increasing the bile salt concentration while holding the cholesterol concentration constant). Increasing the mass of lipid in the micellar phase (by increasing the concentrations of both bile salt and cholesterol) led to only a moderate increase in the initial uptake rate.[2,36]

It is of interest to note that there is substantial evidence to suggest that aqueous monomeric exchange underlies the transfer of cholesterol between cells and lipoproteins (see for example Lange et al.[133]). It should, however, be mentioned that while many studies support the concept of transfer of cholesterol as solvated aqueous monomers,[2] Rampone and Machida[134] could not reproduce the experiments on which this assumption is based in their study of the uptake of cholesterol by everted sacs of rat intestine, and Montet et al. have demonstrated that cholesterol uptake was very much dependent upon the mass of sterol in the micellar phase in their in vitro experiments.[135] Lutton et al.[69] reported that cholesterol absorption in rats fed cholesterol and triacylglycerol (20% of weight of diet) showed no correlation with the amount of sterol found in a micellar phase isolated by centrifugation. Ikeda and Sugano,[99] on the other hand, did find a correlation between the amount of cholesterol present in a clear micellar phase and the rate of absorption in the rat.

The unstirred water layer model and the supporting experimental data contribute significantly to our understanding of the kind of influences we may expect during transport of lipid to the brush border membrane. A great barrier to the complete acceptance of the model as a description of cholesterol absorption in vivo following a fat meal, however, is a body of experimental data suggesting that under actual dietary conditions other factors may be as, or more, important than the unstirred water layer in determining the concentrations of cholesterol at the aqueous-brush border interface.

It is generally accepted that an unstirred water layer surrounds the intestinal villi in vivo as it does in vitro.[2] Although it has been suggested[2] that this layer may actually be thicker in the intact animal than in vitro, Miyamoto and co-workers[136] have argued that the thickness is likely dependent on the net secretion or absorption of water as well as on the rate of flow of gut content through the lumen. On this basis, the unstirred layer was not expected to remain a constant thickness throughout and its influence on the absorption of lipids at each point of absorption may be highly variable.[136]

The form of the lipid aggregates required for absorption is also a point of contention. Patton[37] has pointed out that to obtain micellar phases from human gut content requires long centrifugation periods, and even then only a small proportion of the total volume is present as a clear "micellar" phase that may support the diffusion process. This is amply illustrated in the results of a recent report describing the content of rat small intestine following a lipid meal. After centrifugation of the contents at 100,000 × *g* for 15 hr, no more than 10% of the total cholesterol was found in the clear supernatant.[99] Based on a light microscopic study of the digestion of oil in the presence of human bile and pancreatic juice, Patton[37] suggested that their may be four or more coexisting phases produced during intestinal lipid digestion. The phases represent:

1. *Oil phase* consisting largely of triacylglycerol and diacylglycerol
2. *Calcium soaps* formed from liberated free fatty acids combined with Ca^{2+} in a ratio of 2:1
3. *Viscous isotropic phase* of unknown composition
4. *Fat droplet remnant*
5. *Micellar phase* composed of bile salt and other lipid

The author concluded that the division of duodenal lipids into micellar and oil phases was an oversimplification of the situation. The viscous isotropic phase and the micellar were cited as the most likely substrates for absorption, but they did not rule out absorption from a continuous hydrocarbon domain.[37] In support of this, the author cited experiments on the absorption of octadecane described by Borgstrom.[137] Octadecane fed in triolein carrier was found to be absorbed in large amounts and delivered via the lymph in rats. In view of the negligible aqueous solubility of the hydrocarbon, it appeared unlikely that absorption occurred through an aqueous monomeric phase. Patton further cautioned,[37] however, that a "hydrocarbon continuum" composed of triacylglycerols, monoacylglycerols, and fatty acids would likely not be maintained in the absence of a micellar or liposomal phase.

Evidence that bile salt micelles may not be the only form from which lipid may be absorbed comes from particle size analysis of human bile. Somjen and Gilat[138] have described the existence of large (about 700-Å diameter) particles having the appearance of vesicles in human hepatic bile. They demonstrated that the particles contained low levels of bile salt and were thought to be able to carry more than 80% of the cholesterol present in supersaturated hepatic bile. In this regard, it is of considerable interest that the rate of absorption of cholesterol derived from bile is lower than that fed endogenously in triglyceride in both rat[139] and man.[140] The results suggest that there may not be a thorough mixing of dietary and endogenous cholesterol under in vivo conditions. These results also indicate that while cholesterol absorption from classical bile salt micelles is possible in vivo, they may not be the only substrate from which absorption may occur.

A further determinant of the concentration of cholesterol at the absorption interface is likely to be the composition of the bile salt in which the sterol is solubilized. Reynier and co-workers[79] have studied the absorption of isotopic cholesterol in Swiss female mice in response to diets containing 0.2% (w/w) purified chenodeoxycholic (3α, 7α-dihydroxy), cholic (3α, 7α, 12α-trihydroxy), or ursodeoxycholic (3α, 7β-dihydroxy) acids. The uptake of radioactive label by the mucosa (measured by sterol balance in feces) and the appearance in plasma (measured by a dual isotope ratio method) were highest with cholic acid and lowest with ursodeoxycholic acid (twofold difference). The low absorption from the ursodeoxycholate-supplemented diet was suggested to be the result of the poor ability of this bile acid to solubilize cholesterol. This was subsequently demonstrated in model micelles composed of 10 mM bile acid, oleic acid, and cholesterol. Ursodeoxycholate solubilized the least cholesterol of the three bile acids tested.

The differences between cholic and chenodeoxycholic acids could not be explained on this basis, as the dihydroxy acid solubilized more sterol than the trihydroxy in model micelles containing oleic acid. It was therefore suggested that the degree of saturation (cholate micelles would be expected to be more easily saturated with cholesterol than those composed of chenodeoxycholate) may influence uptake — greater absorption occurring from more saturated micelles.[79] For further information on this aspect of the absorptive process the reader is directed to an excellent article by Montet et al.[135]

Assuming for the moment that cholesterol absorption occurs from the sort of micelles that are produced in the laboratory from bile salt, sterol, and glycerolipid hydrolysis products, there is evidence to suggest that diffusion of these particles may not be, in every case, rate limiting for the initial uptake. In experiments designed to assess the effect of mixed micellar lipid on the uptake of cholesterol, Thompson and co-workers[58] examined the uptake of trace amounts of cholesterol and vitamin D_3 into the intestinal wall of duodenally perfused rats. When uptake was assessed from perfusates containing sodium taurocholate (20 mM) and sterol, it was found that the addition of fatty acids (9.6 mM) and monoacylglycerol (4.8 mM) did not affect the initial uptake of cholesterol into the mucosa. The inclusion of neutral lipid into taurocholate micelles has previously been shown to increase the aggregate radius at least twofold[141] and would thus be expected to reduce the diffusion coefficient of the micelles.

Rampone and Machida,[134] studying cholesterol uptake by evert sacs of rat intestine in vitro, reported that the inclusion of dipalmitoyl PC into bile salt-cholesterol micelles reduced the uptake of cholesterol without significantly increasing the size of the aggregates (assessed by gel filtration). After measuring the diffusion coefficients of the micelles through glass capillary tubes, they concluded that the effects of lecithin could not be attributed to a reduction in the rate of diffusion, but more likely through a direct influence on the permeability of the brush border membrane. The degree of cholesterol saturation of the PC-containing micelles was not considered in this report. To investigate this point, Montet et al.[135] used micelles composed of a variety of bile salts, but saturated with cholesterol and having comparable measured diffusion coefficients. They demonstrated that the uptake of cholesterol by everted intestinal sacs was greatest when the micelles were composed of taurochenodeoxycholate. These structures also solubilized more cholesterol than the other species tested (taurocholate and tauroursodeoxycholate), suggesting that the uptake was proportional to the amount of sterol present in the micellar phase, rather than that in the aqueous phase, as predicted by the unstirred water layer-micellar diffusion concept.

All of these reports demonstrate that under some conditions, the rate of initial cholesterol uptake is not sensitive to changes in the size of the micellar aggregates or their rate of diffusion. It has been pointed out, however, that since the diffusion coefficient for a particle varies with the square root of the diameter, the size of the micelle may have a smaller influence on absorption than initially anticipated.[2]

The above survey of reports demonstrates that although the unstirred water layer can represent a major barrier to cholesterol absorption under some conditions, it is too early to conclude that it is the major barrier in all cases, or under normal dietary conditions, in mammals. A variety of factors may influence the concentrations of cholesterol available for absorption at the aqueous membrane interface and under some conditions the microvillar membranes themselves can provide a rate-limiting barrier to the initial entry of sterol in the cell. A survey of reports dealing with the events occurring within this membrane and others within the villus cell follows.

2. Luminal Events Influencing the Selective Absorption of Cholesterol Analogs

There is a body of evidence suggesting that the solubilization of sterols in the intestinal lumen may be the stage at which the selective absorption of cholesterol over other sterols originates. The yeast sterol, ergosterol, was earlier found to be solubilized to about 1/20th the extent of cholesterol by bile salt micelles containing phospholipid in studies in vitro by Barton and Glover.[142] Similar results were obtained with egg PC liposomes.[143,144] In the study by Bruckdorfer et al.,[143] ergosterol, sitosterol, and 7-dehydrocholesterol were all solubilized to a lesser extent than cholesterol. Kellaway and Saunders[145] further demonstrated that cholest-4-en-3-one, 5-α-cholestan-3-one, and 5-α-cholestan-3-ol are solubilized as readily as cholesterol by egg PC liposomes. A reduced solubility of sitosterol compared to cholesterol in mixed micellar solutions has been reported by Slotta and Ammon.[147] Bhattacharyya[116] has suggested that plant sterols may be less able to partition out of micelles than cholesterol.

Conflicting with these reports are a series of studies on the luminal contents themselves following a sterol-containing meal. McIntyre and co-workers[21] found an identical distribution of campesterol and sitosterol in the micellar and crystalline phases after centrifuging the luminal contents of a rat following a sterol meal. Attempts to recreate the milieu of the intestine in vivo[146] using mixtures of bile salts, sterols, and the hydrolysis products of triacylglycerols confirmed the above reports. The ratio of cholesterol to sitosterol was identical in micellar and emulsion phases separated by gel filtration and Millipore filtration.

There are indications, therefore, that in the intestinal lumen, the solubilization of sterols is a nonselective process. The in vitro data cannot be ignored, but the discrepancy may be rationalized by the demonstration[141] that cholesterol is much more readily solubilized in micellar sodium taurocholate when the lipid components are first mixed together in organic solvent prior to drying down and the addition of water, than when aqueous bile salt is added to crystalline cholesterol. A similar dependence of sterol solubility on the mode of preparation of the samples has been noted by Carey and Small.[148]

3. Luminal Events Bearing on the Hypocholesterolemic Effect of Plant Sterols

The ability of sitosterol and other plant sterols to displace cholesterol from micellar structures has been reviewed in detail by Kritchevsky and Pollack.[3] Reports have since appeared confirming the earlier observations. Ikeda and Sugano[99] centrifugally isolated a clear aqueous phase from the small intestinal content of a rat after feeding a diet containing either cholesterol or cholesterol plus sitosterol. They found that the total sterol concentration was closely similar with both dietary regimens, but that with the inclusion of sitosterol, the cholesterol concentration was approximately one half of that obtained after feeding cholesterol alone. In vitro they found that sitosterol reduced the maximum cholesterol solubility in micelles composed of sodium taurocholate, monoacylglycerol, and fatty acid. Similar in vitro findings have been reported by Slotta et al.[147,149]

In summary, it appears that although the evidence for a selective solubilization of cholesterol in the intestinal lumen is equivocal, there remains convincing evidence that the plant sterols are able, under some conditions, to reduce the available concentration of cholesterol in the luminal aqueous phase.

B. Membrane Events
1. Experiments with Purified Brush Border Membranes

Under some circumstances the brush border membrane of the intestinal epithelial cells can present a significant barrier to the initial entry of sterols into the intestinal wall. A description of the events occurring at or within this membrane structure is still based largely on speculation, but work of the past few years, particularly with purified preparations of brush border, sheds much new light on the subject.

Several reports have provided positive evidence of a major influence of membrane permeability in determining cholesterol uptake under some conditions of absorption. Differences in permeability of the brush border membrane were suggested by Thomson et al. to account for differences in the initial rates of uptake of cholesterol and fatty acids between intestinal slices derived from hamster, rabbit, rat, and guinea pig.[150] The rates of uptake of fatty acids decreased in the order hamster = rabbit > rat > guinea pig and the differences did not correlate with the resistance of the unstirred water layer to micellar diffusion. The rate of uptake of cholesterol from micellar solutions decreased in the order rat > hamster > rabbit > guinea pig and the authors concluded that the trends could only be accounted for by inherent differences in the permeability characteristics of the microvillar membrane, in addition to variations in unstirred water layer resistance. A major role of membrane permeability in the initial rate of cholesterol absorption is also suggested by the finding of a high rate of cholesterol uptake into gut segments derived from animals fed a low cholesterol diet than with those from animals fed high cholesterol.[220] The authors pointed out that the influence may also be mediated through changes in membrane surface area or unstirred layer thickness mediated by the diet.

In a classic study of the brush border membrane during cholesterol absorption, David and co-workers[26] established several important events occurring in this structure. They first determined that the cholesterol content of the brush border region, isolated after feeding, did not change despite a demonstrable uptake of radioactive label provided in a 3-mg dose of cholesterol dispersed in groundnut oil. To account for this observation, the authors proposed a mechanism of uptake whereby an incoming sterol molecule displaces another, forcing it into the cytoplasm of the cell. It should be pointed out that a recent study has reported increased cholesterol/phospholipid ratios in the brush border membranes of rats fed high cholesterol diets.[151] The cholesterol/ phospholipid ratios were not reported in the former study.[26]

The second key observation in the report of David et al.[26] was the complete absence of sterol esters in the brush border fraction, even following dietary administration of radioactive cholesterol ester. The results clearly confirmed a requirement for ester hydrolysis prior to the entry of sterol into the microvillar membrane.

Significant advances in the study of events occurring within the brush border membrane have followed the development of rapid methods for the preparation of large amounts of vesicles derived from the tips of the microvilli.[152] These vesicles have been used in vitro to study aspects of sterol transfer in four recent studies.[127,153-155]

Bloj and Zilversmit[153] studied the exchange of cholesterol between (^{14}C) cholesterol-prelabeled brush border vesicles isolated from rabbits and a large excess of small unilamellar vesicles composed of cholesterol and egg PC. The efflux of label with time generated a curve that was best approximated by the sum of two exponentials and suggested the presence of two pools of cholesterol in the membrane: one rapidly exchangeable (about 35% of the total) and one slowly exchangeable. The two environments persisted despite disruption of the vesicles with deoxycholate. This led the authors to conclude that the heterogeneity was not the result of cholesterol retained in the inner half of the membrane bilayer. Sonicated vesicles prepared from brush border lipids did not appear to possess two distinct pools. The slowly exchangeable sterol, therefore, appeared to be in some way dependent on membrane protein. The rate of cholesterol efflux from the brush border vesicles or the distribution between the pools was not significantly influenced by the level of cholesterol fed to the rabbits prior to the membrane isolations.

Child and Kuksis[127,154] have developed methods for the simultaneous assessment of the rates of uptake of cholesterol, campesterol, and sitosterol by the brush border membrane vesicles in vitro. To allow study of the homologous series (the three sterols are a homologous series with respect to carbon number, but not stereochemistry at C_{24} of the side chain) without interference from endogenous pools of cholesterol, the 7-dehydro analogs of each were prepared and their uptake assessed by high-pressure liquid chromatography (HPLC) following incubations of the intestinal membranes with micellar solutions (composed of sodium taurocholate, egg PC, and egg phosphatidylethanolamine [PE]) containing the test sterols. Representative sterol profiles from these experiments are shown in Figure 4. Panel A shows the composition of the sterol mixture present originally in the micellar mixture. Peak 4 identifies 7-dehydro-cholesterol acetate added subsequently as an internal standard. It is clear from this figure that the brush border vesicles (panel C) and isolated jejunal villus cells (panel B) both preferentially take up the cholesterol analog (peak 1). The inclusion of PC was essential for the expression of the selectivity and a mixture of egg PC and egg PE was found to be optimal.[154,156]

In these experiments, 7-dehydrocholesterol was taken up four times faster than 7-dehydrositosterol by the brush border membrane vesicles. It was pointed out in this study that further correction of the results for nonselectively adsorbed sterol could double the observed selectivity, bringing the ratio close to that found in the lymph in intact animals (see Table 6).

Of interest in the latter study was the capacity of rat erythrocytes to selectively take up the cholesterol analog in a fashion closely similar to that of the brush border mem-

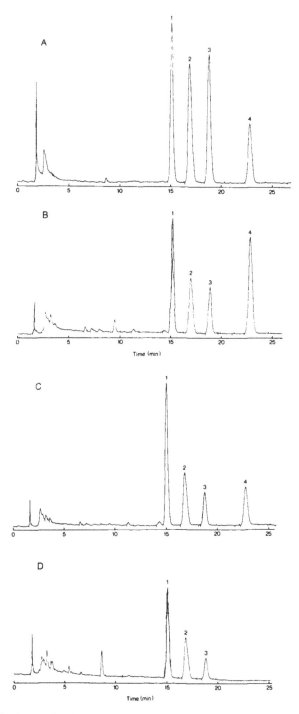

FIGURE 4. The selective uptake of 7-dehydrocholesterol, 7-dehydrocampesterol, and 7-dehydrositosterol by cell and membrane preparations visualized by reversed-phase HPLC. (A) Extract of the micellar sterol mixture; (B) extract of rat jejunal villus cells after incubation with the micellar sterol mixture; (C) purified brush border membranes after incubation; (D) rat erythrocytes after incubation. Peak identification: (1) 7-dehydrocholesterol; (2) 7-dehydrocampesterol; (3) 7-dehydrositosterol; (4) 7-dehydrocholesterol acetate (added subsequently as an internal standard). The cell and membrane preparations were incubated for 30 min at 37°C in a micellar solution containing 6.6 mM sodium taurocholate, 41 μM 7-dehydrocholesterol, 44 μM 7-dehydrocampesterol, 42 μM 7-dehydrositosterol, and 0.6 mM phospholipid containing 40 mol% egg PE and 60 mol% egg PC. The sterols were resolved by HPLC using a 5-μm particle size, C_{18}-reversed phase column eluted with methanol/acetonitrile (1:1). The detection wavelength was 265 mm.

Table 7

IN VITRO UPTAKE OF CHOLESTEROL AND 7-
DEHYDROCHOLESTEROL NORMALIZED TO
MEMBRANE SURFACE AREA AND PC CONTENT

| Membrane type | 7-Dehydrocholesterol[a] | | Cholesterol[b] |
	nmol/m²	pmol/nmol PL	pmol/nmol PL
Jejunal villus cells	2065 ± 130	21[c]	92 ± 37[c]
Brush border vesicles	1603 ± 92	96	698 ± 12
Erythrocytes	38 ± 7	31	86 ± 8

[a] Incubation solution contained sodium taurocholate (7 mM), egg PC + egg
PE, 35 μM each of 7-dehydrocholesterol, 7-dehydrocampesterol, 7-dehydro-
sitosterol.[154] Period of incubation was 30 min.

[b] Incubation solution contained sodium taurocholate (7 mM), egg PC + egg
PE, 200 μM each of cholesterol and sitosterol.

[c] About 4% of the total cellular phospholipid resides in the plasma membrane
of the rat jejunal villus cell (see Child and Kuksis[158]).

branes in the presence of bile salt-containing micellar solutions (panel D of Figure 4).
The ratio of 7-dehydrocampesterol/7-dehydrositosterol in rat erythrocytes incubated
with the micellar carriers was 1.8 after 30 min of incubation.[154] Erythrocytes have
previously been shown to preferentially take up campesterol over sitosterol from liposomes composed of egg PC.[157] In the latter experiments, campesterol entered the cells
2.5 times faster than sitosterol, a ratio that closely approximates that observed in the
intestinal wall following dietary ingestion of the same sterols (Table 6).

The similar degree of selectivity expressed by the erythrocytes and the purified brush
border vesicles clearly demonstrate that the recognition of cholesterol is a membrane
property that is not unique to intestinal membranes. The finding also argues strongly
against the involvement of a specific intestinal cholesterol recognition protein or receptor. The greater selectivity expressed by brush border vesicles compared to the intact
villus cells was suggested to arise from features of membrane structure that the microvilli may have in common with erythrocytes, for example, a high cholesterol/phospho-
lipid ratio.[154]

The actual amounts of sterol that were absorbed by the cell and membrane species
in these experiments were also of interest. Sterol uptake normalized to membrane surface area and membrane phospholipid is given in Table 7. It is evident that with both
7-dehydrocholesterol and cholesterol, the normalized uptake per unit time is much
higher in the intestinal preparations than in the red cells. The apparent uptake per
nanomole phospholipid is the highest in the isolated enterocytes when it is taken into
account that only about 5% of the total cellular phospholipid resides in the membrane
exposed to the micellar solution.[158] It was postulated on the basis of these results[154]
that the intestinal preparations may possess an efficient mechanism for the internali-
zation of sterol that is absent in rat erythrocytes. It is, therefore, of considerable interest that the nonspecific lipid transfer protein from bovine liver increases the rate of
cholesterol efflux from brush border membrane vesicles in vitro.[153] This transfer protein is identical to sterol carrier protein₂ (SCP₂) and an analog has been shown to be
present in large amounts in the intestine of the rat.

Proulx et al.[155] studied the influence of a variety of glycerolipids on the uptake of
cholesterol by purified brush border membranes from rabbit intestine. They were able
to confirm the inhibitory effect of long-chain PCs previously described by Rampone
and Machida,[134] but found that didecanoyl- and dilauryl PCs had no effect. Further-

more, the addition of greater than 0.4 mM palmitic acid to the micelles completely reversed the inhibitory effect of dipalmitoyl PC on the uptake of cholesterol. Preincubation of the purified membranes with dipalmitoyl PC did not influence the subsequent uptake of cholesterol, leading the authors to conclude that the effect of the phosphoglycerides was based on inhibition of the partition of cholesterol out of the micelles rather than an effect upon the membranes themselves.

2. Energy Requirements

The entry of cholesterol into the brush border membrane is generally considered to be a nonenergy-requiring process driven by a large permeability coefficient in favor of the membrane.[2,36] Recent studies[159] of the uptake of cholesterol by loops of intestine in the rat have indicated, however, that there may be a saturable component of the sterol uptake process that is influenced by metabolic inhibitors. The mucosal uptake of cholesterol was reduced to 40 to 60% of the control with the addition of sodium azide, potassium cyanide, 2,4-dinitrophenol, or ouabain. Reduction of body temperature from 37 to 27°C also dramatically reduced the mucosal uptake of cholesterol. On the basis of these observations, the authors suggested that the initial uptake of cholesterol occurs by an active transport system which is saturable and requires energy for its operation.

These findings are reminiscent of those reported by Sylven and Borgstrom[111] in a study of the selective absorption of cholesterol over sitosterol by ligated loops of rat intestine. The absorptive preference for cholesterol present in control rats was abolished by ligation of the intestinal blood supply or by the inclusion of potassium cyanide in the *in situ* loop. The authors concluded that the expression of the selective absorption was an energy-dependent process. A similar dependence of selective absorption on cellular metabolism was reported by Beames et al.,[160] studying the uptake of cholesterol and sitosterol by the midgut of the nematode, *Ascaris sum*. In control nematodes, cholesterol was taken up twice as fast as sitosterol from micellar solutions containing bile salt and mono-oleoylglycerol. The selectivity was abolished when tissue glycolysis was blocked by changing the gas phase from 95% N_2-5% CO_2 to 100% N_2.

The action of metabolic inhibitors on the absorptive process may be consistent with a role of an activated complex formed between cholesterol and membrane protein or mucopolysaccharide during sterol absorption, as suggested by Chow and Hollander.[23] Such a complex was postulated to account for the 20-kcal/mol free energy of cholesterol uptake measured in evert sacs of rat intestine. The similarity between this value and that of the exchange of cholesterol between erythrocytes and plasma lipoproteins (12 to 22 kcal/mol[161]), the efflux of cholesterol from hepatoma cells (12 kcal/mol[162]), and between vesicles of egg PC (16 kcal/mol[162]) suggests, however, that other explanations may also be consistent. Higuchi and co-workers[163] have cautioned against the overinterpretation of results of Arhenius-type plots applied to complex biological systems.

The influence of metabolic inhibitors on sterol absorption is difficult to rationalize with the hypothesis that the major rate-limiting barrier to the entry of cholesterol into the brush border membrane is caused by interfacial resistance at the aqueous-membrane interface. This hypothesis was proposed[163] based on experiments studying the transfer of cholesterol from oil droplets to an aqueous phase containing detergent. As with unstirred water-layer resistance, the interfacial resistance is not expected to be influenced by cellular metabolism or previous dietary cholesterol levels, both of which have been shown to influence sterol absorption in vivo. It is not possible to exclude, however, that subtle changes to the brush border membrane surface occur under these experimental conditions.

Beames and co-workers[160] suggested that the loss of the ability of the intestine of the

Table 8

DISTRIBUTION OF RADIOACTIVITY DUE TO
ABSORBED STEROL AMONGST THE SUBCELLULAR
FRACTIONS OF VILLUS CELLS

Fraction	Total sterol (DPM) (%)	Total phospholipid (nmol) (%)	Total protein (mg) (%)
1,500 × g Pellet	8.5	9.3	13.1
20,000 × g Pellet	73.9	73.3	41.4
160,000 × g Pellet	13.5	13.4	21.7
Cytosol	4.2	4.1	23.8

Note: The villus cells derived from the jejuna of three rats were incubated
for 10 min at 37°C in a micellar solution containing taurocholate (7
mM), oleic acid (1.2 mM), 2-monooleoyl glycerol (0.6 mM), and
sterol [4-^{14}C]cholesterol and [22,23-^{3}H]sitosterol (0.25 mM each).
The results are derived from one determination, carried out on
pooled subcellular fractions.

nematode, *Ascaris sum*, to selectively recognize cholesterol following inhibition of glycolysis arose through a breakdown of the structure and/or functional processes of the
epithelial cell caused by the reduced amount of energy available for cell maintenance.
This remains a logical explanation of the influence of the metabolic inhibitors, as there
is considerable evidence to indicate that membrane damage can severely influence lipid
transport (see Child and Kuksis[154,156] and included references). Membrane integrity
may conceivably be related to the metabolic well-being of the cell. Furthermore, the
chemical gradient acting to draw cholesterol into the cells may be expected to dissipate
in the absence of metabolic energy. Green has reported[164] that intracellular membrane
cholesterol contents equilibrate after administering metabolic poisons to intact tissue
or cells in culture.

3. Membrane Components Involved

In attempting to determine the membrane components most influential in governing
the entry of cholesterol into the intestinal epithelial cells or its subsequent intracellular
movement, it is difficult to avoid the conclusion that the cellular phospholipid is in
some way involved.

Early evidence of this involvement came from subcellular fractionation of the intestinal mucosa of experimental animals after a meal containing radioactive sterols.
Glover and Green[47] reported that the distribution of radioactivity correlated closely
with the phospholipid content of the subcellular fractions, but not their protein content. Similar data derived from in vitro experiments with isolated rat jejunal villus cells
are presented in Table 8. After incubating the cells with a micellar solution containing
radioactive cholesterol and sitosterol, the distribution of radioactivity correlated with
the total phospholipid content of subcellular fractions, but not the protein content. In
this experiment, the total phospholipid is reported and there was no obvious correlation between the distribution of radioactivity and any particular phospholipid class.

A further indication of a role of membrane phospholipid comes from the striking
similarity between the cholesterol exchange characteristics of PC-cholesterol liposomes
and intact cells. Gottlieb[161] has shown that with respect to the rate of cholesterol exchange, sonicated vesicles prepared from erythrocyte total lipid are virtually identical
to the intact cells. Similarly, Lange and co-workers[133] have reported that the rate constants for the desorption of cholesterol from human red cells are nearly the same as

those describing desorption from cholesterol-egg PC vesicles. Phillips et al.,[162] working with mouse LM cells in culture, reached the same conclusion. Based on this sort of result, several authors have concluded that interactions between cholesterol and membrane phospholipid are dominant forces in determining the rate of transfer of sterol through membranes.[127,133,161,162,165,166] There is no reason to believe that the same underlying principles do not apply to intestinal epithelial cells. The work of Bloj and Zilversmit,[153] however, indicates that in the brush border, membrane proteins also play a role in governing cholesterol transfer.

In addition to indicating the types of interaction that may occur within biological membranes, one of the important findings from in vitro work with nonintestinal cell systems is the identification of the desorption of cholesterol from the membrane as the slow step in the exchange process, rather than the entry into the lipid bilayer. This has been shown to be the case in human erythrocytes,[133] mouse LM cells,[162] and also in egg PC-cholesterol liposomes[133,162] and plasma lipoproteins.[167] The underlying reason for this is most probably the energy required to disrupt the interactions between cholesterol and the other membrane components. If this is indeed the case, then the desorption step in the passage of cholesterol through the brush border membranes may be predictably the most sensitive to diet-induced changes in the brush border composition. There have as yet been no direct studies of the influence of diet-induced changes in membrane composition on the exchangeability of cholesterol in brush border membranes, but studies using the trout[168] and rat[169] as experimental models have shown that the composition of the diet can have a dramatic influence on the phospholipid composition of the intestinal brush border membrane and also on the activity of membrane-bound enzymes.

Data from several model systems demonstrate directly that the exchange of cholesterol is influenced by the fatty acyl composition of the membrane phospholipid. Cholesterol exchange has been shown to be more rapid between plasma LDL and erythrocytes when the lipoproteins were isolated from rabbits fed a polyunsaturated fat diet than when fed a saturated diet.[170] Transfer of cholesterol from PC-cholesterol liposomes to erythrocyte ghosts has been shown to be faster when the liposomes were composed of unsaturated or mixed acid PCs than when totally saturated PC was used.[171,172] Similarly, Wattenberg and Silbert[165] have shown that crude endoplasmic reticulum membranes from rat liver take up less cholesterol from donor vesicles than do purified plasma membranes, but more than purified mitochondria. These authors identified high sphingomyelin contents as possible contributors to high membrane cholesterol levels, a concept that was previously suggested by Patton.[173] Child et al.[166] demonstrated that the rate of cholesterol desorption from human erythrocytes depended on the membrane PC composition, which had been systematically modified using the PC-specific transfer protein from beef liver.

From the discussion presented in this subsection it is apparent that a large number of properties of the initial intestinal uptake of sterols can be ascribed to events occurring at the surface of or within the brush border membrane of the jejunal epithelium. Under some circumstances the rate of cholesterol uptake may be determined by this membrane, and it has been directly demonstrated that the preferential uptake of cholesterol over the plant sterols can be brought about during the uptake of sterols by brush border membranes in vitro. It is clear that the possible dependence of intestinal cholesterol absorption on membrane composition must be explored further before a complete understanding of the molecular details of absorption can be achieved.

C. Cytosolic Events

There is little hard evidence concerning the movement of cholesterol within cells of the intestinal mucosa; how, for example, cholesterol gets from its site of entry into the villus cell, the brush border membrane, to the site of its esterification or assembly into lipoproteins. In a general review of the principles involved, Green[164] has suggested that at least three mechanisms are possible: spontaneous transfer between membranes via the aqueous phase, transfer facilitated by soluble carrier proteins, or transfer mediated through membrane vesicles. In the intestine, a fourth, lateral diffusion through continuous membranes should also be included.

The spontaneous exchange of cholesterol between membrane or membrane-like structures has been demonstrated many times using a wide variety of experimental methods (reviewed by Green[164]). Net transfer of cholesterol is known to occur between membrane-like structures containing different amounts of sterol. The most relevant parameter in this regard appears to be the cholesterol/phospholipid ratio of the donor and acceptor structures. Cholesterol flow is from the structure having the higher ratio to that having the lower.[174] How this occurs is not known, but as mentioned earlier, the weight of evidence favors a concept that involves the transfer of aqueous sterol monomers that have been released from the donor membrane to an acceptor membrane.

Applying similar arguments to the intestinal villus cells, transfer of cholesterol from the brush border membrane to the endoplasmic reticulum may be expected to occur from the difference in the cholesterol contents of these membranes. The brush border has a cholesterol/phospholipid ratio reported to be about 0.9, while that of intestinal endoplasmic reticular membranes may be as low a 0.1 (see Child and Kuksis[154] and Ashworth and Green[175] and included citations). Thus, the driving force behind the movement of sterol within intestinal villus cells may be predicted to be the result of a complex combination of influences arising from the structure of the various membranes and the available level of cholesterol at any given time, the latter markedly influenced by ongoing metabolism, particularly sterol esterification.

Tipping and Ketterer[176] have discussed in a theoretical manner how intracellular lipid transfer proteins may be expected to enhance the rate of lipid transfer within cells. They have predicted that the most efficient results would be obtained with a high concentration of protein of low affinity rather than a small amount of a high-affinity transporter. In this case the rate of transfer would be governed by the rate of diffusion of the complex rather than by the desorption from the carrier. In the intestine this role may be fulfilled by the nonspecific lipid transfer protein or sterol carrier protein, which has been shown[177] by enzyme-linked immunoassay to be present at a level of 0.46 g/mg protein in the 105,000 × g supernatant derived from rat intestinal mucosa.

This protein has been shown to increase the conversion of cholesterol to cholesterol esters by rat liver microsomal acyl CoA: cholesterol acyltransferase (ACAT) in vitro[178-180] by enhancing the transport of cholesterol from donor vesicles to the microsomes.[179] A role in an intact cell has yet to be demonstrated conclusively. Van Heusden et al.[181] have shown that cholesterol esterification in Morris hepatomas proceeds as efficiently as in other cell types, despite the fact that these tumor cells contain only low levels of that particular transfer protein. Although an influence of other sterol transfer proteins could not be ruled out, the results suggested that the esterification of cholesterol was not strictly dependent on transfer proteins. A similar conclusion was reached by Suckling et al.[182]

D. Sterol Esterification

1. The Enzymes Involved

There are two known esterases present in the small intestine that may be responsible for the intracellular esterification of cholesterol: cholesterol ester hydrolase (E.C. 3.1.1.13) requiring free fatty acids and bile salts for its activity and ACAT (E.C. 2.3.1.26) requiring fatty acyl CoAs.

Intestinal ACAT has been found in rat,[96,183] guinea pig,[184] rabbit,[97,102,185] and man.[186] The greatest amount ACAT-stimulated cholesterol esterification is found in the human jejunum, with lower amounts in the duodenum and ileum.[186] In rats perfused with micellar bile salt-cholesterol solutions through duodenal cannulae, equal amounts of ACAT-mediated ester formation were found in the proximal and distal halves of the small intestine.[73] In the rat, greater than 90% of the epithelial cell-esterifying activity was found in the villus cell fraction (as opposed to the crypt cell fraction) in both proximal and distal intestine.[96] The enzyme has been further localized to the microsomal fraction following subcellular fractionation of mucosal scrapings,[51,52,96,97,186] and it is generally assumed that these structures are derived from the endoplasmic reticulum. In one study, the enzyme has been further localized to the rough endoplasmic reticulum.[53]

Cholesterol ester hydrolase from pancreas is also associated with the intestinal mucosa. Its presence there was first inferred from studies in which a decreased cholesterol esterifying capability was found in depancreatized rats[187] or those with diverted pancreatic flow.[50] It has subsequently been demonstrated that antibodies to the pancreatic enzymes of rat react extensively with thin sections of rat intestine.[188] In these experiments, the pancreatic enzyme was found to be uniformly distributed throughout the villus cells.

The functional existence of two distinct enzymes in the intestinal mucosa has recently been demonstrated in the rabbit.[97] Using N-(1-oxo-9-octadecenyl)-DL-tryptophan (Z) ethyl ester to inhibit the microsomal ACAT, it was determined that a cytosolic esterifying activity was not inhibited either in vitro or when the compound was fed to rabbits and subcellular fractions subsequently isolated and assayed.

Field[185] has also measured a cytosolic esterifying activity in rabbit intestine and has shown that the cytosolic, but not microsomal, activity can be greatly reduced by washing or wiping the intestinal segments free of adherent mucus before carrying out subcellular fractionation. Based on this finding the author suggested that the presence of the pancreatic esterase in intestinal subcellular fractions occurs as the result of contamination of the preparation with pancreatic fluid.

As there is some difference of opinion over the origins of the intestinal cholesterol esterifying activity, so there is also some dispute over the relative importance of the enzymes in cholesterol absorption. The esterase responsible for the formation of lymphatic esters was initially suggested[50,187] to be pancreatic cholesterol ester hydrolase (E.C. 3.1.1.13) based on experiments in rats with diverted pancreatic flow. Cholesterol esterifying capability was subsequently found to be reduced in intestinal villus cells isolated from the treated animals[189] in addition to the reduced lymphatic cholesterol transport in these animals.

Almost from the outset of this suggestion, reports appeared showing that the diversion of pancreatic fluid from the intestine did not necessarily influence cholesterol absorption. Several workers suggested the existence of an intrinsic intestinal esterifying enzyme. These early papers have been well reviewed by Friedman et al.[190] and a clear and thorough account of the historical developments concerning the two enzymes is given in the introduction to an article by Heider et al.[97]

A lack of agreement over the relative importance of cholesterol ester hydrolase and ACAT to cholesterol absorption continues to be evident in the current literature. Strong evidence in favor of an absorptive role of pancreatic cholesterase in the rat was obtained recently by Gallo et al.[191,193] Rats were perfused with pancreatic fluid that had been previously treated with an antibody prepared against the pancreatic esterase or deprived of pancreatic juice prior to the measurement of cholesterol absorption. In both groups, the esterification of lymph cholesterol was unaffected, but mucosal cholesterol esterase activity and the mass of cholesterol absorbed were greatly reduced. A further indication of a role of the pancreatic enzyme was obtained by the use of inhibitors of this esterase (diisopropyl fluorophosphate and *p*-bromophenylbromic acid). The compounds decreased the accumulation of cholesterol in evert gut sacs prepared from rat intestine.[194] In confirmation of earlier work using jejunal villus cells isolated from the rat,[189] the addition of purified pig pancreatic cholesterol ester hydrolase stimulated the initial cholesterol uptake in these experiments. An active enzyme was required for the stimulatory effect.[194]

In contrast to the above reports, Watt and Simmonds[183] reported that rats with bile fistulae and diverted pancreatic flow absorbed cholesterol and delivered it to the thoracic duct as well as did those with pancreatic flow when duodenally perfused with micellar solutions containing bile salts, cholesterol, and glycerolipids. As an important control in this experiment, the authors confirmed the efficiency of the pancreatic diversion by establishing the absence of proteolytic activity in the intestinal lumen. As in the previously described experiments of Gallo et al.,[191,193] about 80% of the absorbed cholesterol appearing in the lymph was esterified independently of the flow of pancreatic fluid. The authors concluded that ACAT and not pancreatic cholesterol esterase was responsible for the esterification of cholesterol appearing in the lymph.[183]

Gallo et al.[193] responded to this by showing that the depletion of the esterase by pancreatic diversion or immunoprecipitation in their own experiments was not complete. They further suggested that the residual activity was sufficient to esterify the small amount of cholesterol that was transported into the lymph, without assuming a role of ACAT.

A different conclusion was drawn from studies designed to investigate factors influencing the activity of ACAT. In rats duodenally perfused with lecithin-stabilized triolein emulsions, the ACAT-mediated esterification of cholesterol, assayed, subsequently, in whole mucosal homogenates, was depressed compared to controls.[73] Since these same triolein infusions had previously been shown to lead to reduced lymphatic transport of cholesterol with decreased cholesterol ester content,[72] the authors suggested that the secretion of esterified and unesterified cholesterol may be regulated by ACAT. The uptake of cholesterol by the intestinal mucosa was not directly measured in these experiments, but based on the similarity in the luminal cholesterol content in triolein-infused and control-infused rats, differences in the uptake were not considered significant.

Subsequent work by the same investigators[195] showed that the administration of 3-(decyl dimethylsilyl)-*N*-[2(4-methylphenyl)-1-phenylethyl] propanamide, an inhibitor of ACAT in rats, markedly decreased the content of esterified cholesterol in the mesenteric lymph. The effect was evident after a single bolus dose of cholesterol and also under conditions of steady-state absorption. The level of unesterified cholesterol in the lymph was entirely unaffected by the drug. The authors concluded that mucosal ACAT controls the rate of delivery of cholesterol into lymph. They further speculated that ACAT and pancreatic cholesterol esterase may act in series, with the esterase promoting the uptake into the mucosal cells and with ACAT acting as the major esterifying activity within the mucosal cells prior to secretion of cholesterol into the lymph. This hypothesis is consistent with the stimulating effect of pancreatic esterase on cholesterol uptake in vitro[189,194] and with the presence of cholesterol esters in the lymph of animals deprived of pancreatic secretions.

Further attempts to investigate the role of ACAT by the use of chemicals that inhibit the enzyme in cell culture have been carried out by Heider et al.[97] In rabbits fed a diet of 1% cholesterol-supplemented rabbit chow, *N*-(1-oxo-9-octadecenyl)-DL-tryptophan (Z) ethyl ester (compound 57-118) inhibited cholesterol absorption (measured by a dual isotope plasma ratio method) in a dose-dependent manner and also reduced the mucosal microsomal ACAT-mediated sterol esterification rate, measured subsequently.

The authors concluded that the compound exerted its hypocholesterolemic effect through a competitive inhibition of ACAT and suggested that the enzyme played an important role in the absorptive process.

Despite the large number of recent reports suggesting a major role of ACAT as a regulator of the rate of intestinal cholesterol absorption,[73,97,183,185,195] the evidence to support these conclusions is not wholly convincing. In addition to the evidence showing a role of pancreatic esterase, one weakness lies in the difficulty in assaying an enzyme that has not yet been isolated in a pure form, but that carries with it cholesterol substrate in amounts that may vary depending on the treatments used in the experiment or the membrane fraction isolated.

The dependence of the ACAT esterification rate on the available cholesterol supply has been directly studied by varying the levels of cholesterol in membranes containing the enzyme.[179,196] Poorthuis and Wirtz,[179] studying the rat liver microsomal ACAT, were able to achieve a saturating cholesterol substrate supply by incubating the enzyme preparation in the presence of cholesterol-egg PC vesicles and nonspecific lipid transfer protein to facilitate the transfer of cholesterol between the two membrane systems. An approximately linear dependence of the esterification rate on the vesicle cholesterol concentration was found up to the point of saturation, where the rate was no longer influenced by the cholesterol content of the membranes.

Many authors have recognized that the substrate supply may influence the rate of esterification mediated by ACAT when measured during dietary studies in which the membrane cholesterol content may be modified.[96,102,197,198] In this regard it is important to note that earlier reports[52] concerning the ACAT-enzyme cited the percentage of microsomal cholesterol converted to ester per unit time, in this way normalizing the results to the membrane substrate supply. Several recent papers investigating the influence of dietary factors on the esterification of cholesterol in the mucosa have not, however, reported the cholesterol contents of the preparations assayed,[73,97,186] and none, to the author's knowledge, has attempted to ensure saturating substrate cholesterol concentrations.

Comparisons of ACAT esterification rates between animals treated with different dietary fats may be subject to a further complication. This is derived from the observation[196,197,199] that ACAT activity may be influenced by the bulk PC composition of the membranes surrounding it. Spector and co-workers[197] found that in rats fed diets of saturated or polyunsaturated fat, the ACAT-mediated rate of cholesterol esterification in liver microsomes was 70 to 90% higher with polyunsaturated fat. In these experiments, there was no change in the cholesterol/phospholipid ratio or phospholipid head group composition induced by the dietary treatments. Similar observations have been reported for the ACAT from Ehrlich cells, solubilized with Triton® X-100 and reconstituted in PC-cholesterol vesicles,[196] and with rat liver microsomes treated in vitro with pure PCs.[199] Dipalmitoyl PC enrichment decreased the ACAT esterification rate in the former study,[196] as had occurred in rabbit intestinal microsomes,[102] but not in an earlier study with rat liver microsomes.[197] The effect has been suggested to be a result of a modified interaction of the phospholipid with ACAT or an influence of the polar lipid composition on the accessibility of substrate cholesterol to the enzyme.[196] Therefore, while there is little doubt that inhibitors of ACAT can reduce the total lymphatic cholesterol delivery in vivo, it is, in many cases, difficult to interpret the significance of esterification rates measured in vitro.

2. Esterification and the Hypocholesterolemic Effect of Plant Sterols

The potential of the plant sterol, sitosterol, to interfere with the esterification of cholesterol in the intestinal mucosa and thus reduce cholesterol absorption was first presented by Hernandez and co-workers (see Pollack and Kritchevsky[3] for an extensive review). In these experiments, rats were fed ^{14}C-cholesterol in cottonseed oil with or without sitosterol. A marked reduction in the lymphatic delivery of cholesterol was found in the rats treated with the plant sterol.

Two articles have recently appeared describing experiments designed to directly test the influence of plant sterols on the ability of microsomal ACAT to esterify cholesterol.

Tavani and co-workers[200] found that the esterification of exogenously added cholesterol by rat liver microsomal ACAT was reduced by 61% when sitosterol or stigmasterol was added to the reaction mixture dispersed in Triton® WR-1339. Ergosterol and lanosterol reduced the esterification of exogenously added cholesterol by 55%. There was no effect of any of the analogs on the esterification of cholesterol endogenous to the microsomes.

A reduction in the amount of cholesterol esters formed following sitosterol treatment has also been reported to occur with rabbit jejunal ACAT in vivo.[102] The rate of formation of cholesterol esters by microsomal ACAT (assayed in vitro) was significantly decreased in animals fed diets containing sitosterol alone at a level of 1% of the diet. There was no significant difference between the rates of esterification in microsomes derived from the intestines of animals feed 1% cholesterol or 1% cholesterol + 1% sitosterol. The authors concluded that these effects could be accounted for by substrate (cholesterol) depletion, but could not rule out an influence of sitosterol on the enzyme itself. Small amounts of membrane sitosterol introduced in vitro, however, apparently did not influence the rate of esterification of cholesterol.

3. Role of Esterification in the Selective Absorption of Cholesterol Analogs

The suggestion that the esterification of cholesterol may occur preferentially over that of plant sterols and other analogs was first made by Swell and co-workers.[107] This conclusion was based on the finding that the ratio of absorbed radioactive cholesterol to sitosterol was much higher in lymph and blood than it was in the intestinal mucosa. This finding has since been confirmed in a large number of laboratories using a wide variety of experimental designs.[19,54,63,109,113] A good account of the development of ideas concerning this suggested role of ACAT in sterol absorption is given by Field and Mathur.[102]

Several laboratories have attempted to detect a preferential esterification of cholesterol over the plant sterols directly using semipurified enzyme preparations. Early experiments using an acetone powder of intestine demonstrated[201] that cholesterol and sitosterol were esterified to an equal extent when presented to the enzyme suspension dispersed in ethanol with oleic acid. Similar results were found when the sterols were solubilized with bile salts, indicating that the esterification was not the result of ACAT. The ACAT is known to be inhibited by bile salts.[52] In recent studies[200] with the ACAT in rat liver microsomes, cholesterol was esterified preferentially over a wide variety of other sterols when the substrates were solubilized in Triton® WR-1339. Sitosterol, stigmasterol, and 3-epicholesterol were esterified to a level of less than 5% of that of cholesterol. Cholestanol and lathosterol were both efficiently esterified (70 and 41% of the level obtained with cholesterol, respectively). Esterification of exogenously added sitosterol has also been reported[102] to be slow with rabbit intestinal ACAT.

In these studies, the accessibility of the substrate to the enzyme must be considered as a possible influence. In the latter study,[102] for example, a 60-fold greater rate of microsomal esterification of cholesterol (^{14}C) over sitosterol (^{14}C) was reported, but after a 4-hr preincubation with liposomes containing dipalmitoyl-PC and sitosterol (^{14}C), only 6% of the microsomal sterol had been replaced by (^{14}C) plant sterol. Of this, it is not known how much sterol was functionally incorporated into the membrane or accessible to the enzyme. Similarly, in the study of Tavani et al.,[200] no estimate of the amount of functionally accessible sterol was reported.

Finally, it is of interest to mention that a loss of the ability of intestinal ACAT to esterify cholesterol preferentially over sitosterol has been suggested[202,203] to be the basic defect underlying the disease, β-sitosterolemia. Patients displaying this syndrome are characterized by high plasma concentrations of plant sterols arising through a higher-than-normal intestinal absorption of these dietary components.[203]

IV. MECHANISTIC ASPECTS OF INTESTINAL STEROL ABSORPTION

A. Initial Uptake to Lymph Delivery

There is evidence that the transfer of sterols from the carrier particle or aggregate in the intestinal lumen to the brush border surface occurs through monomers, or small aggregates, solvated in the aqueous phase. The absorption need not necessarily occur in this form, however, as there is also evidence to show that an accumulation of lipid at the membrane surface may precede the actual entry into the cell under some conditions.[156,204,205] Partition, in this case, would then occur between a layer of sterol on the membrane surface and the hydrocarbon interior of the membrane.

How this occurs is purely speculative at this point, but it would appear consistent with known facts to suggest that the sterol molecule initially enters the membrane via its side chain rather than its hydroxyl end. This would be compatible with the cone-shaped form of cholesterol (Figure 2) and would also allow for the removal of the hydrocarbon tail from the aqueous environment while allowing the polar end to remain in contact with water molecules. This has previously been suggested[127,154] to account for the inability of plant sterols, having bulkier side chains than cholesterol (Figure 2), to enter the brush border membrane. It was argued that the influence of the side chain would be predictably greater if it entered the membrane first, as it was difficult to rationalize a large effect of side chain alkylation on the further transport of a sterol molecule already embedded in the membrane hydroxyl end first.

The structural features influencing the rate of sterol absorption appear to be different from those required for a condensing interaction of sterol and phospholipid. It has previously been shown that 3-keto cholesterol and vitamin D_3 are efficiently absorbed by intestinal preparations,[58,127] but a 3β-OH group and an intact planar ring system are required for a strong interaction with PC.[206,207] Side chain alkylation, on the other hand, does not prevent strong interaction between sterol and PC,[144,208] but does seriously hamper intestinal absorption. These results indicate that the initial uptake of sterol is not governed by sterol-phospholipid interactions of the type that exists when the two lipids are cosolubilized as in a membrane bilayer.

Following initial entry of the sterol into the brush border membrane, it is reasonable to assume that some sort of "flip-flop" occurs which reorients the molecule at the inner surface of the membrane bilayer. Evidence that the rate of flip-flop is fast compared to the rate of entry and exit of cholesterol in the brush border membrane has been provided by Bloj and Zilversmit.[153] These workers convincingly demonstrated that while there was more than one environment for the cholesterol contained in vesicles derived from the brush border membrane of rabbit intestine, there was no evidence to suggest the existence of a pool of cholesterol sequestered in the inner monolayer of the vesicle membrane. The result suggested that the transbilayer movement was fast compared to the exchange process.

The desorption of sterols from the inner half of the microvillar membrane bilayer may be expected to be a slow event based on evidence derived from experiments using other membrane systems, such as erythrocytes and PC-cholesterol liposomes. The release of a sterol from interactions occurring within an essentially hydrocarbon interior of the membrane into an aqueous medium may by all accounts be considered to be an energetically unfavorable event. A role of cytosolic cholesterol transfer proteins in facilitating this desorption cannot be excluded.

An alternative hypothesis to explain the manner in which cholesterol enters the villus cell has recently been published by Bhat and Brockman.[194] The hypothesis arose from an observed enhancement of cholesterol uptake into evert gut sacs by porcine pancreatic cholesterol esterase. The stimulation required an active enzyme, but did not involve the direct transfer of cholesterol esters into the intestinal structure. To rationalize these apparently contradictory observations, the authors proposed that the enzyme acts catalytically at the lumenal surface of the villus cell membrane to convert a small amount of absorbed sterol into cholesterol ester. It was suggested that the ester may then partition into the membrane and flip-flop within the hydrocarbon interior. An esterase acting at the cytosolic face of the membrane was thought to bring about the hydrolysis of the membrane-bound esters freeing cholesterol to diffuse into the cell.

On entry of the sterol into an intestinal villus cell, the absorbed sterol appears to be rapidly distributed throughout the intracellular membranes in a manner that is in some way related to the phospholipid content of each membrane. There is evidence from in vitro studies with isolated rat jejunal villus cells[158] that these internal membranes can double or triple their total sterol contents during the process of absorption, thus providing a sizable holding capacity for absorbed material. Borgstrom,[209] in his studies of rats on a high cholesterol diet, did not find increased mucosal sterol contents, but Hietanen and Laitinen[210] and Glover and Green[47] found markedly increased levels in their long-term feeding studies.

The concept of the intestinal mucosa as a holding "tank" for cholesterol awaiting packaging into lipoproteins would require that initial entry of sterol into the intestinal wall occurs quickly relative to the subsequent secretion into the lymphatic system. Data plotted in Figure 3 indicates that this is indeed the case. As a corollary to this idea the levels of villus cell sterol achieved during previous dietary regimens would be expected to be a major determinant of the amount of cholesterol that can subsequently be absorbed. There is substantial evidence to indicate that a larger fraction of dietary cholesterol is absorbed by the intestine when the previous dietary level has been low than when it has been high.[2,211,212] It is too early to conclude, however, that the phospholipid content of the villus cell membranes, or the sterol/phospholipid ratio, determines this capacity.

With the accumulation of absorbed sterol within the jejunal or ileal villus cells arises the problem of back diffusion or the dissipation of the concentration gradient attracting sterol into the cell. It is here that intracellular sterol esterification may be expected to play a major role. There is good evidence to show that about one half of the cholesterol newly synthesized in the intestine in a fasting rat is secreted into the lymph, while the other half diffuses into the intestinal lumen unesterified.[213] Esterification may therefore be required for the accumulation of sterol prior to lipoprotein assembly and at the same time for the maintenance of a low intracellular level of free cholesterol and thus, the concentration gradient. Based on the earlier discussion of the action of ACAT, this enzyme may be expected to be stimulated in a concentration-dependent manner to bring about this esterification. In this way, intracellular esterification would determine the bulk of cholesterol secreted into the lymph while not ruling out the possibility of the lymphatic transport of free cholesterol.

The factors influencing, or giving rise to, the levels of cholesterol in intracellular membranes may also be closely related to the intracellular esterification. Wattenbrug and Silbert[165] have studied the exchange of cholesterol between cellular membranes isolated from rat hepatocytes and have concluded that the lipid composition dictates their cholesterol content to some extent, but can in no way account for all of the differences. They suggested that the distribution of cholesterol between membranes was likely further influenced by ongoing cellular metabolism, i.e., a balance between cholesterol synthesis and depletion. Intracellular cholesterol esterification would be one factor acting to reduce the levels of intracellular cholesterol over and above the level dictated by the composition of the membrane itself.

The esterification is generally believed to occur in the endoplasmic reticulum. It has been reported that the accumulation of esters occurs in the smooth surface endoplasmic reticulum, while the ACAT enzyme itself is localized in the rough endoplasmic reticulum. This raises the interesting possibility that ACAT itself acts, not on cholesterol present in membranes directly surrounding it, but that present in smooth-surfaced structures. The idea that smooth membranes act as a "sink" for cholesterol ester was proposed by Hashimoto and Fogelman.[53] It has previously been suggested that cholesterol may be esterified in the smooth endoplasmic reticulum before passing to the apical vesicles where large accumulations of the esters were found to occur.[49]

Throughout the previous discussion in this chapter, an attempt was made to identify rate-limiting steps in the absorption of sterol by the intestine. In Sections III.A and III.B it was established that major barriers to the initial entry of sterol into the intestinal mucosa are presented by the brush border membrane or the layer of still water surrounding it. It is clear that the concentration of sterol reaching that membrane in an absorbable form is of paramount importance in determining the initial uptake rate. From the data in Figure 3, however, we know that the transfer of absorbed sterol to the lymph occurs much more slowly than the initial entry into the cell. Thus, the process that is rate limiting to the delivery into lymph occurs after the initial uptake. This slow step is probably not the esterification of absorbed sterol. Figure 3 shows that sterol can be esterified very soon after entry into mucosal villus cells and much earlier than the beginning of lymph delivery. It therefore appears likely that either the accumulation of a critical mass of sterol, the assembly of lipoproteins containing the absorbed sterol, or the subsequent release from the cell is the slow step limiting the overall rate of absorption. Why the transport of triglyceride in similar lipoproteins should occur faster than that of cholesterol is still unclear, but the slower exchange rate of cholesterol between membranes, compared to triglyceride precursors, may underlie this phenomenon.

B. The Preferential Absorption of Cholesterol over Other Analogs

To aid in the understanding of the selectivity expressed during the intestinal absorption of sterols, it is important to establish where it occurs in the process. From the analysis of the selectivity given earlier (Sections I.D, III.A.2, III.B.1, and III.D.3), it is evident that there are three probable areas in which an absorptive recognition of cholesterol could occur: one during desorption from micelles, the second during entry into the cell, and a third occuring later in the process, perhaps as a result of sterol esterification assembly and secretion of lipoproteins by the villus cells.

Sjostrand and Borgstrom[49] fractionated the intestinal mucosa of rats that had ingested tracer doses of radioactive cholesterol and sitosterol and found that the ratio of the two sterols was the same in the endoplasmic reticulum and apical vesicle fractions as it was in the brush border. This was evidence of a selective process occurring early in the sequence of absorptive events. Bhattacharyya has suggested[116] that this early event may occur during the partition of the sterols out of micelles. There is direct evidence that the micellization process itself may, under some conditions lead to the preferential solubilization of cholesterol. Selectivity occurring at this stage would, therefore, enhance any that may occur during the entry of the sterol into the mucosal wall.

An absorptive preference for cholesterol occurring during uptake by the brush border membrane has been directly demonstrated in vitro[127,154] and has been established to be a general property of a variety of other membrane types. Edwards and Green[157] demonstrated that a more rapid exchange of campesterol, compared to sitosterol, between liposomes and erythrocytes was markedly influenced by the species from which the cells were derived. The greatest differences in the exchange rate of the two sterols were found in ox erythrocytes and the authors suggested that the high content of sphingomyelin in these cells may have been involved. In these experiments, the differences in selectivity could not have occurred during the release of the sterols from the liposomes, which were the same in each case. Child and Kuksis[127,154] have similarly demonstrated that the selectivity expressed during the uptake of 7-dehydrocholesterol, 7-dehydrocampesterol, and 7-dehydrositosterol from micellar solutions is greatly influenced by the type of acceptor membranes used.

Evidence for selective transformations occurring within the absorptive villus cells is based on the finding that the absorptive preference for cholesterol expressed in lymph or blood is generally greater than that found in the intestinal wall (Table 6). The selectivity displayed by membrane preparations in vitro can, however, approach that seen in lymph or blood if nonselectively adsorbed sterol is taken into account. The difference in the ratio of cholesterol/sitosterol measured in the mucosa and in the lymph may, therefore, simply reflect sterol adsorbed to the mucosal surface. This remains to be established in vivo by direct experiment.

Bikhazi and Higuchi[214] postulated that the inability of sitosterol to enter the brush border membrane may be the result of a difficulty in crossing the interfacial barrier existing between the aqueous phase and the membrane surface. These workers were able to demonstrate that sitosterol is transferred more slowly than cholesterol between oil droplets and an aqueous bulk phase. Support for this concept is derived from the wide variety of cell membrane systems that display a similar absorptive preference for cholesterol. This phenomenon has been described in vitro in isolated villus cells,[154] brush border membranes,[154] a wide variety of erythrocyte types,[154,156,157] fibroblasts in culture,[162] the intestine of nematodes,[160] in addition to oil droplets, and it may be justifiably argued that these absorptive structures have little in common except an aqueous-lipid interface.

The ability of metabolic poisons to abolish the selective preference for cholesterol in the rat may argue against this hypothesis unless we assume that the inhibitors in some way disrupt the membrane and, thus, the interface. Some experimental results in vitro also argue against the interfacial barrier hypothesis. In one group of studies[154,156] the absorptive preference for cholesterol over the plant sterols was not evident during the initial accumulation of sterol in rat erythrocytes or brush border membrane vesicles, but increased in a time-dependent fashion during incubation of the membranes in micellar solutions. The initial nonselective uptake of sterol was suggested to represent sterol adsorbed to the outer surface of the membrane. The selectivity was thought to occur during subsequent partition of this material into the membrane interior. The initial approach of the sterols to the membrane could not, in this case, have been selective as required by the interfacial resistance hypothesis.

Other authors have postulated that the selective uptake is initiated in the hydrocarbon interior of the membrane. Barton and co-workers[215] suggested that the inability of sitosterol to enter the intestinal wall was the result of an inability to fit between the acyl chains of the membrane phospholipid. Sylven[216] expanded this concept by postulating the existence of lipid-soluble spaces within the cell membrane into which sterols must fit. He went on to suggest that metabolic energy was required to maintain these spaces and that sitosterol was simply not the correct shape to be accepted into them. This concept was developed further by Child and Kuksis,[127,154] who suggested that the selectivity may occur during entry of the sterols into the membrane, possibly during passage of the side chain through the region of the membrane occupied by the ring systems of membrane cholesterol. This area of the membrane is thought to show the highest degree of acyl chain ordering[217] and was expected to be the most sensitive to variations in side chain bulk or stereochemistry. There is as yet little experimental support for this concept except for the observation that the selective preference for cholesterol does seem to be enhanced in those membranes having high contents of sphingomyelin or high cholesterol/phospholipid ratios.[154,157]

Perhaps one of the newest and most exciting discoveries concerning this problem is the demonstration by Clejan and Bittman[218] that the transbilayer movement of sitosterol in *Acholeplasma laidlawii* is slower than that of cholesterol. The transbilayer "flip-flop" was estimated on the basis of the amount of filipin-reactable sterol in the cells following their initial uptake. The implication was that, after initial entry of sitosterol into the outer membrane of these organisms, an inability to flip into the inner monolayer led to an accumulation in the outer half. The results of these experiments clearly demonstrated a difference in the manner in which sitosterol and cholesterol interact with other membrane components.

The structural features of the sterol molecule itself that are important for the absorptive influence of the side chain have been explored in some detail.[127] It was found that the absorptive preference for sterols with the cholesterol side chain was retained when the hydroxyl group was oxidized, when the ring system was oxidized to the 7-dehydro derivatives, or when ring B was photolytically cleaved to give the vitamin D analogs. Oxidative removal of rings A and B, leaving only a keto group at position 8 of ring C, abolished the absorptive preference for the cholesterol analogs. In this experiment the C_{18}, C_{19}, and C_{20} nor-AB derivatives of cholesterol, campesterol, and sitosterol were absorbed by brush border membrane vesicles at the same rates. It was suggested that the bulk of the intact four-membered ring system forced the side chain into intimate contact with phospholipid acyl chains in the membrane and prevented the plant sterol analogs from shifting into an orientation that would allow their side chains to enter the membrane more easily.

C. Hypocholesterolemic Effects of Plant Sterols

The influence of dietary plant sterols on the absorption of cholesterol may be manifest in any of a variety of transformations, for example, through a competition with the 27-carbon sterol for micellar space, through an influence on the brush border membrane acting to reduce the entry of cholesterol, through the enhanced excretion of mucosal cholesterol, or through a blockage of the esterification of cholesterol or subsequent assembly of lipoproteins. There are undoubtedly others.

There is good evidence to suggest that the presence of sitosterol or other plant sterols or their analogs can reduce the amount of cholesterol that can be accommodated in bile salt micelles. This has been demonstrated in model micelles[99,147,149] and in the intestinal content of animals undergoing sterol absorption.[99] If it is assumed that absorption occurs from a micellar phase in the intestinal lumen, the potential for sitosterol to interfere with cholesterol uptake at this stage is large.

Recent evidence[99] suggests that the absorptive influence of the plant sterols is only manifest under conditions where the micelles are nearly saturated with sterol and the sterols are intimately mixed. The coadministration of sitosterol-containing micelles in loops of rat intestine *in situ* had no effect on the uptake of cholesterol from micelles containing that sterol in these experiments. The situation is further complicated, however, by the possibility that such micelles may not be the only substrates for sterol absorption under normal conditions. The existence of large aggregates of sterol and phospholipid has been detected in human bile[138] and these may have a liposome-like structure. The absorption of cholesterol from structures such as these, if they indeed exist, would be expected to be little affected by the administration of dietary sitosterol in a crystalline form. This factor may underlie the variable success that is evident[3] in reports of the use of phytosterols to reduce plasma cholesterol levels in human patients.

An interesting aspect of the hypocholesterolemic effect of the plant sterols is the apparent independence of this effect and the inability of the phytosterols to be absorbed. This was initially indicated in the work of Sylven and Borgstrom,[63] who demonstrated that although sitosterol was not absorbed by the rat intestine as readily as cholesterol, the plant sterol exerted no influence over the absorption of cholesterol. This has recently been confirmed by Ikeda and Sugano.[99] Vahouny and co-workers[100] found that fucosterol, sitosterol, and stigmasterol were all poorly absorbed by rat intestine, but only sitosterol and stigmasterol exerted any influence over the absorption of cholesterol. Thus, we can conclude with some certainty that the inability of a sterol to be absorbed does not necessarily mean that it will interfere with the absorption of others.

As a corollary to this argument, it is likely that the plant sterols do not interfere with cholesterol absorption by modification of the brush border membrane or by the occupation of sites at the surface of, or within, that structure. Blockage of cholesterol uptake by an accumulation of nonabsorbed sterol at the membrane interface is clearly not compatible with the above-mentioned results. A hypocholesterolemic effect based on the entry of sitosterol into the brush border membrane is, furthermore, not compatible with the results of Sylven and Borgstrom.[63] There was no inhibition of cholesterol absorption in the rat when sitosterol was administered in triolein in a wide range of dietary concentrations, although under these conditions, some absorption of sitosterol was known to occur.

The small amount of sitosterol absorbed by the mammalian intestine may, under some conditions, influence intracellular cholesterol levels in the mucosa. Sitosterol has been shown to accumulate in rat liver and reduce the level of hepatocyte cholesterol.[219] There is similar evidence to suggest that sitosterol may reduce the amounts of microsomal cholesterol in rabbit jejunal villus cells following long-term feeding.[102] This effect was thought to be responsible for a reduced rate of cholesterol esterification in microsomes subsequently isolated from sitosterol-fed rabbits. It is entirely possible, however, that a simple reduction of initial cholesterol uptake by sitosterol may also lead to a depletion of mucosal cholesterol through the normal back-diffusion of cholesterol into the lumen. The entry of sitosterol into the villus cells would therefore not be essential for this effect.

V. SUMMARY AND CONCLUSIONS

The research work of the past few years contains many consistent findings confirming older observations and many new developments contributing to our overall understanding of sterol absorption.

Among the factors influencing the absorption of cholesterol, there is consistent evidence that unsaturated triacylglycerols promote more rapid absorption than do the totally saturated glycerolipids. There is general agreement that cholic and chenodeoxycholic acids added to standard chow diets do not influence the rate of cholesterol uptake or subsequent lymphatic delivery. On the other hand, cholestyramine and colestipol which bind bile acids in the lumen consistently reduce cholesterol absorption. Pectin and cellulose consistently reduce cholesterol uptake in rats, while wheat bran is not effective. The plant sterol, sitosterol has been found to reduce cholesterol uptake in rats, but apparently only when the cholesterol-containing micelles are nearly saturated with sterol. Otherwise, the uptake of the two sterols occurs without mutual interference as has been observed several times in the past. There is no evidence to suggest a direct influence of sitosterol on the brush border membrane or through an accumulation of the plant sterol at the aqueous-membrane interface as a rationale for the hypocholesterolemic effect of the plant sterol.

The absorptive mechanism is extremely sensitive to the structural features of the sterol molecule itself. Alkylation at position 24 of the side chain greatly reduces the absorbability of the sterol in vivo. This has occurred in all recorded cases except those where intestinal epithelial cell metabolism has been disrupted, or in vitro where there has been reason to suspect damage to the absorptive membrane. In the latter case, the selectivity was restored with the adoption of procedures designed to avoid membrane damage, such as the inclusion of PC into the bile salt-containing sterol carrier micelles or the use of phospholipid liposomes as the sterol donor.

Unsaturation in the side chain, in itself, has little influence on the rate of absorption unless it holds an alkylated side chain rigid, as in ergosterol or stigmasterol — two of the most poorly absorbed sterols. The absorptive system is sensitive to reduction of the Δ^5 double bond in ring B, but further oxidation of this ring or complete cleavage, as in the D vitamins, has little effect. There is some evidence to suggest that the disruption of ring B reduces the ability of the absorptive system to recognize alkylation in the side chain. In all cases, cholesterol itself represents the most efficiently absorbed structure, and there has been no structural feature reported to date that increases the rate of its absorption by the mammalian intestine.

Considering the mechanistic aspects of the absorptive process, it is clear that the concentration of sterol at the brush border surface at any one time is an important determinant of the rate of uptake. Some in vitro evidence suggests that the diffusion of micelles across a layer of unstirred water surrounding the villi may be rate limiting to initial uptake. Other evidence, however, indicates that it is too early to conclude that this is a major barrier in all cases, or is a situation which occurs under normal dietary conditions in mammals. The uptake of lipids from other nonmicellar vehicles must also be considered as well as the barrier properties of the brush border surface.

Direct studies of purified brush border membranes have provided evidence that some of the cholesterol contained in this structure is tightly bound to membrane lipid or protein and may not be available for exchange during the absorptive process. Other studies have shown that purified brush border membrane vesicles in vitro are capable of bringing about the absorptive recognition of cholesterol that is a hallmark of absorption in the intact intestine. This capability is, however, also displayed by cell types that are not of intestinal origin, indicating that it may be a general property of membranes. In light of this it appears unlikely that the absorptive recognition of cholesterol requires an intestine-specific cholesterol receptor or transfer protein. It is more likely the result of physicochemical interactions between the incoming sterol molecule and membrane lipids or the water-lipid interface formed by them. This principle may also underlie selectivity occurring during desorption of sterols from micelles and esterification by ACAT, as well as during their passage through membranes like the brush border.

Transfer of the sterol from the brush border membrane to the internal membranes of the intestinal epithelial cells is perhaps the least understood aspect of the absorptive process. Proteins known to have the potential to facilitate this transfer exist in the epithelium of rat intestine, but their involvement has not been demonstrated. The spontaneous exchange or transfer of sterols between phospholipid-containing membranes is a well-documented phenomenon in model systems, but its contribution to intestinal absorption is also unknown.

There is convincing evidence of a role of sterol esterification in the absorptive process. Inhibition or removal of pancreatic cholesterol ester hydrolase can dramatically reduce the lymphatic delivery of cholesterol, possibly by an influence on the initial uptake into the mucosa. Inhibition of the mucosal ACAT enzyme reduces the lymphatic output of esterified cholesterol without affecting the free cholesterol content. The precise role of these enzymes in the process remains unclear, however, and likely will remain so until such time as assay systems for the quantitation of the enzymes themselves are developed and, in the case of ACAT, until the appearance of a method of assay that is not influenced by the cholesterol present in the enzyme-containing microsomal preparations. From the data accumulated to date, it is evident that a certain threshold amount of cholesterol is absorbed and delivered to lymph independently of the action of the two esterifying activities. This raises the possibility that the mucosal esterases function to ensure the temporary localization of large amounts of cholesterol while maintaining a concentration gradient favoring the net influx of sterol into the villus cell. In this view, the esterases would directly influence the amount of cholesterol available for lymphatic delivery without necessarily governing whether a given cholesterol molecule is transferred or not.

The absorption of sterols by the mammalian intestine is an exceedingly complex concert of events that has not yet yielded to attempts to reduce its secrets to a few simple transformations. We still do not fully understand the manner in which other dietary components are able to diminish or enhance the rate of absorption, and we know even less about the exquisite selectivity that allows the intestine to preferentially take up cholesterol over a wide variety of other closely similar analogs.

Despite the remaining gaps in our understanding it is encouraging to note the increase in our knowledge since Schonheimer and colleagues reported,[1] 50 years ago, that the ingestion of cholesterol increased hepatic sterol content in mice. With the advent of a battery of new investigative tools, including methods for the rapid purification of brush border membranes and the development of inhibitors of the mucosal esterases, it is not difficult to imagine that great steps toward an understanding of intestinal sterol absorption will now occur rapidly.

ACKNOWLEDGMENTS

The author wishes to thank Margaret Magee for typing the manuscript. Financial assistance from the Ludwig Foundation and the Medical Research Council of Canada is gratefully acknowledged.

REFERENCES

1. Schonheimer, R., Behring, H. V., and Hummel, R., Uber die spezifitat der Resorption von Sterinen, abhangig von ihrer Konstitution, *Hoppe Seyler's Z. Physiol. Chem.*, 192, 117, 1930.
2. Thomson, A. B. R. and Dietschy, J. M., Intestinal lipid absorption: major extracellular and intracellular events, in *Physiology of the Gastrointestinal Tract*, Vol. 2, Johnson, L. R., Ed., Raven Press, New York, 1981, 1147.
3. Pollack, O. J. and Kritchevsky, D., Sitosterol, in *Monographs on Atherosclerosis*, Vol. 10, Clarkson, T. B., Kritchevsky, D., and Pollack, O. J., Eds., S. Karger, Basel, 1981.
4. Fieser, L. F. and Fieser, M., *Steroids,* Reinhold Publishing, New York, 1969.
5. Nes, W. R., Krewitz, K., and Behzadan, S., Configuration at C-24 of 24-methyl and 24-ethyl cholesterol in tracheophytes, *Lipids,* 11, 118, 1976.
6. Kuksis, A., Intestinal digestion and absorption of fat-soluble environmental agents, in *Intestinal Toxicology,* Schiller, C. M., Ed., Raven Press, New York, 1984, 69.
7. Johnston, J. M., Mechanism of fat absorption, in *Handbook of Physiology,* Sect. 6, Vol. 3, *Intestinal Absorption,* Code, C. F., Ed., American Physiological Society, Washington, D.C., 1968, 1353.
8. Brindley, D. N., The intracellular phase of fat absorption, in *Biomembranes,* Vol. 4B, *Intestinal Absorption,* Smyth, D. H., Ed., Plenum Press, London, 1974, 621.
9. Simmonds, W. J., Absorption of lipids, in *Gastrointestinal Physiology,* Jacobson, E. D. and Shonbour, L. L., Eds., University Park Press, Baltimore, 1974, 343.
10. Rommel, K., Goebell, H., and Bohmer, R., *Lipid Absorption: Biochemical and Clinical Aspects,* University Park Press, Baltimore, 1976.

11. Borgstrom, B., Digestion and absorption of lipids, in *International Review of Physiology, Gastrointestinal Physiology II,* Crane, R. K., Ed., University Park Press, Baltimore, 1977, 305.

12. Johnston, J. M., Esterification reactions in intestinal mucosa, in *Disturbances in Lipid and Lipoprotein Metabolism,* Dietschy, J. M., Gotto, A. M., and Ontko, J. A., Eds., American Physiological Society, Bethesda, Md., 1978, 57.

13. Clement, J., Intestinal absorption of triglycerols, *Reprod. Nutr. Dev.,* 20(4B), 1285, 1980.

14. Friedman, H. Y. and Nylund, B., Intestinal fat digestion, absorption, and transport, *Am. J. Clin. Nutr.,* 33, 1108, 1980.

15. Glickman, R. M., Intestinal fat absorption, in *Current Concepts in Nutrition,* Vol. 9, *Nutrition and Gastroenterology,* Winnick, M., Ed., Wiley-Interscience, New York, 1980, 29.

16. Shiau, Y. F., Mechanisms of intestinal fat absoprtion, *Am. J. Physiol.,* 240, G1, 1981.

17. Trier, J. S. and Madura, J. L., Functional morphology of the mucosa of the small intestine, in *Physiology of the Gastrointestinal Tract,* Vol. 2, Johnson, L. R., Ed., Raven Press, New York, 1981, 925.

18. Borgstrom, B., Dahlqvist, A., Lundh, G., and Sjovall, J., Studies of intestinal digestion and absorption in the human, *J. Clin. Invest.,* 36, 1521, 1957.

19. Schachter, D., Finkelstein, J. D., and Kowarski, S., Metabolism of vitamin D. Preparation of radioactive vitamin D and its intestinal absorption in the rat, *J. Clin. Invest.,* 43, 787, 1964.

20. Sylven, C. and Nordstrom, C., The site of absorption of cholesterol and sitosterol in the rat small intestine, *Scand. J. Gastroenterol.,* 5, 57, 1970.

21. McIntyre, N., Kirsch, K., Orr, J. C., and Isselbacher, K. J., Sterols in the small intestine of the rat, guinea pig and rabbit, *J. Lipid Res.,* 12, 336, 1971.

22. Thompson, A. B. R., Influence of site and unstirred layers on the rate of uptake of cholesterol and fatty acids into rabbit intestine, *J. Lipid Res.,* 21, 1097, 1980.

23. Chow, S. L. and Hollander, D., Initial cholesterol uptake by everted sacs of rat small intestine. Kinetic and thermodynamic aspects, *Lipids,* 13, 239, 1978.

24. Roy, T., Treadwell, C. R., and Vahouny, G. V., Comparative intestinal and colonic absorption of [4-¹⁴C]cholesterol in the rat, *Lipids,* 13, 99, 1978.

25. Shiratori, T. and Goodman, D. S., Complete hydrolysis of dietary cholesterol esters during intestinal absorption, *Biochim. Biophys. Acta,* 106, 625, 1965.

26. David, J. S. K., Malathi, P., and Ganguly, J., Role of the intestinal brush border in the absorption of cholesterol in rats, *Biochem. J.,* 98, 662, 1966.

27. Vahouny, G. V. and Treadwell, C. R., Absolute requirement for free cholesterol for absorption by rat intestinal mucosa, *Proc. Soc. Exp. Biol. Med.,* 116, 496, 1964.

28. Treadwell, C. R., Swell, L., and Vahouny, G. V., Factors in sterol absorption, *Fed. Proc.,* 21, 903, 1962.

29. Borgstrom, B., A note on the intestinal absorption of cholesterol ethers in the rat, *Proc. Soc. Exp. Biol. Med.,* 127, 1120, 1968.

30. Glover, J. and Morton, R. A., The absorption and metabolism of sterols, *Br. Med. Bull.,* 226, 1958.

31. Siperstein, M. D., Chaikoff, I. L., and Reinhard, W. O., Obligatory function of bile in intestinal absorption of cholesterol, *J. Biol. Chem.,* 198, 111, 1952.

32. Swell, L., Trout, E. C., Hopper, J. R., Field, H., and Treadwell, C. R., Specific function of bile salts in cholesterol absorption, *Proc. Soc. Exp. Biol. Med.,* 98, 174, 1958.

33. Vahouny, G. V., Woo, C. H., and Treadwell, C. R., Quantitative effects of bile salt and fatty acid on cholesterol absorption in the rat, *Am. J. Physiol.,* 193, 41, 1958.

34. Mercer, E. I. and Glover, J., The interconversion of cholesterol, 7-dehydrocholesterol and lathosterol in the rat, *Biochem. J.,* 80, 552, 1961.

35. Borja, C. R., Vahouny, G. V., and Treadwell, C. R., Role of bile and pancreatic juice in cholesterol absorption and esterification, *Am. J. Physiol.,* 206, 223, 1964.

36. Dietschy, J. M., The uptake of lipids into the intestinal mucosa, in *Physiology of Membrane Disorders,* Andreoli, T. E., Hoffman, T. F., and Fanestil, D. D., Eds., Plenum Press, New York, 1978, 577.

37. Patton, J. S., Gastrointestinal lipid digestion, in *Physiology of the Gastrointestinal Tract,* Vol. 2, Johnson, L. R., Ed., Raven Press, New York, 1981, 1123.

38. Hofmann, A. F. and Borgstrom, B., Physico-chemical state of lipids in intestinal content during their digestion and absorption, *Fed. Proc.,* 21, 43, 1962.

39. Wilson, F. A. and Dietschy, J. M., Characterization of bile salt absorption across the unstirred water layer and brush border in the rat jejunum, *J. Clin. Invest.,* 51, 3015, 1972.

40. Simmonds, W. J., The role of micellar solubilization in lipid absorption, *Aust. J. Exp. Biol. Med. Sci.,* 50, 403, 1972.

41. Westergaard, H. and Dietschy, J. M., The mechanism whereby bile acid micelles increase the rate of fatty acid and cholesterol uptake into the intestinal mucosal cell, *J. Clin. Invest.,* 58, 97, 1976.

42. Small, D. M., Physiochemical studies of cholesterol gallstone formation, *Gastroenterology*, 52, 607, 1967.

43. Muller, K., Structural dimorphism of bile salt/lecithin mixed micelles. A possible regulatory mechanism for cholesterol solubility in bile? X-ray structure analysis, *Biochemistry*, 20, 404, 1981.

44. Porter, K. R., Independence of fat absorption and pinocytosis, *Fed. Proc.*, 28, 35, 1969.

45. Murthy, S. K., David, J. S. K., and Ganguly, J., Some observations on the mechanism of absorption of cholesterol in rats, *Biochim. Biophys. Acta*, 70, 490, 1963.

46. David, J. S. K., Malathi, P., and Ganguly, J., Role of the intestinal brush border in the absorption of cholesterol in rats, *Biochem. J.*, 98, 662, 1966.

47. Glover, J. and Green, C., Studies on the absorption and metabolism of sterols: mode of absorption, in *Biochemical Problems of Lipids*, Popjak, G. and LeBreton, E., Eds., Butterworths, London, 1955, 359.

48. Borgstrom, B., Lindle, B. A., and Wlodawer, P., Absorption and distribution of cholesterol-4-¹⁴C in the rat, *Proc. Soc. Exp. Biol. Med.*, 99, 365, 1958.

49. Sjostrand, F. S. and Borgstrom, B., The lipid components of the smooth-surfaced membrane-bound vesicles of the columnar cells of the rat intestinal epithelium during fat absorption, *J. Ultrastruct. Res.*, 20, 140, 1967.

50. Swell, L., Byron, J. E., and Treadwell, C. R., Cholesterol esterase of rat intestinal mucosa, *J. Biol. Chem.*, 186, 543, 1960.

51. Haugen, R. and Norum, K. R., Coenzyme-A-dependent esterification of cholesterol in rat intestinal mucosa, *Scand. J. Gastroenterol.*, 11, 615, 1976.

52. Norum, K. R., Lilljeqvist, A. C., Helgerud, P., Normann, E. R., Mo, A., and Selbekk, B., Esterification of cholesterol in human small intestine: the importance of acyl-CoA:cholesterol acyltransferase, *Eur. J. Clin. Invest.*, 9, 55, 1979.

53. Hashimoto, S. and Fogelman, A. M., Smooth microsomes. A trap for cholesterol ester formed in hepatic microsomes, *J. Biol. Chem.*, 255, 8678, 1980.

54. Borgstrom, B., Quantitative aspects of the intestinal absorption and metabolism of cholesterol and beta-sitosterol in the rat, *J. Lipid Res.*, 9, 473, 1968.

55. Klein, R. and Rudel, L., Intestinal regulation of cholesterol transport by lymph lipoproteins during absorption, in Atherosclerosis, Abstr. of the 33rd Annual Meeting, 1979, 21.

56. Borgstrom, B., Quantification of cholesterol absorption in mass by fecal analysis after the feeding of a single isotope-labelled meal, *J. Lipid Res.*, 10, 331, 1969.

57. Kudchodkar, B. J., Sodhi, H. S., and Horlick, L., Absorption of dietary cholesterol in man, *Metabolism*, 22, 155, 1973.

58. Thomson, G. R., Ockner, R. K., and Isselbacher, K. J., Effect of mixed micellar lipid on the absorption of cholesterol and vitamin D₃ into lymph, *J. Clin. Invest.*, 48, 87, 1969.

59. Green, C., The transport of sterols across the mucosal cell, in *Biochemical Problems of Lipids*, Frazer, A. C., Ed., Elsevier, Amsterdam, 1963, 144.

60. Wilson, J. D. and Lindsay, C. A., Studies on the influence of dietary cholesterol on cholesterol metabolism in the isotopic steady state in man, *J. Clin. Invest.*, 44, 1805, 1965.

61. Borgstrom, B., Radner, S., and Werner, B., Lymphatic transport of cholesterol in the human being. Effect of dietary cholesterol, *Scand. J. Clin. Lab. Invest.*, 26, 227, 1970.

62. Swell, L., Trout, E. C., Hopper, J. R., Field, H., and Treadwell, C. R., Mechanism of cholesterol absorption. II. Changes in free esterified cholesterol pools of mucosa after feeding cholesterol-4-C¹⁴, *J. Biol. Chem.*, 223, 49, 1958.

63. Sylven, C. and Borgstrom, B., Absorption and lymphatic transport of cholesterol and sitosterol in the rat, *J. Lipid Res.*, 10, 179, 1969.

64. Glickman, R. M., Perotto, J. L., and Kirsch, K., Intestinal lipoprotein formation: effect of colchicine, *Gastroenterology*, 70, 347, 1976.

65. Arreaza-Plaza, C. A., Bosch, V., and Otayek, M. A., Lipid transport across the intestinal epithelial cell. Effect of colchicine, *Biochim. Biophys. Acta*, 431, 297, 1976.

66. Reaven, E. P. and Reaven, G. M., Distribution and content of microtubules in relation to the transport of lipid, *J. Cell Biol.*, 75, 559, 1977.

67. Dietschy, J. M., Mechanisms for the intestinal absorption of bile acids, *J. Lipid Res.*, 9, 297, 1968.

68. Maislos, M., Silver, J., and Fainaru, M., Intestinal absorption of vitamin D sterols: differential absorption into lymph and portal blood in the rat, *Gastroenterology*, 80, 1528, 1981.

69. Lutton, C., Magot, T., and Chevallier, F., Effect of dietary long chain fatty acids on the rates of cholesterol turnover processes, cholesterol origin and distribution in the rat intestinal lumen, *Reprod. Nutr. Rev.*, 20, 1467, 1980.

70. Rukaj, A. and Serougne, C., Effect of excess dietary cystine on the biodynamics of cholesterol in the rat, *Biochim. Biophys. Acta*, 735, 1, 1983.

71. Feldman, E. B., Russell, B. S., Hawkins, C. B., and Forte, T., Intestinal lymph lipoproteins in rats fed diets enriched in specific fatty acids, *J. Nutr.*, 113, 2323, 1983.

72. Clark, S. B., Chylomicron composition during duodenal triglyceride and lecithin infusion, *Am. J. Physiol.*, 235, E183, 1978.

73. Clark, S. B., Mucosal coenzyme A-dependent cholesterol esterification after intestinal perfusion of lipids in rats, *J. Biol. Chem.*, 254, 1534, 1979.

74. Strandberg, T. E., Tilvis, R. S., and Miettinen, T. A., Regulation of cholesterol synthesis in jejunal absorptive cells of the rat, *Scand. J. Gastroenterol.*, 18, 1017, 1983.

75. Shefer, S., Zaki, F. G., and Salen, G., Early morphologic and enzymatic changes in livers of rats treated with chenodeoxycholic acid and ursodeoxycholic acid, *Hepatology*, 3, 201, 1983.

76. Larusso, N. F. and Thistle, J. L., Effect of litholytic bile acids on cholesterol absorption in gall stone patients, *Gastroenterology*, 84, 265, 1983.

77. Sama, C. and Larusso, N. F., Effect of deoxycholic acid, chenodeoxycholic acid and cholic acid on intestinal absorption of cholesterol in humans, *Mayo Clin. Proc.*, 57, 44, 1982.

78. Marteau, C., Reynier, M. O., Crotte, C., Mathieu, S., Gerolami, A., and Mule, A., Action of cholic acid and chenodeoxycholic acid on biliary secretion of the mouse, effect of the addition of beta sitosterol, *Can. J. Physiol. Pharmacol.*, 58, 1058, 1980.

79. Reynier, M. O., Montet, J. C., Gerolami, A., Marteau, C., Crotte, C., Montet, A. M., and Mathieu, S., Comparative effects of cholic, chenodeoxycholic, and ursodeoxycholic acids on micellar solubilization and intestinal absorption of cholesterol, *J. Lipid Res.*, 22, 467, 1981.

80. Carey, M. C., Armstrong, M. J., Mazer, N. A., Igimi, H., and Salvioli, G., Measurement of hydrophilic-hydrophobic balance of bile salts: correlation with physical chemical interactions between membrane lipids and bile salt micelles, in *Bile Acids and Cholesterol in Health and Disease*, Paugmgartner, G., Stiehl, A., and Gerok, W., Eds., MTP Press, Boston, 1983, 31.

81. Tso, P., Balint, J. A., and Rodgers, J. B., Effect of hydrophobic surfactant pluronic L-81 on lymphatic lipid transport in the rat, *Am. J. Physiol.*, 239, G348, 1980.

82. Vahouny, G. V., Satchithanandam, S., Cassidy, M. M., Lightfoot, F. B., and Furda, I., Comparative effects of chitosan and cholestyramine on lymphatic absorption of lipids in the rat, *Am. J. Clin. Nutr.*, 38, 278, 1983.

83. Fujihashi, T., Munekiyo, K., and Meshi, T., Effects of cholestipol hydrochloride on cholesterol and bile acids. Absorption in the rat intestinal tract, *J. Pharmacobio-Dyn.*, 4, 552, 1981.

84. Vahouny, G. V., Roy, T., Gallo, L. L., Story, J. A., Kritchevsky, D., and Cassidy, M., Dietary fibres. III. Effects of chronic intake on cholesterol absorption and metabolism in the rat, *Am. J. Clin. Nutr.*, 33, 2182, 1980.

85. Change, M. L. W., Effect of dietary pectin on esterification and excretion of exogenous cholesterol in rats, *Nutr. Rep. Int.*, 26, 59, 1982.

86. Malinow, M. R., Connor, W. E., McLaughlin, P., Stafford, C., Lin, D. S., Livingston, A. L., Kohler, G. O., and McNulty, W. P., Cholesterol and bile acid balance in Macaca-Fascicularis. Effects of alfalfa saponins, *J. Clin. Invest.*, 67, 156, 1981.

87. Nagata, Y., Ishiwaki, N., and Sugano, M., The mechanism of antihypercholesterolemic action of soy protein and soy protein type amino acid mixtures in relation to the casein counterparts in rats, *J. Nutr.*, 112, 1614, 1982.

88. Joo, C. N., Cho, Y. D., Koo, J. H., Kim, C. W., and Lee, S. J., Biochemical studies on ginseng saponins. XV. The effect of ginseng saponins on the absorption of fats and cholesterol in rats, *Korean Biochem. J.*, 13, 1, 1980.

89. Knehans, A. W., Johanning, G. L., and O'Dell, B. L., Relationship between fibre-rich foods and serum sterols in guinea pigs, *Nutr. Rep. Int.*, 24, 1075, 1981.

90. Gee, J. M., Blackburn, N. A., and Johnson, I. T., Influence of guar gum on intestinal cholesterol transport in the rat, *Br. J. Nutr.*, 50, 215, 1983.

91. Ferezou, J., Sulpice, J. C., Coste, J., and Chevallier, F., Origin of neutral sterols in human feces studied by stable isotope labelling (deuterium and carbon-13). Effect of phytosterols and calcium, *Digestion*, 25, 164, 1982.

92. Shandilya, L. N. and Clarkson, T. B., Hypolipidemic effects of gossypol in cynomolgus monkeys Macaca-fascicularis, *Lipids*, 17, 285, 1982.

93. Hosokawa, T., Sawada, M., and Tamura, G., Activation of cholesterol metabolism and hypocholesterolemic properties of 4-0 methyl ascochlorin in controlled reverse phase feeding rats, *Agric. Biol. Chem.*, 46, 775, 1982.

94. DiPadora, C., Bosisio, E., Cighetti, G., Rovagnati, P., Mazzocchi, M., Colombo, C., and Tritapepe, R., 3-Hydroxy-3-methyl glutaric acid reduces dietary cholesterol induction of saturated bile in hamsters, *Life Sci.*, 30, 1907, 1982.

95. Sklan, D., Effect of high vitamin A or tocopherol intake on hepatic lipid metabolism and intestinal absorption and secretion of lipids and bile acids in the chick, *Br. J. Nutr.*, 50, 409, 1983.

96. Strange, E. F., Suckling, K. E., and Dietschy, J. M., Synthesis and coenzyme A dependent resterification of cholesterol in rat intestinal epithelium — differences in cellular localization and mechanisms of regulation, *J. Biol. Chem.*, 258, 12868, 1983.

97. Heider, J. G., Pickens, C. E., and Kelly, L. A., Role of acyl coenzyme A cholesterol acyl transferase (EC-2.3.1.26) in cholesterol absorption and its inhibition by 57-118 in the rabbit, *J. Lipid Res.*, 24, 1127, 1983.

98. Glueck, C. J., Jandacek, R. J., Subbiah, M. T. R., Gallon, L., Yunker, R., Allen, C., Hogg, E., and Laskarzerski, P. M., Effect of sucrose polyester on fecal bile acid excretion and composition in normal man, *Am. J. Clin. Nutr.*, 33, 2177, 1980.

99. Ikeda, I. and Sugano, M., Some aspects of mechanism of inhibition of cholesterol absorption by beta sitosterol, *Biochim. Biophys. Acta*, 732, 651, 1983.

100. Vahouny, G. V., Connor, W. E., Subramaniam, S., Lin, D. S., and Gallo, L. L., Comparative lymphatic absorption of sitosterol, stigmasterol and fucosterol and differential inhibition of cholesterol absorption, *Am. J. Clin. Nutr.*, 37, 805, 1983.

101. Shidoji, Y., Watanabe, M., Oku, T., Muto, Y., and Hosoya, N., Inhibition of beta sitosterol on intestinal cholesterol absorption in rat using in vivo dual isotope ratio method, *J. Nutr. Sci. Vitaminol.*, 26, 183, 1980.

102. Field, F. J. and Mathur, S. N., Beta sitosterol esterification by intestinal acyl coenzyme A cholesterol acyl transferase and its effect on cholesterol esterification, *J. Lipid Res.*, 24, 409, 1983.

103. Ikeda, I., Kawasaki, A., Samezima, K., and Sugano, M., Anti-hypercholesterolemic activity of beta sitosterol in rabbits, *J. Nutr. Sci. Vitaminol.*, 27, 243, 1981.

104. Vahouny, G. V., Connor, W. E., Roy, T., Lin, D. S., and Gallo, L. L., Lymphatic absorption of shellfish sterols and their effects on cholesterol absorption, *Am. J. Clin. Nutr.*, 34, 507, 1981.

105. Gould, R. G., Absorbability of beta-sitosterol, *Trans. N.Y. Acad. Sci.*, 18(Sec. II), 129, 1955.

106. Dunham, L. W., Fortner, R. E., Moore, R. D., Calp, H. W., and Rice, C. N., Comparative lymphatic absorption of beta-sitosterol and cholesterol by the rat, *Arch. Biochem. Biophys.*, 82, 50, 1959.

107. Swell, L., Trout, E. C., Field, H., and Treadwell, C. R., Intestinal metabolism of ^{14}C-phytosterols, *J. Biol. Chem.*, 234, 2286, 1959.

108. Hassan, A. S. and Rampone, A. J., Intestinal absorption and lymphatic transport of cholesterol and beta-sitosterol in the rat, *J. Lipid Res.*, 20, 646, 1979.

109. Shidoji, Y., Watanabe, M., Oku, T., Muto, Y., and Hosoya, H., Inhibition of beta-sitosterol on intestinal cholesterol absorption in rat using in vivo dual isotope ratio method, *J. Nutr. Sci. Vitaminol.*, 26, 183, 1980.

110. Roth, M. and Favarger, P., La digestibilite des graisses en presence de certain sterols, *Helv. Physiol. Acta*, 13, 249, 1955.

111. Sylven, C. and Borgstrom, B., Influence of blood supply on lipid uptake from micellar solutions by the rat intestine, *Biochim. Biophys. Acta*, 203, 365, 1970.

112. Kuksis, A. and Huang, T. C., Differential absorption of plant sterols in the dog, *Can. J. Biochem. Physiol.*, 40, 1493, 1962.

113. Subbiah, M. T. R., Kottke, B. A., and Carlo, I. A., Uptake of campesterol in pigeon intestine, *Biochim. Biophys. Acta*, 249, 643, 1971.

114. Konlande, J. E. and Fisher, H., Evidence for a non-absorptive antihypercholesterolemic action of phytosterols, *J. Nutr.*, 98, 435, 1969.

115. Sklan, D., Budowski, P., and Hurwitz, S., Effect of soy sterols on intestinal absorption and secretion of cholesterol and bile acids in the chick, *J. Nutr.*, 104, 1086, 1974.

116. Bhattacharyya, A. K., Uptake and esterification of plant sterols by rat small intestine, *Am. J. Physiol.*, 240, G50, 1981.

117. Ivy, A. C., Tsung-Min, L., and Karvinen, E., Absorption of dihydrocholesterol and soya sterols when fed with oleic acid, *Am. J. Physiol.*, 179, 646, 1954.

118. Ivy, A. C., Tsung-Min, L., and Kavinen, E., Absorption of dihydrocholesterol and soya sterols by the rat intestine, *Am. J. Physiol.*, 183, 79, 1955.

119. Vahouny, G. V., Mayer, R. M., and Treadwell, C. R., Comparison of lymphatic absorption of dihydrocholesterol and cholesterol in the rat, *Arch. Biochem. Biophys.*, 86, 215, 1960.

120. Ikeda, I. and Sugano, M., Comparison of absorption and metabolism of beta-sitosterol and betsitostanol in rats, *Atherosclerosis*, 30, 227, 1978.

121. Bell, N. H., Comparison of intestinal absorption and esterification of 4-^{14}C Vitamin D$_3$ and 4-^{14}-C cholesterol in the rat, *Proc. Soc. Exp. Biol. Med.*, 123, 529, 1966.

122. McIntyre, N., Cholesterol absorption, in *Lipid Absorption: Biochemical and Clinical Aspects*, Rommel, K., Ed., University Park Press, Baltimore, 1976, 73.

123. Salen, G., Ahrens, E. H., Jr., and Grundy, S. M., Metabolism of beta-sitosterol in man, *J. Clin. Invest.*, 49, 952, 1970.

124. Quintao, E., Grundy, S. M., and Ahrens, E. H., Effects of dietary cholesterol on the regulation of total body cholesterol in man, *J. Lipid Res.*, 12, 233, 1971.

125. Hanahan, D. J. and Wakil, S. J., Studies on absorption and metabolism of ergosterol-^{14}C, *Arch. Biochem. Biophys.*, 44, 150, 1953.

126. Glover, J., Leat, W. M. F., and Morton, R. A., The absorption and metabolism of [^{14}C]ergosterol in the guinea pig, *Biochem. J.*, 66, 214, 1957.

127. Child, P. and Kuksis, A., Critical role of ring structure in the differential uptake of cholesterol and plant sterols by membrane preparations *in vitro*, *J. Lipid Res.*, 24, 1196, 1983.

128. Cook, R. P., Kliman, H., and Fieser, L. F., The absorption and metabolism of cholesterol and its main companions in the rabbit with observations on the atherogenic nature of the sterols, *Arch. Biochem. Biophys.*, 52, 439, 1954.

129. Lemmon, R. M., Pierce, R. T., Biggs, M. W., Parsons, M. A., and Kritchevsky, D., The effect of Δ^7-cholesterol feeding on the cholesterol and lipoproteins in rabbit serum, *Arch. Biochem. Biophys.*, 51, 161, 1954.

130. Biggs, M. W., Lemmon, R. M., and Pierce, F. T., Observations on Δ^7-cholesterol metabolism in the rabbit, *Arch. Biochem. Biophys.*, 51, 155, 1954.

131. Glover, J. and Morton, R. A., Provitamin D$_3$ in tissues and the conversion of cholestanol to 7-dehydrocholesterol in vivo, *Biochem. J.*, 51,1, 1952.

132. Connor, W. E. and Lin, D. S., Absorption and transport of shellfish sterols in human subjects, *Gastroenterology*, 81, 276, 1981.

133. Lange, Y., Molinaro, A. L., Chauncey, T. R., and Steck, T. L., On the mechanism of transfer of cholesterol between human erythrocytes and plasma, *J. Biol. Chem.*, 258, 6920, 1983.

134. Rampone, A. J. and Machida, C. M., Mode of action of lecithin in suppressing cholesterol absorption, *J. Lipid Res.*, 22, 744, 1981.

135. Montet, J. C., Lindheimer, M., Gerolami, A., Montet, A. M., Reynier, M. O., Crotte, C., and Brun, B., Intestinal cholesterol uptake from mixed micelles. Effects of micellar size and detergent capacity, in *Bile Acids and Cholesterol in Health and Disease*, Paumgartner, G., Stiehl, A., and Gerok, W., Eds., MTP Press, Boston, 1983, 223.

136. Miyamoto, Y., Hanano, M., Iga, T., and Ishikawa, M., Concentration profile in the intestinal tract and drug absorption model: two dimensional laminar flow in a circular porous tube, *J. Theor. Biol.*, 102, 585, 1983.

137. Borgstrom, B., Fat digestion and absorption, in *Biomembranes*, Vol. 4B, Smyth, D. H., Ed., Plenum Press, New York, 1974, 555.

138. Somjen, G. J. and Gilat, T., A non-micellar mode of cholesterol transport in human bile, *FEBS Lett.*, 156, 265, 1983.

139. Dulery, C. and Reisser, D., Intestinal absorption of biliary and exogenous cholesterol in the rat, *Biochim. Biophys. Acta*, 710, 164, 1982.

140. Samuel, P. and McNamara, D. J., Differential absorption of exogenous and endogenous cholesterol in man, *J. Lipid Res.*, 24, 265, 1983.

141. Woodford, F. P., Enlargement of taurocholate micelles by added cholesterol and monoolein: self-diffusion measurements, *J. Lipid Res.*, 10, 539, 1969.

142. Barton, P. G. and Glover, T., A possible relationship between the specificity of the intestinal absorption of sterols and their capacity to form micelles with phospholipid, *Biochem. J.*, 84, 53p, 1962.

143. Bruckdorfer, K. K., Graham, J. M., and Green, C., The incorporation of steroid molecules into lecithin sols, beta-lipoproteins and cellular membranes, *Eur. J. Biochem.*, 4, 512, 1968.

144. Butler, K. W. and Smith, I. C. P., Sterol ordering effects and permeability regulation in phosphatidylcholine bilayers. A comparison of ESR spin-probe data from oriented multilamellae and dispersions, *Can. J. Biochem.*, 56, 177, 1978.

145. Kellaway, I. W. and Saunders, L., The solubilization of some steroids by phosphatidylcholine and lysophosphatidylcholine, *Biochim. Biophys. Acta*, 144, 145, 1967.

146. Borgstrom, B., Partition of lipids between emulsified oil and micellar phases of glyceride-bile salt dispersions, *J. Lipid Res.*, 8, 598, 1967.

147. Slotta, T. and Ammon, H. V., Effect of beta-sitosterol on oleic acid induced fluid secretion in the human jejunum, *Gastroenterology*, 78(Abstr.), 1264, 1980.

148. Carey, M. C. and Small, D. M., The physical chemistry of cholesterol solubility in bile. Relationship to gallstone formation and dissolution in man, *J. Clin. Invest.*, 61, 998, 1978.

149. Slotta, T., Kozlov, N. A., and Ammon, H. V., Comparison of cholesterol and sitosterol: effects on jejunal fluid secretion induced by oleate, and absorption from mixed micellar solutions, *Gut*, 24, 653, 1983.

150. Thomson, A. B. R., Hotke, C. A., O'Brien, B. D., and Weinstein, W. M., Intestinal uptake of fatty acids and cholesterol in four animal species and man: role of unstirred water layer and bile salt micelle, *Comp. Biochem. Physiol.*, 75A, 221, 1983.

151. Datserko, Z. M., Volkov, G. L., Govseeva, N. N., and Nikiforova, T. N., Lipid composition and activity of certain enzymes in membranes of intestinal epithelium microvilli in rats with experimental hypercholesterolemia, *Ukr. Biokhim. Zh.*, 53, 74, 1981.

152. Kessler, M., Acuto, O., Storelli, C., Murer, H., Muller, M., and Semenza, G., A modified procedure for the rapid preparation of efficiently transporting vesicles from small intestinal brush border membranes. Their use in investigating some properties of D-glucose and choline transport systems, *Biochim. Biophys. Acta,* 506, 136, 1978.

153. Bloj, B. and Zilversmit, D. B., Heterogeneity of rabbit intestine brush border plasma membrane cholesterol, *J. Biol. Chem.,* 257, 7608, 1982.

154. Child, P. and Kuksis, A., Uptake of 7-dehydro derivatives of cholesterol, campesterol, and beta-sitosterol by rat erythrocytes, jejunal villus cells and brush border membrane, *J. Lipid Res.,* 24, 552, 1983.

155. Proulx, P., Aubry, A., Brglez, I., and Williamson, D. G., The effect of phosphoglycerides on the incorporation of cholesterol into isolated brush border membranes from rabbit small intestine, *Biochim. Biophys. Acta,* 775, 341, 1984.

156. Child, P. and Kuksis, A., Differential uptake of cholesterol and plant sterols by rat erythrocytes *in vitro, Lipids,* 17, 748, 1982.

157. Edwards, P. A. and Green, C., Incorporation of plant sterols into membranes and its relation to sterol absorption, *FEBS Lett.,* 20, 97, 1972.

158. Child, P. and Kuksis, A., Uptake and transport of sterols by isolated villus cells of rat jejunum, *Can. J. Biochem.,* 58, 1215, 1980.

159. Watanabe, M., Oku, T., Shidoji, Y., and Aosoya, N., A new aspect on the mechanism of intestinal cholesterol absorption in rat, *J. Nutr. Sci. Vitaminol.,* 27, 209, 1981.

160. Beames, C. G., Bailey, H. H., Rock, C. O., and Schanbacher, L. M., Movement of cholesterol and beta-sitosterol across the intestine of Ascaris Sum, *Comp. Biochem. Physiol.,* 47A, 881, 1974.

161. Gottlieb, M. H., Rates of cholesterol exchange between human erythrocytes and plasma lipoproteins, *Biochim. Biophys. Acta,* 600, 530, 1980.

162. Phillips, M. C., McLean, L. R., Stoudt, G. W., and Rothblat, G. H., Mechanism of cholesterol efflux from cells, *Atherosclerosis,* 36, 409, 1980.

163. Higuchi, W. I., Ghanem, A. H., and Bikhazi, A. B., Mechanistic studies of solute transport rates at oil-water interface, *Fed. Proc.,* 29, 1327, 1970.

164. Green, C., The movement of cholesterol within cells, *Biochem. Soc. Trans.,* 11, 637, 1983.

165. Wattenburg, B. W. and Silbert, D. F., Sterol partitioning among intracellular membranes, *J. Biol. Chem.,* 258, 2284, 1983.

166. Child, P., Op den Kamp, J. A. F., Roelofsen, B., and van Deenen, L. L. M., Molecular species composition of membrane phosphatidylcholine influences the rate of cholesterol efflux from human erythrocytes and vesicles of erythrocyte lipid, *Biochim. Biophys. Acta,* 814, 237, 1985.

167. Smith, L. C. and Scow, R. O., Chylomicrons. Mechanisms of transfer of lipolytic products to cells, *Prog. Biochem. Pharmacol.,* 15, 109, 1979.

168. Di Costanzo, G., Duportail, G., Florentz, A., and Leroy, C., The brush border membrane of trout Salmo-Gairdneri intestine. Influence of its lipid composition on ion permeability, enzyme activity and membrane fluidity, *Mol. Physiol.,* 4, 279, 1983.

169. Datsenko, Z. M., Volkov, G. L., Govseeva, N. N., Nikiforova, T. N., and Palladin, A. V., Lipid composition and activity of certain enzymes in membranes of intestinal epithelium microvilli in rats with experimental hypercholesterolemia, *Ukr. Biokhim. Zh.,* 53, 74, 1981.

170. Spritz, N., Effect of fatty acid saturation on the distribution of the cholesterol moiety of VLDL, *J. Clin. Invest.,* 44, 339, 1965.

171. Poznansky, M. J. and Lange, Y., Transbilayer movement of cholesterol in phospholipid vesicles underequilibrium and non-equilibrium conditions, *Biochim. Biophys. Acta,* 506, 256, 1978.

172. Bloj, B. and Zilversmit, D. B., Transposition and distribution of cholesterol in rat erythrocytes, *Proc. Soc. Exp. Biol. Med.,* 156, 539, 1977.

173. Patton, S., Correlative relationship of cholesterol and sphingomyelin in cell membranes, *J. Theor. Biol.,* 29, 489, 1970.

174. Poznansky, M. J. and Czekanski, S., Cholesterol exchange as a function of cholesterol/phospholipid mole ratios, *Biochem. J.,* 177, 989, 1979.

175. Ashworth, L. A. E. and Green, C., Plasma membranes: phospholipid and sterol content, *Science,* 151, 210, 1966.

176. Tipping, E. and Ketterer, B., The influence of soluble binding proteins on lipophile transport and metabolism in hepatocytes, *Biochem. J.,* 195, 441, 1981.

177. Teerlink, T., van der Krift, T. P., van Heusden, G. P., and Wirtz, K. W. A., Determination of non-specific lipid transfer protein in rat tissues and Morris hepatomas by enzyme immunoassay, *Biochim. Biophys. Acta,* 793, 251, 1984.

178. Gavey, K. L., Noland, B. J., and Scallen, T. J., The participation of sterol carrier proteins in the conversion of cholesterol to cholesterol ester by rat liver microsomes, *J. Biol. Chem.,* 256, 2993, 1981.

179. Poorthuis, B. J. H. M. and Wirtz, K. W. A., Increased cholesterol esterification in rat liver microsomes by purified non-specific phospholipid transfer protein, *Biochim. Biophys. Acta,* 710, 99, 1982.
180. Trzuskos, J. M. and Gaylor, J. L., Cytosolic modulators of activities of microsomal enzymes of cholesterol biosynthesis. Purification and characterization of a non-specific lipid transfer protein, *Biochim. Biophys. Acta,* 751, 52, 1983.
181. van Heusden, G. P. H., van der Krift, T. P., Hostetler, K. Y., and Wirtz, K. W. A., Effect of nonspecific phospholipid transfer protein on cholesterol esterification in microsomes from Morris hepatomas, *Cancer Res.,* 43, 4207, 1983.
182. Suckling, K. E., Boyd, G. S., and Smellie, C. G., Properties of a solubilized and reconstituted preparation of acyl-CoA: cholesterol acyl transferase from rat liver, *Biochim. Biophys. Acta,* 710, 154, 1982.
183. Watt, S. M. and Simmonds, W. J., Effect of pancreatic diversion on lymphatic absorption and esterification of cholesterol in the rat, *J. Lipid Res.,* 22, 157, 1981.
184. Norum, K. R., Lilljeqvist, A. C., and Drevon, C. A., Coenzyme — a dependent esterification of cholesterol in the intestinal mucosa from guinea pig. Influence of diet on the enzyme activity, *Scand. J. Gastroenterol.,* 12, 281, 1977.
185. Field, F. J., Intestinal cholesterol esterase: intracellular enzyme or contamination of cytosol by pancreatic enzymes?, *J. Lipid Res.,* 25, 389, 1984.
186. Helgerud, P., Saarem, K., and Norum, K. R., Acyl coenzyme — a cholesterol acyl transferase (EC-2.3.1.26) in human small intestine. Its activity and some properties of the enzymic reaction, *J. Lipid Res.,* 22, 271, 1981.
187. Swell, L., Byron, J. E., and Treadwell, C. R., Cholesterol esterases. IV. Cholesterol esterase of rat intestinal mucosa, *J. Biol. Chem.,* 186, 543, 1950.
188. Gallo, L. L., Chiang, Y., Vahouny, G. V., and Treadwell, C. R., Localization and origin of rat intestinal cholesterol esterase determined by immunocytochemistry, *J. Lipid Res.,* 21, 537, 1980.
189. Gallo, L. L., Newbill, T., Hyun, J., Vahouny, G. V., and Treadwell, C. R., Role of pancreatic cholesterol esterase in the uptake and esterification of cholesterol by isolated intestinal cells, *Proc. Soc. Exp. Biol. Med.,* 156, 277, 1977.
190. Friedman, M., Byers, S. O., and St. George, S., Cholesterol metabolism, *Ann. Rev. Biochem.,* 25, 613, 1956.
191. Gallo, L. L., Clark, S. B., Myers, S., and Vahouny, G. V., Cholesterol absorption in rat intestine, role of cholesterol esterase and acyl coenzyme A:cholesterol acyl transferase, *Fed. Proc.,* 42, A5647, 1983.
192. Erikson, S. K., Report on the American Society of Biological Chemists Satellite Conference on regulation of intracellular cholesterol esterification, *J. Lipid Res.,* 25, 411, 1984.
193. Gallo, L. L., Bennet Clark, S., Myers, S., and Vahouny, G. V., Cholesterol absorption in rat intestine: role of cholesterol esterase and acyl coenzyme A:cholesterol acyl transferase, *J. Lipid Res.,* 25, 604, 1984.
194. Bhat, S. G. and Brockman, H. L., The role of cholesterol ester hydrolysis and synthesis in cholesterol transport across rat intestinal mucosal membrane. A new concept, *Biochem. Biophys. Res. Commun.,* 109, 486, 1982.
195. Bennet Clark, S. and Tercyak, A. M., Reduced cholesterol transmucosal transport in rats with inhibited mucosal acyl CoA:cholesterol acyl-transferase and normal pancreatic function, *J. Lipid Res.,* 25, 148, 1984.
196. Mathur, S. N. and Spector, A. A., Effect of liposome composition on the activity of detergent-solubilized acyl coenzyme A:cholesterol acyl transferase, *J. Lipid Res.,* 23, 692, 1982.
197. Spector, A. A., Kaduce, T. L., and Dane, R. W., Effect of dietary fat saturation on acyl coenzyme A:cholesterol acyl transferase activity of rat liver microsomes, *J. Lipid Res.,* 21, 169, 1980.
198. Stokke, K. T. and Norum, K. R., Subcellular distribution of acyl-CoA:cholesterol acyl transferase in rat liver cells, *Biochim. Biophys. Acta,* 210, 202, 1970.
199. Hashimoto, S. and Dayton, S., Stimulation of acyl CoA:cholesterol acyl transferase activity in rat liver microsomes by phosphatidylcholine, *Biochem. Biophys. Res. Commun.,* 82, 1111, 1978.
200. Tavani, D. M., Nes, W. R., and Billheimer, J. T., The sterol substrate specificity of acyl CoA:cholesterol acyl transferase from rat liver, *J. Lipid Res.,* 23, 774, 1982.
201. Murthy, S. K. and Ganguly, J., Studies on cholesterol esterases of the small intestine and pancreas of rats, *Biochem. J.,* 83, 460, 1962.
202. Salen, G., Shefer, S., and Berginer, V. M., Familial diseases with storage of sterols other than cholesterol: cerebrotendinous xanthomatosis and sitosterolemia with xanthomatosis, in *Metabolic Basis of Inherited Disease,* 5th ed., Stanbury, J. B., Wyngaarden, J. B., Frederickson, D. S., Goldstein, J. L., and Brown, M. S., Eds., McGraw-Hill, New York, 1983, 713.
203. Bhattacharyya, A. K. and Connor, W. E., Beta-sitosterolemia and xanthomastosis. A newly described lipid storage disease in two sisters, *J. Clin. Invest.,* 53, 1033, 1974.

204. Dermer, G. B., Ultrastructural changes in the microvillous plasma membrane during lipid absorption and the form of absorbed lipid: an in vivo study, *J. Ultrastruct. Res.,* 20, 51, 1967.

205. Dermer, G. B., Ultrastructural changes in the microvillous plasma membrane during lipid absorption: an *in vitro* study at 0°C, *J. Ultrastruct. Res.,* 21, 1, 1968.

206. Demel, R. A., Bruckdorfer, K. R., and van Deenen, L. L. M., Structural requirements of sterols for the interactin with lecithin at the air-water interface, *Biochim. Biophys. Acta,* 255, 311, 1972.

207. Bruckdorfer, K. R., Demel, R. A., de Gier, J., and van Deenen, L. L. M., The effect of partial replacements of membrane cholesterol by other steroids on the osmotic fragility and glycerol permeability of erythrocytes, *Biochim. Biophys. Acta,* 183, 334, 1969.

208. Ghosh, D. and Tinoco, J., Monolayer interactions of individual lecithins with natural sterols, *Biochim. Biophys. Acta,* 266, 41, 1972.

209. Borgstrom, B., Quantitative aspects of the intestinal absorption and metabolism of cholesterol and beta-sitosterol in the rat, *J. Lipid Res.,* 9, 473, 1968.

210. Hietanen, E. and Laitinen, M., Dependence of intestinal biotransformation on dietary cholesterol, *Biochem. Pharmacol.,* 27, 1095, 1978.

211. Sklan, D., Dahan, M., Budowski, P., and Hurwitz, S., Differential absorption of endogenous and exogenous cholesterol in the chick as affected by dietary oil level and phytosterols, *J. Nutr.,* 107, 1996, 1977.

212. Quintao, E., Grundy, S. M., and Ahrens, E. A., Effects of dietary cholesterol on the regulation of total body cholesterol in man, *J. Lipid Res.,* 12, 233, 1971.

213. Wilson, J. D. and Reinke, R. T., Transfer of locally synthesized cholesterol from intestinal wall to intestinal lymph, *J. Lipid Res.,* 9, 86, 1968.

214. Bikhazi, A. B. and Higuchi, W. I., Interfacial barriers to the transport of sterols and other organic compounds at the aqueous polysorbate 80-hexadecane interface, *Biochim. Biophys. Acta,* 233, 676, 1971.

215. Barton, P. G., Booth, R., Glover, J., and Thompson, D. G., Factors concerned in the specificity of absorption and transport of lipids, *Biochem. J.,* 96, 28p, 1965.

216. Sylven, C., Intestinal Cholesterol and Sitosterol Absorption in the Rat, Ph.D. dissertation, University of Lund, Sweden, 1970.

217. Smith, I. C. P., Organization and dynamics of membrane lipids as determined by magnetic resonance spectroscopy, *Can. J. Biochem.,* 57, 1, 1979.

218. Clejan, S. and Bittman, R., Distribution and movement of sterols with different side chain structures between the two leaflets of the membrane bilayer of mycoplasma cells, *J. Biol. Chem.,* 259, 449, 1984.

219. Sugano, M., Morioka, H., Kida, Y., and Ikeda, I., The distribution of dietary plant sterols in serum lipoproteins and liver subcellular fractions of rats, *Lipids,* 13, 427, 1978.

220. Thomson, A. B. R., unpublished; as cited in Thomson, A. B. R. and Dietschy, J. M., *Physiology of the Gastrointestinal Tract,* Vol. 2, Johnson, L. R., Ed., Raven Press, New York, 1981, 1147.

Chapter 2

INTESTINAL ABSORPTION OF BILE ACIDS

Joseph D. Fondacaro

TABLE OF CONTENTS

I. INTRODUCTION

The role of bile acids in the digestive process has been well defined over the last two decades. These detergent-like steroid derivatives serve to emulsify dietary lipids within the intestinal lumen in order to facilitate lipolysis and are required for the optimal activity of pancreatic lipase. Also, bile acids serve to disperse the end products of lipid digestion into micelles. These mixed micelles then facilitate the absorption of lipids, mainly, long-chain fatty acids and 2-monoglycerides. The efficient absorption of dietary cholesterol and the fat-soluble vitamins A, D, E, and K is also dependent upon the presence of bile acids in the upper small intestine. The absorption of dietary lipids and other lipophilic substrates is suppressed as much as 50 to 80% in some disease states in the absence of bile acids. In addition, bile acids play an essential role in solubilizing cholesterol during secretion and storage of bile in the gall bladder so as to prevent the precipitation of gallstones.

To perform these functions, which in total ensure the proper digestion and absorption of dietary lipids, an adequate intraluminal concentration of bile acids is essential. Maintenance of this concentration, in turn, depends upon the efficient absorption of these sterols from the small bowel and subsequent return to the liver via the portal circulation and sequestration by the hepatocytes. This now well-defined enterohepatic circulation thus allows recycling and conservation of these physiologically important substances.

Thus, the importance of bile acids in the spectrum of human health is apparent. Furthermore, the daily and usually immediate demand for these sterols underscores the importance of the enterohepatic circulation. As stated earlier, the intestinal absorption of bile acids is an integral part of this total process and highly efficient absorptive mechanisms are involved, resulting in a 95 to 98% recovery. This chapter will review the essential features of intestinal bile acid absorption, highlighting some of the intraluminal events which influence this process and some pathophysiology which has been described in certain clinical settings.

II. PROCESS OF ABSORPTION

The intestinal absorption of bile acids in mammals occurs predominantly in the small intestine, in particular, the more distal portion, the ileum. This absorptive process has characteristics similar to that of other substrates, as well as some features which are unique to this chemical class of molecules. Absorption takes place by active transport and passive diffusion with necessity for various requirements and with varying rates, depending upon the intestinal region under study.

A. Active Transport

It is now well established that an active transport process, uptake against a concentration gradient, contributes significantly to bile acid absorption. Several authors have contributed to and confirmed this finding.[1-7] This active transport process has been characterized in many different mammalian species, including man, and with several well-established in vitro and in vivo experimental techniques. Early experiments by Lack and Weiner[1] using the everted gut sac preparation demonstrated that active transport of taurocholic and glycocholic acid was limited to the distal ileum in rats and guinea pigs. Furthermore, this transport process could be inhibited by anoxia and the metabolic poisons dinitrophenol and sodium azide. Holt[3] reported that the active transport of taurocholate and cholate by slices of rat ileum obeyed Michaelis-Menton kinetics. This property was absent in other portions of the gut. Also, the presence of endogenous bile acids resulted in competitive inhibition of ileal active absorption. Hea-

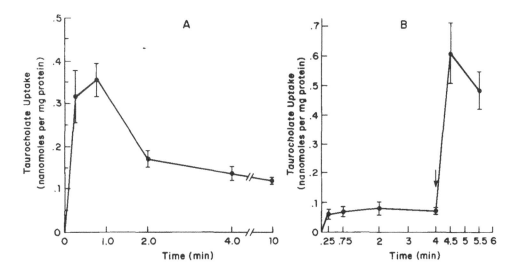

FIGURE 1.   Uptake of ^{14}C-taurocholic acid by brush border membrane vesicles prepared from distal ileum of the guinea pig. Panel A: the reaction was initiated by adding an equal volume of the taurocholate (15 nmol) in 50 mM NaCl to a suspension of vesicles (2 mg protein per milliliter) in 100 mM mannitol (0 time). Panel B: between 0 time and 4 min vesicles were incubated in 15 nmol of taurocholate in 100 mM mannitol. At point indicated by the arrow, an equal volume of 50 mM NaCl was added. (Reprinted with permission from *Life Sci.,* 20, Lack, L., Walker, J. T., and Hsu, C.-Y. H., Taurocholate uptake by membrane vesicles prepared from ileal brush borders, Copyright 1977, Pergamon Press, Ltd.)

ton and Lack[8] also demonstrated this inhibition in an in vivo perfusion system in the guinea pig. Later, Wilson and Treanor[6] demonstrated the active transport mechanism for bile acid uptake into epithelial cells isolated from rat intestine. Active bile acid transport following Michaelis-Menton kinetics has also been demonstrated in villi isolated from hamster ileum.[7]

Perhaps the most pronounced feature of this process is the requirement for the presence of luminal Na$^+$. This co-transport system is not unlike that of glucose[9] and most of the amino acids,[10] in that operation of the basolateral Na$^+$K$^+$-ATPase-driven Na$^+$ pump provides the electrochemical driving force for Na$^+$ movement inward. Since the active transport of bile acids is linked to Na$^+$, translocation of bile acids from lumen to enterocyte occurs in a co-transport fashion. Bile acids are then free to diffuse down their concentration gradient out of the cell via the basolateral route and to enter the bloodstream, thus completing the active transcellular transport and absorption of bile acids. Playout and Isselbacher[4] reported that in the absence of mucosal Na$^+$, taurocholic acid transport in everted sacs of rat and hamster was markedly inhibited, even to a greater extent than that caused by anoxia or by dinitrophenol. Also, with graded media concentrations of Na$^+$, Gallagher et al.[11] demonstrated a direct relationship with the degree of bile acid transport inhibition. However, in that same study, even with 84% replacement of Na$^+$ with choline chloride, there was still 69% of the active transport observed in control-everted ileal gut sacs. Also, in vivo perfusion studies demonstrated that the inhibition of bile acid absorption following the removal of Na$^+$ is reversible.[11]

The Na$^+$-bile acid co-transport process is clearly demonstrable in studies utilizing the brush border membrane vesicles to measure uptake. Lack et al.,[12] using brush border membrane vesicles prepared from guinea pig ileum, demonstrated enhanced vesicular uptake of taurocholate in incubations where an inwardly directed Na$^+$ gradient was present (Figure 1, panel A). The initially rapid uptake has now become known as the "overshoot" phenomenon and is not present in vesicles prepared from more proximal

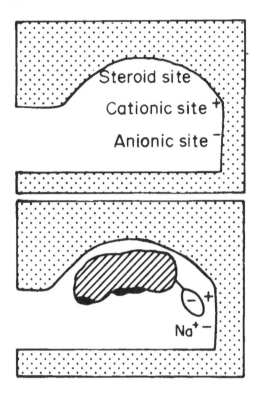

FIGURE 2. Hypothetical scheme for the interaction of a bile salt
with its recognition site in the membrane. Three sites interact co-
operatively. All this takes place in a hydrophobic space or cleft.
(From Lack, L., *Environ. Health Perspect.*, 33, 79, 1979. With
permission.)

regions of the small bowel. With dissipation of the Na^+ gradient, there was a decline in
the intravesicular taurocholate concentration. When isosmotic mannitol was substituted for NaCl, no overshoot was observed until subsequent addition of NaCl to the
incubation mixture (Figure 1, panel B). Similar results have subsequently been reported.[13,14] Later studies have also demonstrated the presence of the overshoot phenomenon for taurocholate and the conjugated, dihydroxy bile acid taurochenodeoxycholate in vesicles prepared from rat ileum.[15]

While this review of works which have demonstrated an active transport component
to intestinal bile acid absorption is brief, they nonetheless point out several important
features of this process. First, the active transport of bile acids occurs exclusively in
the distal portions of the small intestine, namely the ileum. Second, Na^+ is required for
this active co-transport process. Third, there appears to be mutual inhibition of this
transport process between various species of bile acids.

With regard to the luminal membrane compartment or "carrier" responsible for the
translocation of Na^+ and bile acid into the enterocyte, Bundy et al.[16] and later Lack,[17]
have proposed the model depicted in Figure 2. The model suggests that the carrier
possesses three distinct interactive sites: a steroid site which accommodates the molecular structure of the bile acid; a cationic site, a positively charged component of the
carrier which allows the coulombic interaction of the negatively charged side chain of
bile acids; and an anionic site permitting the interaction of the sodium ion. When all
three sites are occupied, inward flux of the substrates occurs. Other features of the
carrier are (1) the hydrophobicity of the carrier locus and (2) the affinities of the reactive sites are cooperative in that one enhances the binding at the other. Lack[17] proposes

that although complex, this model does provide predictions which are testable regarding active Na⁺-bile acid co-transport. Thus, Na⁺ has a physiological role in active intestinal bile acid absorption.

B. Passive Diffusion

While considerable research attention has been given to the active bile acid transport process, passive intestinal transport of these sterols continues to be of significant physiologic importance. In contrast to the regional specificity of active transport, passive diffusional absorption of bile acids can take place at all levels of the small intestine and in most regions of the large bowel. Studies by Dietschy et al.[18] demonstrated that in proximal small bowel, bile acids move passively across the intestinal wall in the presence of a concentration or activity gradient between mucosal and serosal solutions in vitro. This mucosal-to-serosal flux increased proportionally with an increase in the mucosal concentration of bile acid. This has also been demonstrated for passive uptake of taurocholic acid into villi isolated from hamster jejunum.[7] Dietschy et al.[18] also reported that solvent drag (obligatory movement of a solute which follows the movement of water) had a slight but measurable effect on passive bile acid absorption.

Besides the permeability characteristics of the intestinal mucosal membrane to specific bile acids species, the most important variables affecting the rate of passive absorption of bile acids across the intestine are centered around the physiochemical state of these molecules in solution. Specifically, the pK_a and the critical micellar concentration (CMC) will influence how rapidly each bile acid passively traverses the intestinal mucosa. The normal pH range of the small intestine is usually well above the pK_a for most bile acids, with unconjugated forms having a dissociation constant much higher than conjugated forms and closer to the pH of the intestinal contents.[19] Thus, under normal physiologic conditions, the bulk of the bile acid content in the small bowel is in ionized form. Dietschy et al.[18] demonstrated that the passive diffusion of cholic acid (pK_a 6.4) was markedly enhanced when the pH of the mucosal solution was lowered from 8.0 to 6.0. Under these conditions, passive diffusion is the sum of ionic and nonionic diffusion down their respective concentration gradients. As these authors showed, the magnitude of passive *ionic* flux of cholic acid dropped 75% at pH 6.0 compared to that at pH 8.0. By subtracting the value for ionic flux from total flux, these investigators demonstrated a marked increase in passive nonionic diffusion of cholic acid. Thus, the nonionized form of bile acid is more readily diffusible through the intestinal mucosa. In contrast, the pK_a of the taurine conjugate of cholic acid is about 1.5, hence this experiment could not be performed without damaging the intestinal mucosa. One can assume, however, that very little, if any, nonionic diffusion of taurocholic acid occurs in vivo; and, since diffusion of ionic species is a much more delayed process, passive diffusion contributes little to the total absorptive process for conjugated bile acids in the gut. However, this in turn provides for the maintenance of an adequate intraluminal concentration of bile acids for the purpose of fat digestion and absorption in the upper and middle small bowel.

As the intraluminal concentration of bile acids increases, it approaches a level at which micelles are formed, that is, the critical micellar concentration or CMC. Bile acids are only absorbed in monomeric form,[20,21] and since the monomer concentration approaches a constant at and above the CMC, the value for passive diffusion of bile acid should plateau. Consequently, as pointed out by Schiff et al.,[5] the presence of micelles in the intestinal lumen is particularly relevant to the physiological setting during digestion. Micelle size is also important in this scheme and will be discussed in some detail later as a factor which alters absorption of bile acids.

Thus, we see that passive diffusional absorption contributes quantitatively little to the total bile acid absorptive process in the normal functioning small intestine. It nevertheless plays a key role in the processes of intestinal fat digestion and absorption. This limitation of passive bile acid absorption has been considered a physiological adaptation[22] and provides for adequate solubilization of dietary lipids in the water environment of the small bowel.

C. Structure-Activity Relationship

Lack and Weiner[23] published an extensive study defining some of the structure-activity relationships involved in active bile acid transport in the guinea pig ileum. Using everted intestinal sacs, they reported active transport for trihydroxy, dihydroxy, and triketo bile acids. The taurine and glycine conjugates of cholic acid were transported more efficiently than the unconjugated trihydroxycholanic acid form. Substitution of the amino acid with other ligands produced variable effects. The *d*-aminovaleric acid conjugate was transported somewhat better than cholic acid, but roughly 2.5 times less avidly than either the taurine or glycine conjugate. Aspartic acid and aminoethylphosphonic acid conjugates were only marginally transported, while phosphorylethanolamine and trimethylethylenediamine substitutes were not actively transported. Substitution of ketonic groups (=O) for the hydroxy groups (−OH) at positions 3, 7, and 12 reduced active transport of the glycine conjugate by nearly 67%.

Studies with dihydroxy derivatives showed different results. Taurine conjugates of deoxycholate (3,12-dihydroxy) and chenodeoxycholate (3,7-dihydroxy) were transported more efficiently than the glycine conjugates of the same derivatives. However, taurine and glycine conjugates of hyodeoxycholic acid (3,6-dihydroxy) were transported equally well and at a rate comparable to the taurine conjugates of the other dihydroxy forms.

With respect to charge, a positively charged trihydroxy derivative, N-cholyl-N^2-trimethylethylenediamine, was not actively transported by the guinea pig ileum. These authors concluded that in general, the ability of the system to actively transport a bile acid derivative was not greatly influenced by changing the conjugating moiety unless the change involved altering the number of potential negative charges. Later, Bundy et al.[16] reported that removal of the anionic charge as well as alteration of the steroid moiety increases the Na^+ requirement for active transport.

D. Postnatal Development

Because of the important role of bile acids in intestinal absorption of dietary lipids, the neonatal development of the enterohepatic circulation and, in particular, the active transport process for bile acids play key roles in growth and development. The ontogenesis of intestinal bile acid transport has been studied in several species including the human. Lester et al.[24] and Little and Lester[25] have shown in the dog and rat, at 2 weeks of age, everted rings of ileal mucosa can concentrate taurocholic acid. This concentrating ability increases with age as demonstrated by increasing mucosal-to-media ratios. More recently, Heubi and Fondacaro[26] have described the postnatal development of ileal bile acid transport in the guinea pig. Using the villus technique,[27] taurochate uptake into villous cells was correlated with developmental age and media substrate concentration. As shown in Figure 3, ileal transport was linear with concentration in fetal and 1-day-old animals. By 5 days, a hyperbolic curve for ileal uptake was manifest indicating the presence of an active transport mechanism. This system continued to develop to age 21 days when it was nearly identical to that shown for the adult. Villous cell uptake by jejunal samples remained linear with initial taurocholate concentration throughout the developmental period.

In the normal human newborn, it is known that fat absorption is impaired, and Koldovsky[28] has stated that the competency of fat digestion in infants is a function of postnatal age and birth weight. This correlates with the findings of de Belle et al.[29] who

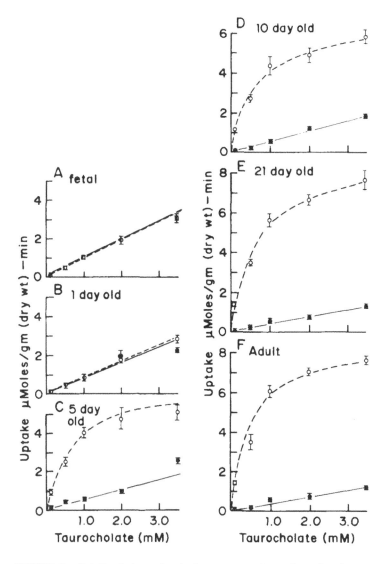

FIGURE 3. Relation between incubation concentration and uptake of tauro-cholate in jejunal (solid line) and ileal (dashed line) villus cells from fetal (A), 1-day-old (B), 5-day-old (C), 10-day-old (D), 21-day-old (E), and adult (F) guinea pig. (From Heubi, J. E. and Fondacaro, J. D., *Am. J. Physiol.*, 243(Gastrointest. Liver Physiol. 6), G189, 1982. With permission.)

have shown that human active bile acid transport in the ileum develops postnatally and appears by 8 months of age.

Information has been provided that suggests some factors which may influence development of the active transport system for bile acids in the ileum. It is known that hormonal influence from the adrenal cortex and the thyroid gland accounts for changes in cellular enzyme systems required for active transport.[3] Likewise, glucocorticoid administered to suckling rats hastens the normal development of ileal bile acid active transport.[25] Nutritional factors also are thought to play a role in this development, but further studies are required to elucidate their impact.

E. Role of the Large Bowel

Of the considerable amount of bile acid introduced into the duodenum of mammals every day, only a small fraction of this amount enters the large intestine. Although bile acid absorption does occur in the colon, reports have varied as to the amount of bile acid that can be absorbed by passive diffusion from the colon of humans and of other mammals.[30,31] Conjugated bile acids are generally not absorbed as such by the large bowel. Intestinal bacteria modify bile acids as they pass through the distal ileum and proximal colon. Hence, bile acids absorbed via the colonic mucosa are usually the deconjugated and dehydroxylated forms of those present in the upper and middle small bowel.

Quantitation of the amount of bile acid absorbed by the colon is difficult, but attempts have been made. Olivecrona and Sjovall[32] determined that 15% of the bile acid in rat portal blood is unconjugated and thus may be of colonic origin. However, this may be an overestimate if some deconjugation occurs in the distal small bowel. In another study, Norman and Sjovall[33] found that 50% of labeled unconjugated bile acid injected into the rat colon was recovered in 24 hr.

While the absorptive capacity of the colon is less than that of the ileum, it is clear that other factors influence the rate of bile acid absorption from this portion of the gastrointestinal tract. For example, since conversion of cholic acid to deoxycholic acid will cause partial precipitation of bile acid from solution (making this form unavailable for absorption), the rate of bacterial dehydroxylation will influence the amount and rate of bile acid absorbed from the colon.[34] Also, the number of enterohepatic cycles per day will vary and so the amount of bile acid discharged into the intestine is variable. Thus, the frequency and amount of feeding can profoundly affect colonic absorption of bile acids.[31] Therefore, while it is generally agreed that the permeability of the colon permits some bile acid to diffuse from its lumen to the portal blood, estimations of this amount have not been readily available and are perhaps only marginally significant in the overall scheme of the enterhepatic circulation.

III. FACTORS INFLUENCING BILE ACID ABSORPTION

A variety of in vivo and in vitro test systems has been used to characterize the intestinal bile acid absorption in man and in most common experimentally used species. The physiology and biochemistry of this process is thus well defined and the unique features of bile acid transport and diffusion are well understood. As with any physiological process, there are factors which alter the normal absorption of these acid sterols, and several of these naturally occurring factors have been described using the same basic techniques that are used to characterize the process itself. The following discussion will present some of these rate-limiting factors and how they influence intestinal bile acid absorption.

A. Unstirred Water Layer

In the absorptive state, the contents of the intestine are found to be dissolved, suspended, or otherwise contained within either the bulk water phase or the unstirred water layer of the intestinal lumen. The bulk water phase is generally considered to be that fraction of the intestinal contents which is kinetic and is under the influence of intestinal motor activity. Contents of this phase are for the most part in a partially or nearly complete digestive state and are generally moving slowly through the small bowel in a caudal direction, except for brief orally directed movements set up by segmentation. In contrast, the unstirred water layer is visualized as a stagnant "lining" on the luminal surface of the intestinal mucosa which usually contains end products of digestion that are making their way to the absorbing enterocyte. Hydrophilic, polar substrates can traverse this unstirred layer more rapidly than hydrophobic, less polar molecules. Thus, for many substances to be absorbed by the intestine, the unstirred water layer represents a resistance and a rate-limiting factor influencing absorption.

Wilson and Dietschy[35] have characterized the relationship of bile acid absorption and the unstirred water layer in the rat intestine in vitro. As could be predicted, a more polar bile acid such as taurocholic acid traversed the unstirred water layer quite readily and stirring the incubation mixture did not enhance the uptake of taurocholate. A less polar bile acid, taurodeoxycholic acid, did not permeate this unstirred layer as readily. These investigators also found that the unstirred water layer was more rate limiting to uptake of bile acids from micellar solutions and that stirring enhanced uptake from these solutions.[35] Therefore, it is apparent that the unstirred water layer lining the intestinal mucosa does present a rate-limiting diffusional barrier for the absorption of many bile acid forms and can adversely influence the total amount of bile acid transported by the intestine.

B. Critical Micellar Concentration (CMC)

The critical micellar concentration or CMC of a bile acid is the concentration at which, in solution, bile acid monomers aggregate to form micelles. These simple micelles are optically clear and because bile acids are amphipathic molecules, their orientation around the micelle is such that the lipophilic portions of these molecules face inward. This creates the lipid environment which holds fatty acids, monoglycerides, and fat-soluble vitamins and cholesterol in solution, as this now mixed micelle traverses the unstirred water layer to the absorptive surface of the enterocyte. Monomeric bile acid and micellar bile acid are in equilibrium.[35] Since bile acids are only absorbed in the monomeric form,[20,21,35,36] the intraluminal concentration of a bile acid and any alteration of the monomer-to-micelle equilibrium can influence the rate of intestinal bile acid absorption.

Wilson and Dietschy[35] demonstrated the effect of the CMC on absorption of the taurine conjugates of deoxycholate and cholate. In the rat jejunum, uptake of taurodeoxycholic acid was linear from 0.1 to about 1.0 mM, at which point a shift occurred in the slope of the uptake curve, suggesting a slower absorption rate. The change in slope occurred at the CMC for this bile acid (about 1.0 mM). Similar results were obtained with taurocholic acid uptake; i.e., a slowing of the rate, as indicated by a downward shift in the uptake curve, took place at about 4.0 mM, the CMC for taurocholic acid.[35] These observed reductions in rate of uptake occurring at the CMC are explained by a decrease in the monomer concentration as bile acid molecules are incorporated into micelles. This essentially reduces the driving force for passive diffusion (i.e., effective substrate concentration) and thus a slower rate of diffusion is seen. It is interesting to note that the passive uptake curve of taurodehydrocholic acid did not change slope over an entire concentration range (1.0 to 8.0 mM). This triketo-cholic acid derivative does not form micelles,[19,35,36] therefore, its uptake rate is unaltered and increased uniformly with substrate concentration.[35]

Inasmuch as the CMC, when reached, will alter bile acid absorption, then logically, anything which influences the CMC will also affect bile acid absorption. If a factor increases the CMC of a given bile acid, one would presume that the established steady-state absorption of that bile acid would continue unaltered until the new CMC is reached by the addition of more monomers of that bile acid. Conversely, if a factor lowers the CMC for a particular bile acid, one could predict a deceleration of the absorption rate to perhaps a new steady state which would be governed in part by the rate of conversion of micelles to monomers for that species of bile acid. Certain in vivo conditions are known to cause a reduction in the CMC and have been shown experimentally to inhibit the absorption of specific bile acids.[36] These will be examined in the following section.

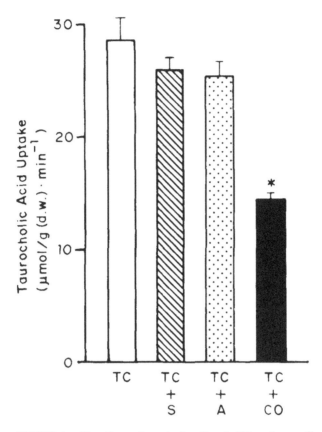

FIGURE 4.   The effects of starch (S), albumin (A), and corn oil (CO) on taurocholic acid (TC) uptake by villi isolated from hamster ileum. * = Significantly less than TC alone. (From Fondacaro, J. D. and Wolcott, R. H., *Proc. Soc. Exp. Biol. Med.*, 168, 276, 1981. With permission.)

C. Dietary Nutrients

In the recent past, our laboratory has concerned itself with the influence of dietary nutrients, in particular, dietary lipids, on intestinal bile acid absorption. This interaction is important not only to the proper digestion and absorption of these lipids, but to the overall scheme of bile acid metabolism in the body. This relationship also becomes important in certain clinical conditions. For example, in diseases of the intestine such as inflammatory bowel disease of the ileal mucosa, active and/or passive bile acid transport (and thus the enterohepatic circulation) may be severely compromised and a reduction in the total bile acid pool is possible. Under these conditions the influence of the intraluminal environment, mainly, dietary nutrients, becomes extremely important in an already reduced bile acid absorption state. Recently, Koga and colleagues[37] have shown that dietary fat increases fecal bile acid loss in patients with established Crohn's disease and conclude the dietary fat represents an important consideration in the evaluation of bile acid malabsorption in Crohn's disease. Likewise, in diseases such as cystic fibrosis and chronic pancreatitis, adverse changes in the elaboration of major digestive enzymes by the pancreas create conditions where the interaction of undigested dietary substrates may and do influence absorption of bile acids. Certain clinical implications will be addressed later in this chapter.

The effects of a representative group of unhydrolyzed dietary components on taurocholic acid absorption were examined in vivo and in vitro.[38] Results of in vitro experiments, shown in Figure 4, indicate that of the major food classes, unhydrolyzed

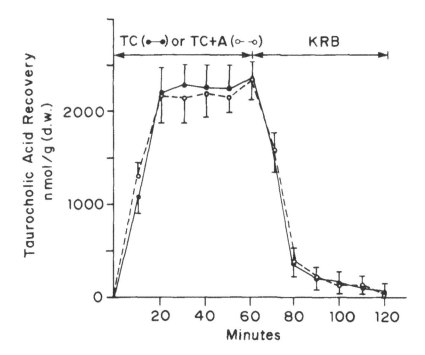

FIGURE 5.   Biliary recovery of taurocholic acid (TC) during perfusion of the distal ileum of rats with solutions of TC alone or TC plus albumin (A). (From Fondacaro, J. D. and Wolcott, R. H., *Proc. Soc. Exp. Biol. Med.*, 168, 276, 1981. With permission.)

carbohydrate (starch) and protein (albumin), did not alter absorption of taurocholic acid by villi isolated from hamster ileum. However, corn oil (as a representative unhydrolyzed triglyceride) markedly suppressed taurocholate uptake in these villus preparations (Figure 4). Furthermore, in the same study it was shown that when various combinations of undigested nutrients were added to the absorptive media, only those combinations containing triglyceride inhibited bile acid uptake.[38] Similar results were also seen with the uptake of unconjugated cholic acid[39] and with uptake of taurocholic acid and cholic acid in the presence of phospholipid,[7] using the same villus technique.

In in vivo studies,[38] similar results were noted. Using the ileal perfusion-biliary recovery technique in anesthetized rats, both albumin (Figure 5) and starch (Figure 6) failed to alter the recovery of ^{14}C-labeled taurocholic acid in bile during concomitant ileal perfusion. When triglyceride (triolein) was included into the perfusion buffer, significant reductions in biliary recovery of radiolabeled bile acid were observed throughout most of the test period (Figure 7). Harries et al.[40] have demonstrated similar results with albumin and taurocholate absorption in the rat in vivo. Conversely, Sklan et al.[41] reported that casein inhibited taurocholic acid uptake in isolated duodenal loops of chick small bowel. This inhibition, however, may have been produced by the presence of oleic acid in the medium, which has been shown by Roy et al.[42] and others[36] to inhibit bile acid absorption.

The effects of these unhydrolyzed lipids and the mechanism by which they cause inhibition of bile acid absorption were further examined in vitro.[36] A long-, medium-, and short-chain triglyceride (represented by triolein, tricaprylin, and tributyrin) all produced similar inhibition of taurocholic acid uptake by villi isolated from hamster ileum (Figure 8). The same was found to be true for glycocholic acid uptake in the same experimental model.[50] It is curious to note, however, that the inhibition only occurred

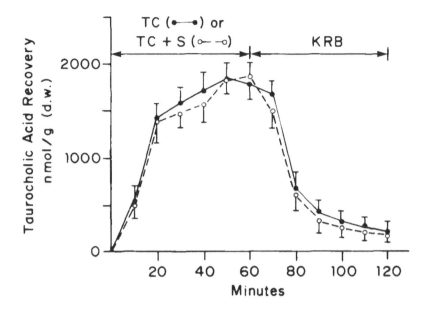

FIGURE 6.   Biliary recovery of taurocholic acid (TC) during perfusion of the distal ileum of rats with solutions of TC alone or TC plus starch (S). (From Fondacaro, J. D. and Wolcott, R. H., *Proc. Soc. Exp. Biol. Med.*, 168, 276, 1981. With permission.)

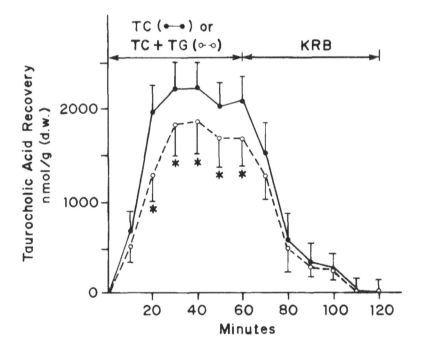

FIGURE 7.   Biliary recovery of taurocholic acid (TC) during perfusion of the distal ileum of rats with solutions of TC alone or TC plus triglyceride (TG). * = Significantly less than TC alone. (From Fondacaro, J. D. and Wolcott, R. H., *Proc. Soc. Exp. Biol. Med.*, 168, 276, 1981. With permission.)

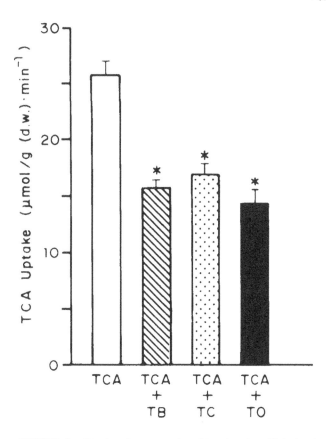

FIGURE 8.   Uptake of taurocholic acid (TCA) by villi isolated from hamster ileum and the effects of tributyrin (TB), tricaprylin (TC), and triolein (TO) on this process. * = Significantly different from TCA alone. (From Fondacaro, J. D., *Proc. Soc. Exp. Biol. Med.*, 173, 118, 1983. With permission.)

when the initial concentration of bile acid was above the CMC and presumably when micelles were present. No inhibition was seen with a bile acid concentration of 0.1 mM. It is also of interest to note that these representative triglycerides failed to alter absorption of taurodehydrocholic acid, the triketo cholic acid derivative which does not form micelles.[38] This suggests that the mechanism of inhibition involves the bile acid micelle, either micellar size or micelle formation. By increasing micellar size, more monomolecular bile acid would be incorporated into these enlarged micelles. This would reduce transport by shifting the monomer-micelle equilibrium to the micelle phase and lowering the concentration of monomers, thus, providing less driving force for passive absorption and less substrate for active transport. However, it is well established from the work of Carey and Small[43] that triglycerides partition poorly into micelles and therefore would tend not to swell these spheres. Oleic acid, a long-chain fatty acid, also inhibits bile acid uptake,[36,39] but these substances do partition into micelles and are thought to enlarge their structure. These facts were ultimately verified by equilibrium dialysis studies.[36]

Another method by which triglycerides may affect the micelle is by enhancing its formation, i.e., by lowering the CMC. This is known to occur with other lipids.[43] At any given bile acid concentration addition of triglyceride may lower the CMC such that the monomer-micelle equilibrium is again shifted toward the micelle phase and less monomer is available for absorption. This hypothesis, however, remains to be exam-

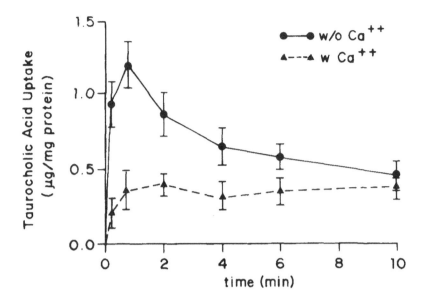

FIGURE 9.   The effects of intravesicular Ca^{2+} on Na^+-coupled taurocholic acid uptake
into brush border membrane vesicles prepared from rat ileum. Initial taurocholic acid
concentration was 1.0 mM in both series. (Reprinted with permission from *Life Sci.,*
35, Fondacaro, J. D. and Madden, T. B., Inhibition of Na^+-coupled solute transport
by calcium in brush border membrane vesicles, Copyright 1984, Pergamon Press, Ltd.)

ined and is still open to question. Nonetheless, it has been shown that a variety of lipids
*inhibit intestinal bile acid absorption and in the in vivo state following a meal and in
certain clinical settings this may be an important influence on the enterohepatic circulation and the bile acid pool.*

As discussed earlier, the requirement for luminal Na^+ for active bile acid transport
and the effects of varying Na^+ concentration on this process are well established. However, other cations and anions are present both intraluminally and intracellularly which
may influence intestinal bile acid absorption. Their effects, if any, on this process are
less well understood. Considerable interest of late has been drawn to the role of calcium in the regulation of cellular function and, in particular, in the regulation of transport function in the enterocyte.[44] Therefore, the role of Ca^{2+} and its influence on bile
acid absorption were recently investigated in vitro. Na^+-coupled taurocholic acid uptake by brush border membrane vesicles prepared from guinea pig ileum was unaltered
in the presence of 2.5 mM $CaCl_2$ in the extravesicular fluid.[14] Likewise, passive vesicular uptake of this bile acid, measured in the absence of Na^+, was also unaffected in
the presence of external Ca^{2+}. However, in the presence of intravesicular Ca^{2+}, taurocholic acid uptake into brush border membrane vesicles was markedly reduced (Figure
9). This is evidenced by a disappearance of the typical "overshoot" characteristic of
Na^+-coupled vesicular transport. Similar results were also expressed for vesicular uptake of the taurine conjugate of chenodeoxycholate.[15] In fact, in the presence of intravesicular Ca^{2+}, the uptake of both bile acids by these vesicles resembled a passive uptake process even in the presence of adequate concentrations of Na^+. It is worth noting
also that in the same report, Na^+-coupled vesicular uptake of glucose and valine and
the uptake of Na^+ were also significantly inhibited by intravesicular Ca^{2+}. This suggests
that since the common feature of these uptake processes inhibited by internal Ca^{2+} was
the requirement for Na^+, internal Ca^{2+} appears to, in some way, exert its inhibitory
influence on the Na^+-sensitive component of the carrier systems.[15]

While it is not possible from these studies to determine the mechanism by which internal Ca^{2+} exerts its inhibitory influence on Na^+-coupled transport, Chase and Al-Awqati[45] suggest that perhaps Ca^{2+} alters surface charge of the membrane and thus indirectly influences the Na^+ permeability. Also, Ca^{2+} on the inner surface of the apical membrane may interact with the carrier proteins and alter their affinity for Na^+. While these initial studies indicate that intravesicular Ca^{2+} inhibits Na^+-coupled bile acid uptake, the role of intracellular Ca^{2+} in the regulation of this and other Na^+-coupled transport processes requires further investigation.

IV. CLINICAL SIGNIFICANCE

The importance of an intact enterohepatic circulation in the spectrum of human health has been emphasized and clinical investigations have contributed significantly to this concept. Interruption of the enterohepatic circulation of bile acids has been documented in humans with surgical resection of the ileum for regional enteritis, volvulus, or ischemia, or with ileal by-pass by construction of a jejunocolic shunt.[46] Likewise, similar disruption of bile acid absorption has been reported in patients with intact ilea, but who suffer from inflammatory bowel disease of the ileal region of the intestine, radiation damage, or from amyloid infiltration of the ileum. Thus, as reported by Garbutt et al.,[46] it is clear that functional loss of the distal one half to one third of the small bowel causes a significant reduction in the enterohepatic circulation and the presence of serious ileal disease results in the inability to maintain bile acids, in particular, conjugated trihydroxy forms, in the enterohepatic circulation.

Recently, two new clinical entities have been identified which involve a compromised intestinal bile acid absorption. In the first, two boys with congenital diarrhea, steatorrhea, and growth failure were identified as having an interrupted enterohepatic circulation.[47] Radiographically, the ileal structure was normal and ileal function as assessed by vitamin B_{12} absorption appeared normal. Bile acid absorptive capability of the distal ileum was determined to confirm or deny the clinical suspicion of bile acid malabsorption. Heubi and colleagues[47] recovered ileal mucosa from these patients by peroral ileal biopsy. Using the villus technique, they determined that taurocholic acid absorption by these mucosal samples was markedly reduced at every concentration examined as compared to control ileal mucosa taken from seven otherwise normal ileostomy patients within the same age group. Furthermore, examination of these biopsies by electron microscopy failed to show significant ultrastructural abnormalities. These investigators thus reported a previously undescribed, congenital transport defect called primary bile acid malabsorption.[47] This primary defect represents an absence of active ileal bile acid transport presenting as diarrhea from infancy and accompanied by clinical and biochemical findings of an interrupted enterohepatic circulation.

The second recent clinical finding involving bile acid absorption centers around cystic fibrosis (CF). It was known for some time that clinically significant bile acid malabsorption consistently accompanied CF. The cause of this malabsorption was unknown, but was believed to be due to some intraluminal event associated with pancreatic insufficiency.[48] In our laboratory, a study was undertaken to determine the bile acid absorptive capability of ileal mucosae from CF patients. Taurocholic acid uptake was measured at three initial concentrations in ileal mucosa taken by peroral biopsy from CF patients.[49] These values were compared to those of normal control ileal and normal control jejunal mucosa obtained by biopsy and studied under identical experimental conditions. At all three initial taurocholic acid concentrations, 0.1, 1.0, and 10.0 mM, significant reduction of bile acid uptake was observed in tissue from every CF patient (n = 7). Table 1 presents the findings of this study. Furthermore, the uptake rate of taurocholate by CF ileal mucosa resembled that of jejunal mucosa from healthy volunteers. Thus, since bile acid absorption occurs only by passive diffusion in

Table 1
TAUROCHOLIC ACID UPTAKE BY INTESTINAL
MUCOSA FROM NORMAL AND CYSTIC
FIBROSIS PATIENTS

$[TC]_i$ (mM)	C(Il) (n = 7)	CF(Il) (n = 7)	C(Je) (n = 5)
0.1	1.42 ± 0.28	0.34* ± 0.06	0.29* ± 0.05
1.0	6.36 ± 1.33	2.41* ± 0.34	3.58* ± 0.75
10.0	76.21 ± 19.30	22.17* ± 3.04	27.90* ± 5.42

Note: All values are expressed in μmol/g dry weight/min as mean ± SEM. $[TC]_i$ is the initial taurocholic acid concentration in mM. C(Il) is control ileum, CF(Il) is cystic fibrosis ileum, and C(Je) is control jejunum. * = Significant difference from C(Il) to $p < 0.05$. The uptake values for CF(Il) and C(Je) at each concentration of taurocholate are not statistically different from each other.

Data taken from Fondacaro et al.[49] and converted to tabular form.

jejunum in vivo, this suggests that passive, diffusional absorption is occurring in tissue ordinarily possessing active bile acid transport capability (Table 1). Ultrastructurally, the CF ileal mucosa was indistinguishable from normal ileal mucosa and elevated fecal bile acid levels confirmed malabsorption in CF patients included in this study. The results of this study suggest that bile acid malabsorption in CF may be a primary mucosal cell defect, since CF ileal mucosa, when placed in physiological medium, failed to absorb taurocholate as well as control ileal tissue in the same setting. Therefore, this study has, likewise, described a cellular defect of CF ileal mucosa suggestive of a function lesion in the bile acid transport mechanism.[49]

V. SUMMARY

The purpose of this chapter was to present an accurate and concise description of the process of intestinal bile acid absorption. We have attempted, also, to provide some information regarding those factors which influence this process and to point out some clinical situations where bile acid malabsorption is present as a primary pathophysiologic entity. A final scenario is appropriate.

Considering the events of intestinal lipid digestion and absorption and from descriptions of the enterohepatic circulation of bile acids, certain biological and biochemical adaptations can be described for the maintenance of optimal bile acid concentrations in the upper small bowel. As first described by Tyor et al.[22] and later by Lack,[17] the enterohepatic circulation allows for the daily circulation and use of bile acids far in excess of the total body pool. Likewise, the delivery of bile acids to the upper small bowel (via the common bile duct) and the restriction of active transport to the distal small intestine provide for an optimal intraluminal concentration of bile acid in the region where lipid digestion and absorption occur. Finally, the influence of dietary lipids on micelle size and formation insures that minimal loss of bile acid to absorption takes place in these proximal regions of the intestine. One can visualize, therefore, as food moves down the small bowel and lypolysis continues, end products of this process are absorbed with the aid of bile acid micelles. Eventually, micellar size and number are reduced and the equilibrium between micelles and monomers shifts toward the monomer phase. Coincident with intraluminal lipid concentrations approaching a minimum in the ileum, the more efficient bile acid absorptive mechanisms are encountered and monomeric bile acid is now rapidly recaptured by the mucosa and the portal circulation. In the norm, this process is likely repeated six to ten times daily. If for some reason an interruption of the enterohepatic circulation occurs, the liver increases its conversion of cholesterol to bile acids in an attempt to maintain bile acid pool homeostasis. Thus, we see how the combination of the unique physicochemical nature of bile acids, the biochemical interaction of bile acids with lipids, and the biological features of intestinal absorption together provide for the physiological process of lipid digestion and absorption.

REFERENCES

1. Lack, L. and Weiner, I. M., *In vitro* absorption of bile salts by small intestine of rats and guinea pigs, *Am. J. Physiol.,* 200, 313, 1961.
2. Weiner, I. M. and Lack, L., Absorption of bile salts from the small intestine *in vivo, Am. J. Physiol.,* 202, 155, 1962.
3. Holt, P. R., Intestinal absorption of bile salts in the rat, *Am. J. Physiol.,* 207, 1, 1964.
4. Playoust, M. R. and Isselbacher, K. J., Studies on the transport and metabolism of conjugated bile salts by intestinal mucosa, *J. Clin. Invest.,* 43, 467, 1964.
5. Schiff, E. R., Small, N. C., and Dietschy, J. M., Characterization of the kinetics of the passive and active transport mechanisms for bile acid absorption in the small intestine and colon of the rat, *J. Clin. Invest.,* 51, 1351, 1972.
6. Wilson, F. A. and Treanor, L. L., Characterization of the passive and active transport mechanisms for bile acid uptake into rat isolated intestinal epithelial cells, *Biochim. Biophys. Acta,* 406, 280, 1975.
7. Fondacaro, J. D. and Rodgers, J. B., Characterization and effects of phospholipid on bile acid absorption by villi isolated from hamster small intestine, *Am. J. Dig. Dis.,* 23, 12, 1978.
8. Heaton, K. W. and Lack, L., Ileal bile salt transport: mutual inhibition in an *in vivo* system, *Am. J. Physiol.,* 214, 585, 1968.
9. Kimmich, G. A., Coupling between Na⁺ and sugar transport in small intestine, *Biochim. Biophys. Acta,* 300, 31, 1973.
10. Schultz, S. G., Frizzell, R. A., and Nellans, H. N., Ion transport by mammalian small intestine, *Ann. Rev. Physiol.,* 36, 51, 1974.
11. Gallagher, K., Mauskopf, J., Walker, J. T., and Lack, L., Ionic requirements for the active ileal bile salt transport system, *J. Lipid Res.,* 17, 572, 1976.
12. Lack, L., Walker, J. T., and Hsu, C.-Y. H., Taurocholate uptake by membrane vesicles prepared from ileal brush borders, *Life Sci.,* 20, 1607, 1977.
13. Rouse, D. J. and Lack, L., Ion requirements for taurocholate transport by ileal brush border membrane vesicles, *Life Sci.,* 25, 45, 1979.
14. Fondacaro, J. D. and Garvey, J. L., Bile acid uptake and calcium flux in brush border membrane vesicles, *Life Sci.,* 32, 1449, 1983.
15. Fondacaro, J. D. and Madden, T. B., Inhibition of Na⁺-coupled solute transport by calcium in brush border membrane vesicles, *Life Sci.,* 35, 1431, 1984.
16. Bundy, R., Mauskopf, J., Walker, J. T., and Lack, L., Interaction of uncharged bile salt derivatives with the ileal bile salt transport system, *J. Lipid Res.,* 18, 389, 1977.
17. Lack, L., Properties and biological significance of the ileal bile salt transport system, *Environ. Health Perspect.,* 33, 79, 1979.
18. Dietschy, J. M., Salomon, H. S., and Siperstein, M. D., Bile acid metabolism. I. Studies on the mechanisms of intestinal transport, *J. Clin. Invest.,* 45, 832, 1966.
19. Small, D. M., The physical chemistry of cholenic acid, in *The Bile Acids,* Vol. 1, Nair, P. P. and Kritchevsky, D., Eds., Plenum Press, New York, 1971, 249.
20. Playoust, M. R., Lack, L., and Weiner, I. M., Effects of intestinal resection on bile acid absorption in dogs, *Am. J. Physiol.,* 208, 363, 1965.
21. Wilson, F. A., Intestinal transport of bile acids, *Am. J. Physiol.,* 241(Gastrointest. Liver Physiol. 4), G38, 1981.
22. Tyor, M. P., Garbutt, J. T., and Lack, L., Metabolism and transport of bile salts in the intestine, *Am. J. Med.,* 51, 614, 1971.

23. Lack, L. and Weiner, I. M., Intestinal bile salt transport: structure-activity relationships and other properties, *Am. J. Physiol.,* 210, 1142, 1966.
24. Lester, R., Smallwood, R. A., Little, J. M., Brown, A. S., Piasecki, G. J., and Jackson, B. T., Fetal bile salt metabolism. The intestinal absorption of bile salts, *J. Clin. Invest.,* 59, 1009, 1977.
25. Little, J. M. and Lester, R., Ontogenesis of intestinal bile salt absorption in the neonatal rat, *Am. J. Physiol.,* 239(Gastrointest. Liver Physiol. 2), G319, 1980.
26. Heubi, J. E. and Fondacaro, J. D., Post-natal development of intestinal bile salt transport in the guinea pig, *Am. J. Physiol.,* 243(Gastrointest. Liver Physiol. 6), G189, 1982.
27. Fondacaro, J. D., Nathan, P., and Wright, W. E., Methionine accumulation in villi isolated from maturing hamster intestine, *J. Physiol. (London),* 241, 751, 1974.
28. Koldovsky, O., Digestion and absorption, in *Perinatal Physiology,* Stave, U., Ed., Plenum Press, New York, 1978, 337.
29. de Belle, T. C., Vaupstras, V., Vitullo, B. B., Haber, L. R., Shaffer, E., Mackie, G. G., Owen, H., Little, J. M., and Lester, R., Intestinal absorption of bile salts: immature development in the neonate, *J. Pediatr.,* 94, 472, 1979.
30. Henning, S. J., Biochemistry of intestinal development, *Environ. Health Perspect.,* 33, 9, 1979.
31. Lack, L. and Weiner, I. M., Role of the intestine during the enterohepatic circulation of bile salts, *Gastroenterology,* 52, 282, 1967.
32. Olivecrona, T. and Sjovall, J., Bile acids in rat portal blood, *Acta Physiol. Scand.,* 46, 284, 1959.
33. Norman, A. and Sjovall, J., On the transformation and enterohepatic circulation of cholic acid in the rat, bile acids and steroids 68, *J. Biol. Chem.,* 233, 872, 1958.
34. Thompson, A. B. R. and Dietschy, J. M., Intestinal lipid absorption: major extracellular and intracellular events, in *Physiology of the Gastrointestinal Tract,* Vol. 2, Johnson, L. R., Ed., Raven Press, New York, 1981, 1147.
35. Wilson, F. A. and Dietschy, J. M., Characterization of bile acid absorption across the unstirred water layer and brush border of the rat jejunum, *J. Clin. Invest.,* 51, 3015, 1972.
36. Fondacaro, J. D., Influence of dietary lipids on intestinal bile acid absorption, *Proc. Soc. Exp. Biol. Med.,* 173, 118, 1983.
37. Koga, T., Nishida, T., Miwa, H., Yamamoto, M., Kaku, K., Yao, T., and Okumura, M., Effects of dietary butter fat on fecal bile acid excretion in patients with Crohn's disease on elemental diet, *Dig. Dis. Sci.,* 29, 994, 1984.
38. Fondacaro, J. D. and Wolcott, R. H., Effects of dietary nutrients on intestinal taurocholic acid absorption, *Proc. Soc. Exp. Biol. Med.,* 168, 276, 1981.
39. Wolcott, R. H. and Fondacaro, J. D., The effects of undigested nutrients on ileal bile acid absorption, *Fed. Proc.,* 38, 952, 1979.
40. Harries, J. T., Muller, P. R., McCullum, K., Lipson, A., Roma, E., and Norman, A. P., Intestinal bile salts in cystic fibrosis, *Arch. Dis. Child.,* 54, 19, 1979.
41. Sklan, D., Budowski, P., and Hurwitz, S., Absorption of oleic and taurocholic acids from the intestine of the chick, *Biochim. Biophys. Acta,* 573, 31, 1979.
42. Roy, C. C., Lefebvre, D., Belanger, G., Chartrand, L., Lepage, G., and Weber, A., The effects of lipids on taurocholic acid absorption from the rat small intestine, *Proc. Soc. Exp. Biol. Med.,* 161, 105, 1979.
43. Carey, M. C. and Small, D. M., The characteristics of mixed micellar solutions with particular reference to bile, *Am. J. Med.,* 49, 590, 1970.
44. Donowitz, M., Ca++ in the control of active intestinal Na+ and Cl- transport: involvement in neurohumoral action, *Am. J. Physiol.,* 245(Gastrointest. Liver Physiol. 8), G165, 1983.
45. Chase, H. S., Jr. and Al-Awqati, Q., Calcium reduces the sodium permeability of luminal membrane vesicles from toad bladder, *J. Gen. Physiol.,* 81, 643, 1983.
46. Garbutt, J. T., Lack, L., and Tyor, M. P., The enterohepatic circulation of bile salts in gastrointestinal disorders, *Am. J. Med.,* 51, 627, 1971.
47. Heubi, J. E., Balistreri, W. F., Fondacaro, J. D., Partin, J. C., and Schubert, W. K., Primary bile acid malabsorption: defective *in vitro* ileal active bile acid transport, *Gastroenterology,* 83, 804, 1982.
48. Weber, A., Roy, C. C., Chartrand, L., Lepage, G., DuFour, O. L., Morin, C. L., and Lasalle, R., Relationship between bile acid malabsorption and pancreatic insufficiency in cystic fibrosis, *Gut,* 17, 295, 1976.
49. Fondacaro, J. D., Heubi, J. E., and Kellogg, F. W., Intestinal bile acid malabsorption in cystic fibrosis: a primary mucosal cell defect, *Pediatr. Res.,* 16, 494, 1982.
50. Fondacaro, J. D., unpublished observation.

Chapter 3

ABSORPTION OF FAT-SOLUBLE VITAMINS

A. Kuksis

TABLE OF CONTENTS

I. INTRODUCTION

The fat-soluble vitamins include vitamins A, D, E, and K. Although all these compounds are originally derived from isoprenoid building blocks, they possess differing chemical structures and biological functions. They are all nonpolar lipids with extremely low solubilities in aqueous media. It is customary to consider them together when examining the mechanisms involved in their absorption.[1,2] The elucidation of the absorption of the fat-soluble vitamins has contributed significantly to the understanding of the absorption of fatty acids, monoacylglycerols, cholesterol, and other lipid-soluble compounds.

II. STRUCTURE AND NOMENCLATURE

The molecular structures of the four fat-soluble vitamins are shown in Figures 1 to 4. The free hydroxyl groups of natural vitamin A, D, and E may be esterified to the common fatty acids, while commercial preparations of these vitamins may be acetylated.

The term "vitamin A" is used generically for all β-ionone derivatives, other than the provitamin A carotenoids, which exhibit the biological activity of all-*trans* retinol[3] (Figure 1). Of the more than 400 carotenoids that have been well characterized, only about 30 have provitamin A activity. For humans and experimental animals, biologically active carotenoids invariably contain at least one unsubstituted β-ionone ring. The most active carotenoid is the all-*trans* β-carotene.[4] In addition to naturally occurring forms of carotenoids and vitamin A, a large number of analogs, termed the retinoids, has been synthesized. Many of these retinoids are therapeutic against some forms of cancer.[4,5] Only a few of these compounds have been tested directly for their intestinal absorptivity. Seven *cis-trans* isomers of vitamin A have been identified, namely, the all-*trans*, 13-*cis*, 11-*cis*, 9-*cis*, 7-*cis*, and 9,13-*cis* forms. Although the all-*trans* isomer predominates in nature, the 13-*cis* and 9-*cis* forms are found in foods and liver oils, and the stereochemically hindered 11-*cis* isomer predominates in the dark-adapted eye. Isomerases exist for the conversion of the 11-*cis* and 9-*cis* isomers of retinol to the all-*trans* form, and some mechanism must exist in the eye for the energy-dependent conversion of all-*trans* retinaldehyde or retinol to the 11-*cis* form. The 13-*cis* retinoic acid, which is used therapeutically against some forms of cancer, is only slowly converted, if at all, to the all-*trans* form.[5] A significant proportion of biliary retinoyl β-glucuronide, and probably of other more polar derivatives, is reabsorbed in the gut and transported back to the liver. Thus, an enterohepatic circulation of vitamin A derivatives occurs. Vitamin A provides the chromophore of the visual pigments and is involved in the biosynthesis of glycoproteins and thus plays a role in many other biological functions.[6]

Vitamin D is unique among the vitamins because it can be synthesized endogenously.[7] The parent molecule is generated by the UV light-induced cleavage of the 9,10-bond of the B ring of 7-dehydrocholesterol, yielding precholecalciferol, which undergoes isomerization to cholecalciferol known as vitamin D_3[8] (Figure 2). Cholecalciferol hydroxylation in the liver produces 25-hydroxycholecalciferol [25-OH-D_3], which may be transformed further in the kidney to a 1-α,25-dihydroxycholecalciferol [1,25-OH)$_2D_3$] or 24R,25-dihydroxycholecalciferol [24,25-(OH)$_2D_3$]. Another metabolite, 25,26-dihydroxycholecalciferol, has been described and it has been suggested that there are other, as yet unidentified, derivatives of vitamin D in plasma and tissues. The metabolite 1,25-(OH)$_2D_3$ mediates the enhanced intestinal absorption of calcium, and it is known as the active metabolite of vitamin D. Much recent research has been directed at studying its formation in the kidney and its action in the intestinal mucosa.[9]

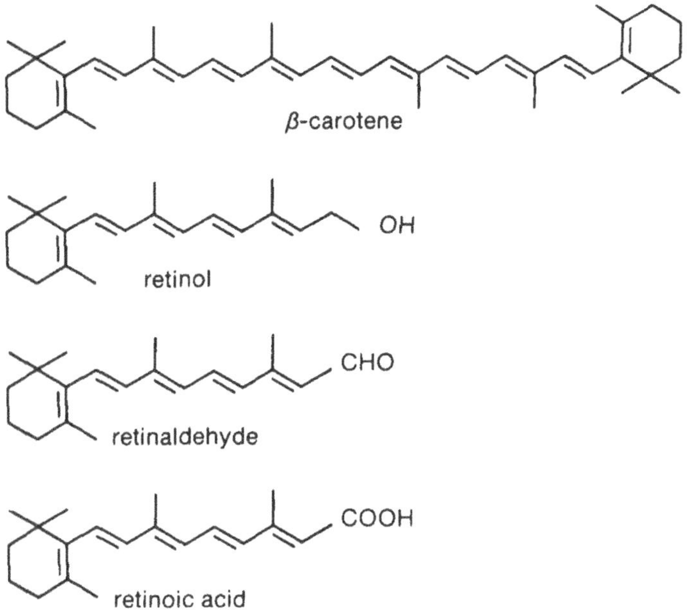

FIGURE 1. Structure of all *trans* β-carotene, all-*trans* retinol, all-*trans* retinaldehyde or retinal, and all-*trans* retinoic acid.

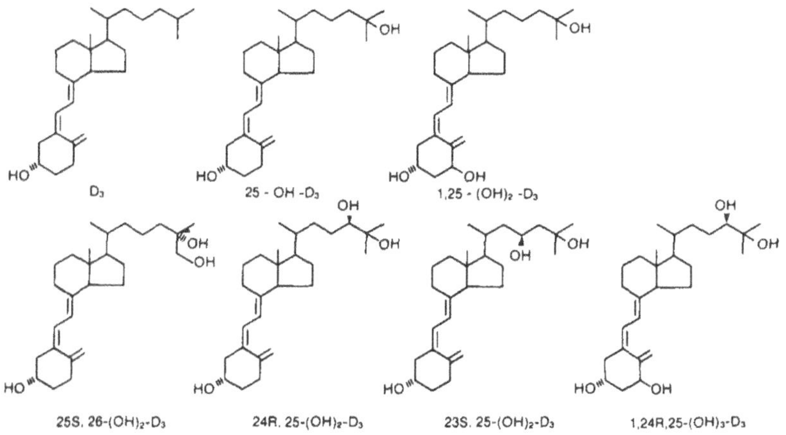

FIGURE 2. Structure of vitamin D_3 and its major metabolites: D_3, vitamin D_3 or cholecalciferol; 25-OH-D_3, 25-hydroxy-cholecalciferol; 1,25-$(OH)_2$-D_3, 1-α,25-dihydroxy cholecalciferol; 24R,25-$(OH)_2$-D_3, 24R,25-dihydroxy cholecalciferol; 25S,26-$(OH)_2$-D_3, 25S,26-dihydroxy cholecalciferol; 1,24R,25-$(OH)_3$-D_3, 1,24R,25-trihydroxy cholecalciferol.

Vitamin E represents a class of eight naturally occurring phenols, known as the tocopherols and tocotrienols (Figure 3). The four tocopherols (α, β, γ, and δ) possess a saturated isoprenoid side chain and differ from each other in the number and position of aromatic methyl groups.[10] Because these compounds possess varying degrees of biological activity, it is important to differentiate among them. The tocotrienols are lesser components of food and have the same structure as the tocopherols, except for one additional double bond per isoprene unit in the side chain. The dietary RRR-α-tocopherol equivalents are[11] mg RRR-α-tocopherol × 1.0; mg RRR-β-tocopherol × 0.4;

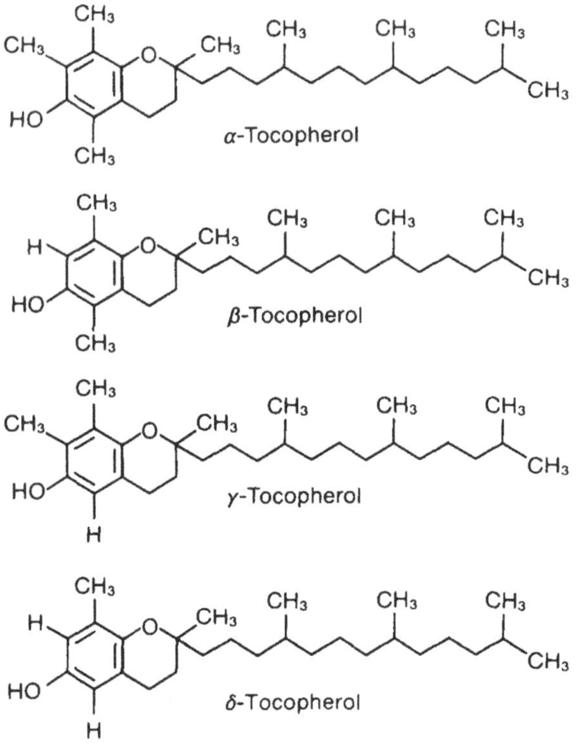

CH₃ labels...

α-Tocopherol

β-Tocopherol

γ-Tocopherol

δ-Tocopherol

FIGURE 3.   Structures of members of the vitamin E complex: α-tocopherol, 2,5,7,8-tetramethyl-2-(4′,8′,12′-trimethyltridecyl)-6-chromane, also 2R, 4′R, 8′R-α-tocopherol or RRR-α-tocopherol; β-tocopherol, 2,5,8-trimethyl-2-(4,8,12-trimethyl tridecyl)-6-chromanol; γ-tocopherol, 2,7,8-trimethyl-2-(4,8,12-trimethyltridecyl)-6-chromanol; δ-tocopherol, 2,8-dimethyl-2-(4,8,12-trimethyltridecyl)-6-chromanol.

mg RRR-γ-tocopherol × 0.1; mg RRR-α-tocotrienol × 0.25; mg all-*rac*-α-tocopherol acetate × 0.67; and IU all-*rac*-α-tocopherol acetate (formerly designated DL-α-tocopherol) × 0.67. Vitamin E is thought to act as a biological antioxidant, which protects tissues against oxidation. It may also play a more specific biochemical role.[12]

Vitamin K occurs in two forms, which differ in the degree of saturation of the isoprenoid side chain attached to the benzoquinoid nucleus (Figure 4). Phylloquinone (vitamin K₁) has only a single double bond, which occurs in the isoprenoid side chain, while menaquinone (vitamin K₂ has double bonds in each of the isoprene units of the side chain). The number of isoprene residues in the side chains of some of the menaquinones also differs, and thus there are subtle differences in the total interaction of the K vitamins with any protein or membrane surface. Vitamin K is needed for the synthesis and release of clotting factors from the liver.[13] Vitamin K₁ is of plant origin and is especially concentrated in green leafy vegetables. The vitamin K₂ group is synthesized by specific strains of bacteria in the terminal ileum and large bowel. Normally, its absorption in this region of the bowel is sufficient to prevent vitamin K deficiency in man.[2]

Vitamin A is mainly ingested in the form of long-chain fatty acid esters, predominantly retinyl palmitate, whereas the fatty acid esters of vitamin D₃ occur in minor amounts in the diet. Foods fortified with vitamin E contain chiefly tocopheryl acetate or other synthetic esters of this vitamin.

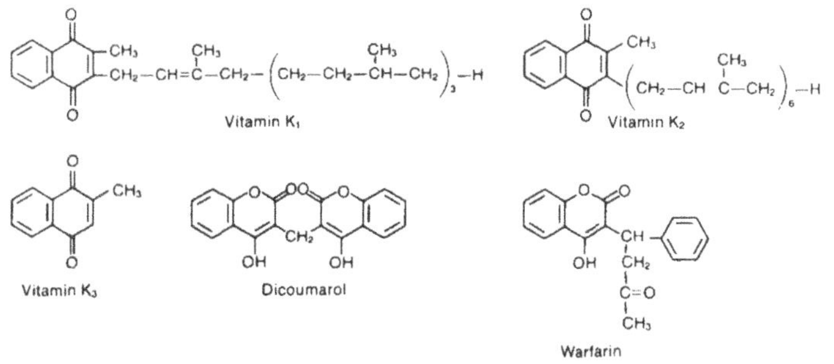

FIGURE 4. Structures of vitamin K and two antagonists: vitamin K₁, 3-methyl-3-phytyl-1,4-naphthoquinone or phylloquinone; vitamin K₂, 2-methyl-3-difarnesyl-1,4-naphthoquinone or menaquinone, also other multi-prenylmenaquinones; vitamin K₃, 2-methyl-1,4-naphthoquinone or menadione; dicoumarol, 3,3′-methylene-bis-(4-hydroxycoumarin) or bis-hydroxy-coumarin; warfarin, 3-(α-acetonylbenzyl)-4-hydroxycoumarin.

III. METHODS OF EXTRACTION AND ANALYSIS

Until recently, retinol and other fat-soluble vitamins were determined by spectrometric methods, colorimetric assays, or fluorescence assays with a minimum of preliminary work-up of the samples. However, all three of these methods are relatively time consuming and nonselective, even though they do achieve adequate sensitivity.[14] Many of the methods require extraction of the fat-soluble vitamins from the tissues into an organic solvent, which is then evaporated to dryness.[15,16] This evaporation step, even when performed under nitrogen, can lead to excessive loss of retinol and other fat-soluble vitamins.[17,18] Thus, extraction methods which do not require this evaporation step would be preferable for certain measurements.

A. Solvent Extraction

The fat-soluble vitamins are readily extracted with the conventional fat solvents, such as chloroform-methanol 2:1[14] or ethanol-diethyl ether[19] in the presence of antioxidants. These solvents are eventually evaporated to dryness and the lipid residues are taken up in other solvents (e.g., hexane) for further chromatographic separation or for a direct spectrometric examination. However, for rapid analyses[17,18] the solvent evaporation step is undesirable and it may result in losses of material due to autoxidation. In many instances it has proven convenient to saponify the sample and to recover the vitamins in the unsaponifiable matter by hexane extraction.[20] Obviously, this leads to a loss of information about the nature of combined forms of the vitamins in the tissues or foods.

For various reasons it is desirable to remove any proteins by precipitation. Goodman et al.[21] have proposed the extraction of retinol after precipitation of the protein by 5% perchloric acid. Nierenberg[17,18] has found that this routine produces inconsistent results, because retinol and retinoic acid are both rapidly destroyed when exposed to 0.83% perchloric acid. Perchloric acid causes less destruction of β-carotene, retinol, and retinoic acid when they are in a serum or plasma matrix. Similar difficulties may be encountered when using trifluoroacetic acid as a means of protein precipitation.

Zaspel and Csallary[22] have proposed a rapid procedure for specifically measuring vitamin E in tissue and plasma using acetone. Burton et al.[23] have proposed a new, general method for lipid extraction and measurement of vitamin E and other fat-soluble vitamins. The new extraction procedure uses a combination of sodium dodecylsulfate, ethanol, and *n*-heptane. It is mild and efficient and allows the completion of the extraction in less than 5 min. This method also results in cleaner samples for high-pressure liquid chromatography (HPLC) analyses, at least for vitamin E, which is the only fat-soluble vitamin thus far tested. However, the extraction and analysis of β-carotene and vitamins A, D, and K should also be satisfactory. The extraction of the fat-soluble vitamins into *n*-butanol-ethyl acetate 1:1 has given excellent recoveries and these extracts have been found to be suitable for high-pressure liquid chromatography without evaporation.[17]

The measurement of vitamin K_1 in infant formulas has created some controversy due to its presence in coated beadlet form. The discrepancies in the analytical measurements have precipitated an exchange of letters,[24,25] which are instructive to review in regard to general difficulties of accurate measurement of the fat-soluble vitamins. It appears that an important result of this work is that the purity of the chromatographic peaks measured as vitamin K_1 be supported by mass spectrometric (MS) data.[26]

The lipid-soluble vitamin extracts may be freed from nonlipid components by chromatography on DEAE-Sephadex A-25 columns in the hydroxyl form equilibrated with methanol. The sample is applied to the column in methanol (50 mℓ) and an additional 100 mℓ of methanol is passed through the column to elute retinol and its esters[27] and other fat-soluble vitamins.

B. Chromatography

An accurate determination of the absorption of the fat-soluble vitamins requires effective methods for their isolation, precise identification, and quantitation. Although the methods differ somewhat for each vitamin, the most recent methods include as a basic routine HPLC, which can usually analyze more than one fat-soluble vitamin at a time. Definitive methods of quantitation of fat-soluble vitamins require the use of mass spectrometry with selected ion monitoring and stable isotope-labeled internal standards. Likewise, definitive characterization and identification of the metabolites of the fat-soluble vitamins require combined chromatographic and MS approaches.[28]

The separation of retinol and its fatty acid esters may be effected on columns of alumina, which have been deactivated with 10% water. The lipid residue is applied to the column in hexane and the retinyl esters are eluted with 2% diethyl ether in hexane, while free retinol is recovered with 50% diethyl ether in hexane.[29] When the efficiency of the separation was tested less than 0.5% [³H]retinol was eluted in the retinyl ester fraction and about 96% in the retinol fraction, giving a total recovery of more than 96%. This routine is commonly employed for the determination of the ratio of esterified to free radioactive retinol.[30]

Lipid extracts containing fat-soluble vitamins may be applied to thin-layer adsorbent plates in hexane and the plates developed in light petroleum (bp 60 to 70)-diethyl ether 85:15 as ascending solvent.[31] In this system retinyl palmitate is recovered in the front fraction and retinol in fractions about halfway down the plate. Imai et al.[32] have used thin-layer chromatography (TLC) on silica gel 60 in a dark place to isolate vitamin E from various aqueous solutions, which, prior to plating, were diluted approximately 1:1 with ethanol. The plates were developed with chloroform. The vitamin E was located as a fluorescent band at approximately $R_f = 0.5$.

The application of gas-liquid chromatography (GLC) to the separation of fat-soluble vitamins has been reviewed by Croteau and Ronald.[33] Only vitamins D and E have been readily measured by this method, while vitamins A and K have presented problems.

Reversed-phase HPLC for separation of fat-soluble vitamins has recently been described by several laboratories and it offers major advantages for high resolution and quantitative recoveries with little or no production of artefacts.[17,18,20,22,30,34-36] Its adaptation to the routine determinations provides a valuable supplement to the earlier methods of adsorption chromatography on alumina columns or silica gel plates. An isocratic HPLC method specifically developed to allow simple and rapid determination of retinol concentrations in serum and plasma has been reported by Nierenberg.[17] Rettinol and retinol acetate (the internal standard) are extracted into butanol-ethyl acetate. Separation is achieved on a reversed-phase C_{18}-column, with a mobile phase consisting of acetonitrile-1% ammonium acetate (89:11) and UV detection at 313 nm. Recoveries of both retinol and the internal standard were 100%, and both compounds were stable in the extraction solvent for at least 2.5 hr. Effective separations of esterified and free retinol by HPLC have been obtained by Helgerud et al.[30] on adsorption columns using 100% methanol as the eluting solvent. Napoli et al.[27] have used reversed-phase HPLC columns to resolve retinol, retinal, and the linoleate, palmitate, plus oleate and stearate esters of retinol. The column was eluted with water-methanol-2-propanol 1:3:1 followed by a linear gradient (over 5 min) to methanol-2-propanol 3:1.

IV. GENERAL FEATURES OF ABSORPTION OF FAT-SOLUBLE VITAMINS

The fat-soluble vitamins in the diet are believed to be dissolved in the dietary fats, which consist largely of triacylglycerols and glycerophospholipids, with smaller amounts of free fatty acids, steryl esters, and free sterols. During the digestion, micellarization, and absorption of the dietary fats the fat-soluble vitamins are believed to undergo several transformations in common, although the exact chronological sequence of the events varies depending on whether the vitamin is present in a free or bound form. The first part of this review considers the common features and differences in the digestion and absorption of the fat-soluble vitamins, while in the second part, some of the factors which influence absorption of lipid-soluble vitamins are considered in greater detail.

A. Digestion

To be absorbed, the dietary fats must first be converted to the free acids, monoacylglycerols, lysophospholipids, and free fatty alcohols, by processes that take place primarily in the lumen of the small intestine under the influence of pancreatic lipases and bile salts. The dietary fats are first acted upon by the enzymes of the mouth. The lingual lipase secreted by the lingual serous glands of the tongue is a potent enzyme that hydrolyzes long-chain triacylglycerols.[37] It is not known whether or not this enzyme also attacks the acyl esters of the fat-soluble vitamins. The enzymes involved in the hydrolysis of vitamin A esters have been studied in considerable detail by Ganguly,[38] who has proposed the existence of both pancreatic and brush border esterases. The bile salt-activated milk lipase also hydrolyzes such esters as p-nitrophenylacetate and other simple esters.[37] The acylglycerol and glycerophospholipid hydrolysis is completed in the small intestine by the action of pancreatic lipase and phospholipases in the presence of bile salts.[37] The pancreatic lipase does not attack esters of secondary alcohols directly, while the phospholipases do attack.

The pancreatic juice also contains a nonspecific carboxylic ester hydrolase, which is responsible for the hydrolysis of the cholesteryl esters not absorbed in the ester form.[39] In addition, this enzyme catalyzes the hydrolysis of a variety of other esters including p-nitrophenylacetate, β-naphthyl acetate and laurate, and the long-chain monoacylglycerols,[40,41] including the 2-monoacylglycerols.[42] It requires bile salts for activation, except for the more water-soluble substrates which are also more rapidly hydrolyzed in the presence of the bile salts.[43] It is believed that this enzyme is responsible for the hydrolysis of the esters of vitamin A, D, and E.[44,45] The nonspecific lipase from human[46] and porcine[47] pancreatic juice has been purified and characterized. It hydrolyzes cholesteryl esters, esters of vitamin A, D_3, and E, and glyceryl esters solubilized by bile salts. Its activity against vitamin E acetate can only be demonstrated in the presence of taurocholate. Two sites of bile salt recognition have been proposed for the human enzyme:[46] one site, specific for the 3-α,7-α-hydroxyl group of cholanic acid, is thought to induce dimerization and activation of the enzyme; the other, unspecific toward bile salt hydroxylation, is considered to be located at the active center and presumably is responsible for substrate recognition.[46]

All vitamin esters are hydrolyzed by the pancreatic ester hydrolase, as shown in studies with human pancreatic juice carried out by Lombardo and Guy[46] and by Mathias et al.[48,49] Hydrolysis takes place in the intestinal lumen during solubilization of the vitamin esters in the mixed micelles.[50] Any vitamin esters not hydrolyzed in the micelle phase are subject to hydrolysis on the surface of the mucosal brush border membrane or inside the enterocytes.[51,52] Recently, Mathias et al.[48,49] have demonstrated in rats the existence of a mucosal ester hydrolase, which is distinct from pancreatic carboxylic ester hydrolase and capable of splitting tocopheryl acetate. It is not known whether or not this enzyme is also present in the human jejunum or is active against other esters of fat-soluble vitamins. The studies of Mathias et al.[49] suggest that micellar solubilization of tocopheryl acetate without prior luminal hydrolysis may be sufficient for the uptake of the vitamin by the jejunal enterocyte.

B. Micellarization and the "Hydrophobic Continuum"

The lipolysis products of glycerolipids along with the fat-soluble vitamins and other fat-soluble substances, such as cholesterol, are subject to the process of micelle formation, in which the lipid components interact with bile salts to form mixed aggregates or micelles. According to Hofmann and Borgstrom,[53] the lypolytic products are dispersed as if there were two phases in the intestinal lumen: one an emulsion phase of lipid particles in the 2000- to 50,000-Å range suspended in an aqueous medium; the other a mixed micellar phase made up of particles 20 to 30 Å, which is formed in the presence of bile salts. The mixed micelles are polymolecular aggregates with a completely lipophilic core where water-insoluble substances, such as fat-soluble vitamins, are dissolved, whereas the outer surface of the mixed micelles is hydrated by the surrounding water phase.[54] In the past it has been assumed that the formation of mixed micelles is a prerequisite for removing the lipid-soluble vitamins from the oily phase and for their solubilization in the aqueous micellar phase, from which the fat-soluble vitamins can be taken up by the intestinal mucosal cells. Imai et al.[32] have determined the intestinal absorption of *dl-α*-tocopherol from various micellar solutions by the *in situ* recirculating perfusion technique in rat small intestine. The relative absorption ratio of the vitamin was sodium taurocholate > sodium taurodeoxycholate > polysorbate 80 micellar solutions. The addition of egg lecithin to all micellar solutions caused a decrease in the absorption. The absorption order did not necessarily have a simple correlation with the vitamin solubilization in these micellar solutions, but rather correlated with the micellar size and with net water flux in the intestinal lumen. It was suggested that vitamin E is absorbed together with water as micelles, i.e., vitamin E in micelles is absorbed by solvent drag. Gallo-Torres et al.[55] had earlier discussed the effect of dietary fat on the micelle formation and the appearance of vitamin E in the lymph. In connection with the micellar solubilization of vitamin E it is necessary to mention the formation of α-tocopherol complexes with fatty acids, which Erin et al.[56] have regarded as a mechanism of membrane stabilization by vitamin E against the damaging action of free fatty acids. The interaction between α-tocopherol and fatty acid is noncovalent, since the complex formation occurs almost instantaneously after the mixing of the reagents. The values of equilibrium constants for unsaturated fatty acids are much higher than those for saturated fatty acids and within a given range linearly dependent on the number of double bonds in the fatty acid molecule. Vitamin E can form complexes with the free fatty acids incorporated into phosphatidylcholine liposomes or into skeletal muscle sarcoplasmic reticulum membranes.

According to Patton,[57] the division of duodenal lipids into micellar and oil phases is an oversimplification based on an artifact of centrifugation. There is evidence of the presence of other types of liquid phases that occur in the duodenum during fat digestion, depending on the chemical composition of the dietary fat and the physiological state of the animal. Obviously, there is often far too much lipid in duodenal content to be solubilized by the available bile salt.[54,58,59] On the basis of a microscopic observation of direct transfer of colored or fluorescent compounds from the oil phase to the enzyme product phase, Patton and Carey[60] have concluded that there is a continuous hydrocarbon domain in triacylglycerols, which remains intact during the enzymatic conversion of the triacylglycerols to fatty acids and monoacylglycerols. The preservation of this hydrophobic domain during fat digestion allows nonpolar molecules (trace lipids) that are dissolved within it to flow from nondispersible triacylglycerols to the products of fat digestion, which are dispersed by bile salts and absorbed.[61-63] Such a concept of an uninterrupted hydrophobic domain that can carry many different molecules during fat digestion has profound significance for the absorption of fat-soluble vitamins. Patton[57] has proposed a "zipper model" of fat digestion to illustrate the hydrocarbon continuum. The active site of the lipase is depicted as the zipper. The movement of the nonpolar lipid among the hydrocarbon chains is based on the experiments with the hydrocarbon probe molecules. Also, within this region are lipid and bile salt monomers. It is assumed that the fatty acid chains that make up the hydrocarbon domain are in the liquid state. The idea that lipids can flow en masse through biological systems with an intact hydrophobic continuum is not new. Scow et al.[64,65] and Smith and Scow[66] have suggested that "rivers of fatty acids" flow in membranes between chylomicrons undergoing lipolysis at the endothelial cell surface and sites of deposition within parenchymal cells. In connection with absorption of the fat-soluble vitamins it is significant that this hypothesis does away with the concept of the unstirred water layer.

In the past the penetration of the unstirred water layer has been believed to be the first step associated with the uptake of the lipids by the villus cell. The unstirred water layer has been claimed[67,68] to exist as an aqueous phase immediately adjacent to the microvillus surface and to remain relatively undisturbed in spite of vigorous mixing of the lumenal fluid. The fat-soluble vitamins would have to be moved toward the cell membrane in the form of mixed micelles. However, the mechanism by which the fat-soluble vitamins and the lipids penetrate from the micellar phase into the mucosal cells has never been clarified. It has been assumed[2,50] that after collision of micellar particles with the membrane of the villus cell the components of the mixed micelles are partitioned into the monomeric phase and that lipids including the fat-soluble vitamins, but not bile salts, pass through the lipid phase of the mucosal cell membranes to reach the cytoplasm. Comparative intestinal perfusion studies with vitamin A, β-carotene, 25-hydroxyvitamin D_3, and vitamin K_1 have shown[2] that the rate of absorption increases two- to fourfold over control values with increased speed of perfusion, while that of vitamin D_3 and vitamin K_2 changes minimally. The concept of the hydrophobic continuum avoids the necessity of solving this problem, but leaves us with some unexplained kinetic delays in the absorption of the lipids, which, however, may be artefacts of the experimental systems based on deliberate introduction of micellar solutions. These conditions may not exist in vivo.

C. Mechanism of Uptake

According to Hollander,[2] the absorption of vitamin A in the range of physiological intraluminal concentrations takes place by a saturable, but energy-independent, mechanism. Therefore, it was concluded that a carrier-mediated, passive absorption process is involved. The retinol may penetrate the brush border membrane by passive diffusion, but after penetration the vitamin is transported by retinol-binding protein present on the cytosol side of the absorptive cell membrane. Recently, Ong[69] has isolated a novel retinol binding protein, which is present in the intestine of the adult rat at 500-fold higher levels than in any other tissue examined, with decreasing gradient from jejunum to colon. The protein is distinct from other known retinol-binding proteins, but may have similarities to known lipid carrier proteins, such as Z protein.[70] It is a single polypeptide chain with a molecular weight of about 16,000. The protein binds all-*trans* retinol as an endogenous ligand. At high intraluminal concentrations of vitamin A, transport by the gut takes place by simple passive diffusion. It appears that vitamin A can be equally well absorbed by the proximal and distal portion of the small bowel of the rat.[2] There is evidence that small amounts of unesterified vitamin A can be absorbed directly into the portal vein.[71] According to extensive in vivo and in vitro studies with rats,[2] it is likely that vitamins D and E are taken up by the gut by means of nonsaturable passive diffusion processes, whereas the absorption of vitamin K_1 is saturable and dependent on energy mediation for transport. In contrast to vitamin K_1, which is taken up by the proximal jejunum, vitamin K_2, which originates from intestinal bacterial sources, is absorbed by the distal portion of the ileum and the colon and follows nonsaturable passive diffusion. In the case of vitamin D, it is still uncertain whether the vitamin is taken up mainly by the jejunum or the ileum.[72,73] The intestinal site of optimum absorption of fat-soluble vitamins in man has not been specifically determined.

Following the passage through the lipid membrane of the villus cell the fate of the fat-soluble vitamins diverges to some extent. Retinol is esterified with long-chain fatty acids, primarily palmitic acid.[74] Negligible amounts of vitamin D may also be esterified,[75,76] but tocopherol and vitamin K remain largely in the free form. The enzymatic systems involved in this esterification of vitamin A and D have not been identified, but the cholesterol esterase of the villus cell may be suggested as a potential source of this activity. The results of immunological studies suggest that cholesterol ester digestion in the lumen and cholesterol ester synthesis in the rat small intestine may be mediated by the same enzyme.[77,78] Since the lumenal cholesterol esterase is known to hydrolyze the esters of vitamins A and D, it is possible that it may also perform the resynthesis of the vitamin esters inside the villus cell. It is known that the action of pancreatic lipase is also reversible,[79,80] but thus far there is no evidence for the intracellular function of this enzyme. The smooth endoplasmic reticulum is the primary intracellular site for activation of dietary and endogenous fatty acids and for their transfer to 2-monoacylglycerols and sn-glycerol-3-phosphate.[81] Furthermore, the smooth endoplasmic reticulum also forms lipid droplets within the confines of bulbous expansions of its membranes, which, after reaching a certain size, are released as discrete lipid-containing vesicles and then accumulate in the Golgi complex.[50] It is here that the synthesis of chylomicrons and other triacylglycerol-rich lipoproteins is completed by assembly of the lipid aggregates with apoproteins. The lipoproteins render the water-insoluble vitamins and lipids transportable in the aqueous medium of the extracellular space.[50,82] It is not known what mechanisms are involved in packaging the fat-soluble vitamins in the lipid droplets of the smooth endoplasmic reticulum. In view of the isolation and identification of a retinol binding protein,[83] it may be assumed that vitamin A would be transported by the intracellular retinol binding protein to the membranes of the smooth endoplasmic reticulum to be esterified and incorporated into the lipid droplets. Helgerud et al.[31,84] have shown that rat and human intestinal microsomes catalyze the formation of retinyl esters by an acyl-CoA:retinol acyltransferase with several properties in common with ACAT located in the same subcellular fraction. This activity was not due simply to a reversal of the hydrolase reaction. The activity of the microsomal ARAT described by Helgerud et al.[30] was estimated to account for all retinyl esters recovered in lymph of rat, suggesting that the enzyme has physiological importance. It was speculated that the ARAT from human intestine could account for the formation of all the retinyl esters appearing in the chylomicrons.[84] Thus, practically no esterification was produced by palmitate itself, whereas a marked stimulation was observed when an acyl-CoA generating system or preformed acyl-CoA was used. In addition to the similarities already discussed, the ARAT and ACAT have several other properties in common. The inhibition of ARAT activity by DTNB suggests that the enzyme is dependent on reduced thiol groups as previously shown for intestinal ACAT.[30] Inhibition by the detergent taurocholate is also a common feature as is the pH optimum. Furthermore, the specific activity of the two enzymes with oleoyl-CoA as the acyl donor was almost the same and highly correlated during feeding and fasting. Helgerud et al.[31,84] have shown that the in vitro incorporation of activated palmitate, oleate, and stearate was of the same relative order as the composition of the predominating retinyl esters in rat and human lymph.[86] Most rat tissues including intestinal mucosa contain a cellular retinol binding protein,[87] and it is possible that the esterification of retinol occurs while it is still associated with this protein.[88] Specific cytosolic binding proteins may also exist for the other fat-soluble vitamins, e.g., α-tocopherol[89] and the fatty acid[90] binding protein, and participate in the translocation of the fat-soluble vitamins.

Studies with radioactive fat-soluble vitamins in man[91-96] have demonstrated that chylomicrons in the lymph represent the major route of transport for lipid-soluble vitamins. These vitamins are eventually taken up by the liver from either the chylomicrons or the chylomicron remnants. Blomhoff et al.[97] have presented data which suggest that the absorption of the vitamins A and D_3 differ to some extent in the rat intestine. When the two labeled vitamins were given simultaneously, [^{14}C]-labeled vitamin D_3 appeared later in the intestinal lymph than [^{3}H]retinol and the rate of absorption of vitamin D_3 was still maximal at a time when that of retinol had declined. Hollander[2] has indicated that retinol and vitamin D_3 are absorbed by facilitated diffusion and passive diffusion, respectively. However, the different time course of absorption may also reflect differences in intraluminal events, rather than differences in absorption mechanisms or intracellular processing. No esterified vitamin D_3 was found in the lymph. Most of the radioactivity was recovered as free vitamin D_3, but a significant amount of [^{14}C]radioactivity recovered in the protein fraction co-migrated with authentic 25-hydroxyvitamin D_3. 25-Hydroxylase activity against vitamin D has been detected in the intestinal mucosa of chicken.[98] The latter observation indicates that vitamin D_3, which is absorbed into the lymph, is not exclusively transported in chylomicrons and their remnants. The retinyl ester carried by the chylomicrons is relatively nonexchangeable, while significant amounts of vitamin D_3 are transferred from chylomicrons to other lymph or plasma fractions. This is apparently due to the location of the vitamin A esters in the interior of the chylomicrons, while the vitamin D_3 may be carried in the surface monolayer. Barua and Olson[36] have studied the metabolism of the radioactive analog of 4,4-difluororetinyl acetate and its biological activity on growth in rats. After oral administration of the analog in oil to rats, difluororetinol, difluororetinyl palmitate and related esters, 4-oxoretinol, 4-oxoretinoic acid, and polar conjugated derivatives, were identified in the intestine, liver, kidney, and/or blood. All-*trans* difluororetinyl acetate showed about 26% of the biological activity of all-

trans retinyl acetate in the rat growth assay. Clearly, difluororetinyl acetate is hydrolyzed in the intestinal tract, is esterified with long-chain fatty acids, and is stored as the ester in significant amounts in the liver.

The absorption of phylloquinone and the menaquinones requires bile and pancreatic juice for maximum effectiveness.[99] Dietary vitamin K is absorbed in the small bowel, is incorporated into chylomicrons, and appears in the lymph.[93] It apparently undergoes extensive transformation in tissues, since Rietz et al.[100] have shown that the human liver contains menaquinone-7, menaquinone-8, menaquinone-9(2H), menaquinone-9(4H), menaquinone-10, and menaquinone-11, in addition to phylloquinone.

D. Efficiency of Uptake

An interesting aspect of intestinal uptake of lipid-soluble vitamins concerns the absorption efficiency. According to the evidence obtained with radiolabeled lipid-soluble vitamins administered in physiological doses in a formula meal, the absorption appears to be incomplete.[92-95] For vitamins A, D, and K, an average absorption rate of about 50 to 80% was found, but only 20 to 30% of vitamin E was absorbed. The rat can absorb 6 mg retinol per day,[101] corresponding to about 17 nmol/min, with most of it leaving the intestine as retinyl palmitate. Using another method of measuring the absorption efficiency of vitamin E,[73] it was observed that in man and rats the absorption rates were 30 to 80%, when given in tracer amounts (micrograms) and clearly decreased with increased doses (milligrams). On the average, about one third of dietary vitamin E seems to be absorbed into the body.[102] There is insufficient information available to distinguish among the relative absorbabilities of the α-, β-, γ-, and δ-tocopherols as well as the corresponding tocotrienols. However, there is evidence to indicate that the absorption of α-tocopherol is higher than that of the β-, γ-, or δ-tocopherols[102] despite its relatively lower polarity. Recent results in man suggest that intestinal uptake and/or plasma transport is more efficient for α-tocopherol.[103] The absorption efficiency of vitamin E depends on the maturity of infants. Jansson et al.[104] have found that 25 mg of water-soluble *dl*-α-tocopherol acetate is adequate to increase serum levels of α-tocopherol in VLWB-infants during the first week of life. These serum levels are comparable to those achieved using larger doses of vitamin E. It is suggested that there is an upper limit to the enteric absorption of vitamin E, owing to a limited plasma transport capacity for vitamin E in such infants.

The efficiency of absorption of vitamin K has been measured at 40 to 80%, depending upon the vehicle in which the vitamin is administered and the extent of the enterohepatic circulation. When isotopically labeled phylloquinone was given orally to animals[105] and humans[106] in doses ranging from the physiological to the pharmacological, the vitamin appeared in the plasma within 20 min and reached a peak in the plasma in 2 hr. When 1 mg of phylloquinone was given intravenously, peak plasma levels of about 100 ng/mℓ were observed. The concentration of the vitamin then declined exponentially to low values over a period of 48 to 72 hr to reach postabsorption values of 1 to 5 ng/mℓ.[107,108] During this period it appeared to be transferred from the chylomicron remnants to the liver to be incorporated into very low-density lipoproteins and ultimately distributed to tissues via low-density lipoproteins. This transfer system also applies to dietary sterols and carotenoids. Thus far, a specific plasma carrier protein for vitamin K has not been identified. The turnover of vitamin K in the animal body is rapid, as the total body pool size is very small. From average daily intakes of vitamin K, body pool sizes were estimated to be 50 to 100 μg, which is less than the body pool size for vitamin B_{12} and extraordinarily low for a fat-soluble vitamin.[109]

V. REGULATION OF FAT-SOLUBLE VITAMIN ABSORPTION

In view of the discussion of the mechanism of absorption, it can be easily appreciated that absorption of lipid-soluble vitamins may be influenced by various alimentary factors or other circumstances. Some of the influences may be positive, because they promote the absorption of the vitamins, while others may be negative and constitute an interference. These influences may be exerted at the different levels of fat digestion and absorption and may affect the vitamin assimilation directly or indirectly.

A. Solubilization of the Vitamins

In regard to the micellarization of the vitamins in the intestinal lumen, both bile salts and pancreatic juice are essential components for the solubilization of lipid-soluble vitamins prior to intestinal absorption. Pancreatic lipolysis of dietary fat is a prerequisite for the formation of 2-monoacylglycerols and fatty acids, which generate the mixed micellar phase.[110] An incomplete hydrolysis of dietary triacylglycerols impairs absorption of the lipid-soluble vitamins, because their partitioning is greatly favored into the oily phase rather than into the micellar phase. In addition, fat contained in a meal delays gastric emptying. The bile acids are the major intestinal detergents and are obligatory for the formation of mixed micelles. The addition of fatty acids or lecithin to micellar solutions containing bile acids expands the micelles and thereby increases the solubilization of the fat-soluble vitamins within the micellar particles. In addition, the bile acids stimulate the activity of pancreatic lipase and are important co-factors for optimal activities of carboxylic ester hydrolase and phospholipase A_2 from pancreatic juice.[50] It is therefore obvious that the absorption of the fat-soluble vitamins is markedly decreased in patients with intestinal disorders, especially biliary or pancreatic insufficiencies. This implication has been confirmed in studies with labeled vitamins.[95,96] When the bile acid concentration is increased above the critical micellar concentration, there is no further increase in the absorption of the fat-soluble vitamins.[2]

The physicochemical properties of the fat-soluble vitamins, especially their lipid solubility or polarity, may play an important role in the absorption of these vitamins. As can be demonstrated by reversed-phase liquid chromatography,[72,73] among the fat-soluble vitamins retinol is relatively the most polar. This fact is consistent with the higher absorption of retinol in comparison to the other vitamins, as noted above. The fact that the relative polarity of vitamin E is lower than that of vitamin D_3 or retinol may account for the lower solubilization of vitamin E either in the mixed micelles or during the uptake into the villus cells, thus accounting for the relatively less efficient intestinal absorption. Whether β-, γ-, and δ-tocopherols or the tocotrienols are competing with α-tocopherol for sites of absorption and for transport is not known. Peake et al.[111] found no marked difference in the absorption of α- and γ-tocopherols in rats with a lymphatic cannula, but they concluded that γ-tocopherol disappeared faster from tissues. It is, therefore, not possible to conclude with certainty at the present time which barriers may be responsible for the inefficient absorption of vitamin E. It is not known whether the bottleneck is related to a limited packing capacity of mixed micelles for vitamin E, to a limited partitioning of the vitamin in the lipid membrane of the villus cells, or to limitations in the incorporation of α-tocopherol into chylomicrons and in transport of the vitamin.

B. Effect of Dietary Fat

The content and the nature of the dietary fat constitute other important aspects of the solubilization and the overall absorption of lipid-soluble vitamins when administered orally to humans or animals for experimental or therapeutic purposes. Animal studies have revealed a better absorption of vitamins A and E from water-miscible emulsions than from oily solutions,[112,113] but only when given in relatively high doses of milligram amounts. When α-tocopherol was administered *in the dose levels of micrograms*, intestinal absorption was quantitatively the same from either arachis oil, Tween® emulsion, or alcoholic suspension,[114] and vitamins A and E were assumed to be taken up directly from the water-miscible emulsions by the mucosal cells, by-passing the luminal mixed micelles, and to be transported both by the lymphatic and portal pathways.[114-116]

Conflicting results are reported in the literature with regard to the effects of medium-chain triacylglycerols on the intestinal absorption of vitamins A and E. In studies with rats, an enhancement of α-tocopherol absorption was observed when administered in an emulsion containing medium-chain triacylglycerols. Therefore, it was concluded that the medium-chain triacylglycerols facilitate the solubilization and, accordingly, increase the concentration of α-tocopherol in the lumenal mixed micelles.[110] However, in other investigations on the absorption of vitamin A in rats, the intestinal absorption was similar whether the vitamin A emulsion used contained medium-chain or long-chain triacylglycerols.[117] Tests with healthy subjects and patients suffering from gastrointestinal diseases did not show any difference in the absorption rate of vitamin A when dissolved either in arachis oil or in medium-chain triacylglycerols.[118] Moreover, the vitamin E status of infants fed a formula in which a significant percentage of the fat was in the form of medium-chain triacylglycerols was absolutely comparable with infants receiving fats consisting exclusively of long-chain triacylglycerols.[119] These studies indicate the general difficulties which arise in carrying out investigations on the intestinal uptake of fat-soluble vitamins and in comparing the various experimental absorption studies with each other. The use of different oils or the procedure for the preparation of the dosage form, and the concentration of the vitamins in these preparations make the comparisons very difficult. Furthermore, whether or not the corresponding vitamin preparations are administered to fasting or nonfasting humans or animals may also be of significance. Likewise, it matters whether or not the vitamins are given together with or without a meal. In the case of in vitro studies with isolated intestinal preparations, a number of other problems, such as altered influx and efflux of the vitamins within the intestinal tissue and the absence of the natural environment at the mucosal and basal sites of the intestinal wall, may become limiting factors for absorption processes.

C. Effect of Polyunsaturated Fatty Acids

An interesting problem is the apparent adverse effect of unsaturated fatty acids on the absorption of lipid-soluble vitamins. Weber et al.[120] and Weber and Wiss[121] have demonstrated that the amount of radioactive vitamin E recovered in various tissues was significantly lowered in animals when linoleic acid or an oil rich in polyunsaturated fatty acids was included in the diet compared to coconut oil with a low content of polyunsaturated fatty acids. Furthermore, an increase of linoleic acid in the diet considerably decreased the time period within which intense red blood cell hemolysis appeared as a sign of vitamin E deficiency.[55] Since the fecal excretion of orally administered tocopherol was significantly higher in the animals fed linoleic acid, the intestinal uptake of vitamin E was concluded to have been reduced by the unsaturated fatty acid. This conclusion was confirmed by measuring the lymphatic appearance of tocopherol in rats receiving linoleic acid, clearly showing that linoleic acid depresses the absorption of vitamin E. Similar conclusions have been reached by Hollander[2] from intestinal perfusion studies. When linoleic and linolenic acids were added to the infusate, a pronounced decrease in the rate of absorption of the fat-soluble vitamins occurred. Dose-response measurements indicated that the absorption of the vitamins was progressively decreased in proportion to the concentration of the added polyunsaturated fatty acid.

The mechanism by which the polyunsaturated fatty acids inhibit the intestinal absorption of lipid-soluble vitamins is not known. It has been suggested[2] that the lower absorption rate observed could be due to an increase in the physical size of the lumenal micelles caused by the polyunsaturated fatty acids and, therefore, to a lower diffusion rate of such micelles across the unstirred water layer. Alternatively, the polyunsaturated fatty acids could have shifted the partitioning of the vitamins between mixed micelles and mucosal cell membranes in favor of the micelles, which would impair the intestinal uptake of the fat-soluble vitamins. Weber[1] has noted the possibility that the polyunsaturated fatty acids could influence the synthesis of chylomicrons or very low-density lipoproteins within the mucosal cells and affect, in this way, the lymphatic transport of lipid-soluble vitamins within carrier lipoproteins.[55,110] Thomson et al.[122] have proposed that small changes in the percentage of total dietary lipids composed of essential and nonessential fatty acids (without concurrent alterations in dietary total fat, carbohydrate, or protein) influence active and passive intestinal transport processes in the gut. Rats fed a diet high in saturated fatty acids demonstrated enhanced in vitro jejunal uptake of fatty acids and cholesterol when compared with uptake in animals fed a diet high in polyunsaturated fatty acids, but equivalent in total content of fat and other nutrients. The mechanism by which these changes occur remains unknown.

D. Interactions of the Fat-Soluble Vitamins

There is evidence of a synergistic effect between the *d*- and *l*-α-tocopherols during absorption. It has been demonstrated[123,124] that the *l*-enantiomer is absorbed more readily than the *d*-form of the vitamin and that there is synergism between the two enantiomers when administered as a mixture. Behrens and Madera[125] have studied the interrelationships of α- and γ-tocopherols during intestinal absorption, plasma transport, and liver uptake. It was observed that these processes are specific for α-tocopherol. Only when the concentration of α-tocopherol is low can γ-tocopherol successfully compete for binding sites at these three stages. Studies with rats[126] and with chicks[127] have indicated a marked decrease of vitamin E levels in liver or plasma, respectively, when the animals were supplemented concomitantly with high amounts of vitamin A. This suggests a real competition of vitamin A with vitamin E for sites of intestinal absorption and transport. High amounts of dietary vitamin A have also been described as impairing intestinal vitamin K utilization in rats.[128] Weber has emphasized that this antagonism of vitamin A toward vitamins E and K is most probably not relevant in humans under normal dietary conditions, since it occurs only at vitamin A doses many times the amounts normally present in food. In contrast, vitamin E does not appear to antagonize vitamin A absorption, since relatively large supplements of vitamin E to rats did not diminish vitamin A absorption as determined by its hepatic deposition rate.[129] In fact, high amounts of vitamin E seemed to enhance the intestinal absorption of vitamin A, even when given in massive doses.[130,131] This synergistic interaction between vitamins E and A needs further clarification.

Napoli et al.[27] have shown that α-tocopherol inhibited retinyl palmitate hydrolase in vitro in the intestine, liver, and kidney, had minimal effect on the testes hydrolase, and stimulated the lung hydrolase. Vitamin K_1 (phylloquinone) inhibited the retinyl palmitate hydrolase in vitro in all tissues tested and was about fivefold more potent than α-tocopherol. The effects of phylloquinone and α-tocopherol on the liver hydrolase were additive, not synergistic. The antioxidant *N,N′*-diphenyl-*p*-phenylenediamine, the most effective synthetic vitamin E substitute known, had little effect on the hydrolase. Prystowsky et al.[132] have also concluded that α-tocopherol, introduced as antioxidant, inhibited retinyl palmitate hydrolysis in liver homogenates. According to Napoli et al.,[27] α-tocopheryl acetate was as potent an inhibitor as unesterified α-tocopherol, but α-tocopheryl succinate was weaker. Menadione (vitamin K_3) also inhibited retinyl palmitate hydrolysis in liver, but seemed to be less potent than phylloquinone. Intestinal hydrolysis was inhibited by α-tocopherol, but was relatively insensitive to menadione. The qualitative effects of α-tocopherol were similar when the hydrolase activity of liver, intestine, and lung was solubilized. Interestingly, the antioxidant, butylated hydroxy toluene (BHT), a poor α-tocopherol substitute, inhibited the hydrolase significantly. Napoli et al.[27] point out that lipophilicity, an aromatic ring, and at least one heteroatom contribute to, but are not sufficient for, interaction with hydrolases, and that inhibition of retinyl palmitate hydrolase activity does not occur simply by an antioxidant mechanism. Furthermore, the inhibition is not likely to be a nonspecific lipid effect, because of the pronounced tissue specificity. Interestingly, Sklan[133] has reported that in chick plasma, transport and clearance of retinol were enhanced by feeding high levels of vitamin A and further enhanced when tocopherol at high concentrations was present in the diet. Dietary tocopherol had no effect on absorption, but increased hepatic vitamin A stores and in vitro retinol esterification, and decreased retinyl glucuronide flow through the duodenum.

Hollander[2] observed no changes in the absorption rate of vitamin K_1 when vitamins K_2 or K_3 were added to the perfusate used in the in vivo intestinal perfusion method with unanesthetized, restrained rats. However, the inclusion of retinoic acid in the perfusate, together with vitamin A in equimolar concentrations, resulted in a lower rate of appearance of vitamin A in the lymph and bile. This suggests that retinoic acid is competing with retinol for transport by the intracellular retinol binding protein from the absorptive cell membrane to the smooth endoplasmic reticulum of the villus cells. Earlier, retinoic acid added to the diet had been shown to interfere with the absorption of vitamin K in rats.[130] More recently, Bieri et al.[131] have reported a marked reduction in the intestinal absorption of vitamin E by low dietary levels of retinoic acid in rats and chicks when compared with animals receiving similar amounts of retinyl palmitate instead of retinoic acid. The mechanisms by which retinoic acid impairs intestinal absorption of vitamins E and K have not been established. It may be speculated that the more polar retinoic acid alters in some manner the solubilization process of vitamins E and K in the micellar phase of the intestinal lumen, as noted above. Since about 15% of the total radioactivity recovered in the lymph collected from rats receiving radioactive α-tocopherol is associated with α-tocopherylquinone and 80% with unchanged α-tocopherol, Weber[1] has speculated that some oxidation of vitamin E was caused by the retinoic acid either in the intestinal lumen or inside the mucosal cells, and that it was responsible for the apparent impairment of vitamin E absorption by retinoic acid.

The fat-soluble vitamins A and E are known to antagonize vitamin K. This has been attributed to the interference of vitamin A with the absorption of vitamin K.[134] Olson and Jones[135] observed that high dietary intakes of vitamin E in rats antagonize vitamin K activity and increase vitamin K requirement. The question of whether the antagonism between α-tocopherol and phylloquinone is exerted at the level of absorption or metabolism has not been settled. Vitamin E does not appear to affect vitamin K uptake and transport from the gut, nor directly the vitamin K-dependent carboxylase in vitro. Bettger et al.[136] found that α-tocopherolquinone is a more potent antagonist of vitamin K activity in the rat than α-tocopherol. Neither compound appeared to affect the absorption or distribution of vitamin K.

E. Effect of Proteins and Carbohydrates

There are conflicting reports in the literature regarding the possible effect of proteins and carbohydrates on the intraluminal dispersion of fat-soluble vitamins. Thus, DeLuca et al.[71] have assumed that soluble proteins and peptides, having surface-active properties, would favor the emulsification of lipids. In contrast, protein malnutrition might impair the functional activity of the pancreas, the proper secretion of bile, and the structural integrity of the intestinal mucosa, leading to malabsorption of lipid-soluble vitamins.[137-139] However, Silvakumar and Reddy[140] did not observe significant alterations in the absorption of vitamin A in children with kwashiorkor (a severe protein deficiency).

VI. SUMMARY AND CONCLUSIONS

This brief review clearly demonstrates that our understanding of the intestinal digestion and absorption of lipid-soluble vitamins is incomplete. The physicochemical events and the biochemical mechanisms of the absorption processes at the molecular level are complex and incompletely investigated. Furthermore, the overall process of absorption becomes further confounded, because the intestinal uptake of the fat-soluble vitamins is affected by a number of interactions with other constituents of ingested food. Physiological factors, such as the rate of gastric emptying, intestinal motility, and biliary and pancreatic secretions, influence the absorption of fat-soluble vitamins. In order to better understand all phenomena involved in intestinal uptake of fat-soluble vitamins, further information is needed with regard to such basic questions as the partitioning equilibrium of the lipid-soluble vitamins in the gut between oily and micellar phases, extramicellar fluid, and absorptive microvillus membranes. More information is also needed with respect to the more complex problems of intracellular processes taking place in the intestinal wall, such as binding of lipid-soluble vitamins to specific intracellular proteins and the mechanisms of their secretion into lymphatic or portal circulation. Finally, a better understanding of the effects of food constituents on absorption of lipid-soluble vitamins is desirable. A fuller knowledge of the mechanisms involved in the digestion and absorption processes of these vitamins may contribute considerably to development of supplementary or therapeutic treatments of human subjects with primary or secondary nutritional deficiency states.

ACKNOWLEDGMENTS

The studies by the author and collaborators referred to in this review were supported by the Medical Research Council of Canada and by the Ontario Heart Foundation, Toronto, Ontario, Canada.

REFERENCES

1. Weber, F., Absorption mechanisms for fat-soluble vitamins and the effect of other food constituents, in *Nutrition in Health and Disease and International Development: Symposia from the XII International Congress of Nutrition,* Harper A. E. and Davis, G. K., Eds., Alan R. Liss, New York, 1981, 119.
2. Hollander, D., Intestinal absorption of vitamins A, E, D, and K, *J. Lab. Clin. Med.,* 97, 449, 1981.
3. Frickel, F., Chemical and physical properties of retinoids, in *The Retinoids,* Vol. 1, Sporn, M. B., Roberts, A. B., and Goodman, D. S., Eds., Academic Press, New York, 1984, 7.
4. Olson, J. A., Vitamin A, in *Nutrition Reviews' Present Knowledge in Nutrition,* Nutrition Foundation, Washington, D.C., 1984, 176.
5. Moon, R. C. and Itri, L. M., Retinoids and cancer, in *The Retinoids,* Vol. 2, Sporn, M. B., Roberts, A. B., and Goodman, D. S., Eds., Academic Press, New York, 1984, 327.
6. Weber, F., Biochemical mechanisms of vitamin A action, *Proc. Nutr. Soc.,* 42, 31, 1983.
7. Fraser, D. R., Vitamin D, in *Nutrition Reviews' Present Knowledge in Nutrition,* 5th ed., Nutrition Foundation, Washington, D.C., 1984, 209.

8. Holick, M. F., Richtand, N. M., McNeil, S. C., Holick, S. A., Frommer, J. E., Henley, J. W., and Potts, J. T., Isolation and identification of previtamin D3 from the skin of rats exposed to ultraviolet irradiation, *Biochemistry,* 18, 1003, 1979.

9. DeLuca, H. F. and Schnoes, H. K., Vitamin D: recent advances, *Ann. Rev. Biochem.,* 52, 411, 1983.

10. Bieri, J. G., Vitamin E, in *Nutrition Reviews' Present Knowledge in Nutrition,* 5th ed., Nutrition Foundation, Washington, D.C., 1984, 226.

11. Machlin, L. J., Ed., *Vitamin E, A Comprehensive Treatise,* Marcel Dekker, New York, 1980.

12. Burton, G. W. and Ingold, K. U., Beta-carotene: an unusual type of lipid antioxidant, *Science,* 224, 569, 1984.

13. Olson, R. E., The function and metabolism of vitamin K, *Ann. Rev. Nutr.,* 4, 281, 1984.

14. International Vitamin A Consultative Group, *Biochemical Methodology for the Assessment of Vitamin A Status,* The Nutrition Foundation, Washington, D.C., 1982.

15. McCormick, A. M., Napoli, J. L., and DeLuca, H. F., High pressure liquid chromatography of vitamin A metabolites and analogs, *Methods Enzymol.,* 67, 220, 1980.

16. Napoli, J. L. and McCormick, A. M., Tissue dependence of retinoic acid metabolism *in vivo, Biochim. Biophys. Acta,* 666, 165, 1981.

17. Nierenberg, D. W., Determination of serum and plasma concentrations of retinol using high performance liquid chromatography, *J. Chromatogr. Biomed. Appl.,* 311, 239, 1984.

18. Nierenberg, D. W., Serum and plasma beta-carotene levels measured with an improved method of high performance liquid chromatography, *J. Chromatogr. Biomed. Appl.,* 339, 273, 1985.

19. Nakamura, T., Aoyama, Y., Fujita, T., and Katsui, G., Studies on tocopherol derivatives. V. Intestinal absorption of several d,l-3,4-3H2-alpha-tocopheryl esters in the rat, *Lipids,* 10, 627, 1975.

20. Zonta, F., Stancher, B., and Bielawny, J., High performance liquid chromatography of fat-soluble vitamins. Separation and identification of vitamins D2 and D3 and their isomers in food samples in the presence of vitamin A, vitamin E and carotene, *J. Chromatogr.,* 246, 105, 1982.

21. Goodman, G. E., Einspahr, J. G., Alberts, D. S., Davis, T. P., Leigh, S. A., Chen, H. S., and Meyskens, F. L., Pharmacokinetics of 13-cis-retinoic acid in patients with advanced cancer, *Cancer Res.,* 42, 2087, 1982.

22. Zaspel, B. J. and Csallary, A. S., Determination of alpha-tocopherol in tissues and plasma by high performance liquid chromatography, *Anal. Biochem.,* 130, 146, 1983.

23. Burton, G. W., Webb, A., and Ingold, K. U., A mild, rapid, and efficient method of lipid extraction for use in determining vitamin E/lipid ratios, *Lipids,* 20, 29, 1985.

24. Sarett, H. P., Manes, J. D., and Barnett, S. A., Measurement of vitamin K1 in infant formulas, *J. Nutr.,* 113, 470, 1983.

25. Shearer, M. J., Haroon, Y., and Barkhan, P., Response to Dr. Sarett's letter, *J. Nutr.,* 113, 471, 1983.

26. Haroon, Y., Shearer, M. J., Rahim, S., Gunn, W. G., McEnery, G., and Barhan, P., The content of phylloquinone (vitamin K1) in human milk, cows' milk and infant formula foods determined by high performance liquid chromatography, *J. Nutr.,* 112, 1105, 1982.

27. Napoli, J. L., McCormick, A. M., O'Meara, B., and Dratz, E. A., Vitamin A metabolism: alpha-tocopherol modulates tissue retinol levels in vivo, and retinyl palmitate hydrolysis in vitro, *Arch. Biochem. Biophys.,* 230, 194, 1984.

28. Chiku, S., Hamamura, K., and Nakamura, T., Novel urinary metabolite of d-delta-tocopherol in rats, *J. Lipid Res.,* 25, 40, 1984.

29. Harrison, E. H., Smith, J. E., and Goodman, D. S., Unusual properties of retinyl palmitate hydrolase activity in rat liver, *J. Lipid Res.,* 20, 760, 1979.

30. Helgerud, P., Petersen, L. B., and Norum, K. R., Retinol esterification by microsomes from the mucosa of human small intestine, *J. Clin. Invest.,* 71, 747, 1983.

31. Helgerud, P., Petersen, L. B., and Norum, K. R., Acyl CoA:retinol acyltransferase in rat small intestine: its activity and some properties of the enzymic reaction, *J. Lipid Res.,* 23, 609, 1982.

32. Imai, J., Hayashi, M., Awazu, S., and Hanano, M., Intestinal absorption of dl-alpha-tocopherol from bile salts and polysorbate 80 micellar solutions in rat, *J. Pharm. Dyn.,* 6, 897, 1983.

33. Croteau, R. and Ronald, R. C., Terpenoids, in *Chromatography, Part B, Applications,* Heftman, E., Ed., Elsevier, Amsterdam, 1983, B147.

34. Lefevere, M. F., DeLeenheer, A. P., Claeys, A. E., Claeys, I. V., and Styaert, H., Multidimensional liquid chromatography: a breakthrough in the assessment of physiological vitamin K levels, *J. Lipid Res.,* 23, 1068, 1982.

35. Stancher, B. and Zonta, F., High performance liquid chromatography of fat-soluble vitamins. Simulataneous quantitative analysis of vitamins D2, D3 and E. Studies of percentage recoveries of vitamins from cod liver oil, *J. Chromatogr.,* 256, 93, 1983.

36. Barua, A. B. and Olson, J. A., Metabolism and biological activity of all-trans-4,4-difluororetinyl acetate, *Biochim. Biophys. Acta,* 799, 128, 1984.

37. Wang, C.-S., Digestion of dietary glycerides and phosphoglycerides, in *Fat Absorption*, Kuksis, A., Ed., CRC Press, Boca Raton, Fla., 1987.
38. Ganguly, J., Absorption of vitamin A, *Am. J. Clin. Nutr.*, 22, 923, 1969.
39. Hyun, S. A., Kothari, H., Herm, E., Mortenson, J., Treadwell, C. R., and Vahouny, G. V., Purification and properties of pancreatic juice cholesterol ester hydrolase, *J. Biol. Chem.*, 244, 1937, 1969.
40. Mattson, F. H. and Volpenhein, R. A., Hydrolysis of primary and secondary esters of glycerol by pancreatic juice, *J. Lipid Res.*, 9, 79, 1968.
41. Morgan, R. G. H., Barrowman, J., Filipek-Wender, H., and Borgstrom, B., The lipolytic enzymes of rat pancreatic juice, *Biochim. Biophys. Acta*, 167, 355, 1968.
42. Shakir, K. M. M., Gabriel, L., Sundaram, S. G., and Margolis, S., Intestinal phospholipase A and triglyceride lipase: localization and effect of fasting, *Am. J. Physiol.*, 242, G168, 1982.
43. Erlanson, C. and Borgstrom, B., Purification and further characterization of co-lipase from porcine pancreas, *Biochim. Biophys. Acta*, 271, 400, 1970.
44. Bell, N. H. and Ryan, P., Absorption of vitamin D oleate in the rat, *Am. J. Clin. Nutr.*, 22, 425, 1969.
45. Gallo-Torres, H. E., Obligatory role of bile for the intestinal absorption of vitamin E, *Lipids*, 5, 379, 1970.
46. Lombardo, D. and Guy, O., Studies on the substrate specificity of a carboxyl ester hydrolase from human pancreatic juice. II. Action on cholesterol esters and lipid soluble vitamin esters, *Biochim. Biophys. Acta*, 611, 147, 1980.
47. Momsen, W. E. and Brockman, H. L., Purification and characterization of cholesterol esterase from porcine pancreas, *Biochim. Biophys. Acta*, 486, 103, 1977.
48. Mathias, P. M., Harries, J. T., Peters, T. J., and Muller, D. P. R., Optimization and validation of assays to estimate pancreatic esterase activity using well characterized micellar solutions of cholesterol oleate and tocopheryl acetate, *J. Lipid Res.*, 22, 177, 1981.
49. Mathias, P. M., Harries, J. T., Peters, T. J., and Muller, D. P. R., Studies on the *in vivo* absorption of micellar solutions of tocopherol and tocopheryl acetate in the rat. Demonstration and partial characterization of a mucosal esterase located to the endoplasmic reticulum of the enterocyte, *J. Lipid Res.*, 22, 829, 1981.
50. Friedman, H. I. and Nylund, B., Intestinal fat digestion, absorption, and transport; a review, *Am. J. Clin. Nutr.*, 33, 1108, 1980.
51. Wasserman, R. H. and Carradino, R. A., Metabolic role of vitamins A and D, *Ann. Rev. Biochem.*, 40, 501, 1971.
52. Mathur, S. N., Murthy, S. K., and Ganguly, J., Studies on the intracellular distribution of retinyl acetate hydrolase in rat and chicken intestine, *Int. J. Vitam. Nutr. Res.*, 42, 115, 1972.
53. Hofmann, A. F. and Borgstrom, B., The intraluminal phase of fat digestion in man: the lipid content of the micellar and oil phases of intestinal content obtained during fat digestion and absorption, *J. Clin. Invest.*, 43, 247, 1964.
54. Carey, M. C. and Small, D. M., The characteristics of mixed micellar solutions with particular reference to bile, *Am. J. Med.*, 49, 590, 1970.
55. Gallo-Torres, H. E., Weber, F., and Wiss, O., The effect of different dietary lipids on the lymphatic appearance of vitamin E, *Int. J. Vitam. Nutr. Res.*, 41, 504, 1971.
56. Erin, A. N., Spirin, M. M., Tabidze, L. V., and Kagan, V. E., Formation of alpha-tocopherol complexes with fatty acids. A hyopthetical mechanism for stabilization of biomembranes by vitamin E, *Biochim. Biophys. Acta*, 774, 96, 1984.
57. Patton, J. S., Gastrointestinal lipid digestion, in *Physiology of the Gastrointestinal Tract*, Johnston, L. R., Ed., Raven Press, New York, 1981, 1123.
58. Miettinen, T. A. and Siurala, M., Bile salts, sterols, sterol esters, glycerides and fatty acids in micellar and oil phases of intestinal contents during fat digestion in man, *Z. Klin. Chem. Klin. Biochem.*, 9, 47, 1971.
59. Miettinen, T. A. and Siurala, M., Micellar solubilization of intestinal lipids and sterols in gluton enteropathy and liver cirrhosis, *Scand. J. Gastroenterol.*, 6, 527, 1971.
60. Patton, J. S. and Carey, M. C., Watching fat digestion, *Science*, 204, 145, 1979.
61. Borgstrom, B., Fat digestion and absorption, in *Biomembranes*, Vol. 4B, Smyth, D. H., Ed., Plenum Press, New York, 1974, 555.
62. Carey, M. C., Role of lecithin in the absorption of dietary fat, in *Phospholipids and Atherosclerosis*, Avogaro, P., Mancini, M., Ricci, G., and Paoletti, R., Eds., Raven Press, New York, 1983, 33.
63. Borgstrom, B., The micellar hypothesis of fat absorption: must it be revisited?, *Scand. J. Gastroenterol.*, 20, 389, 1985.
64. Scow, R. O., Blanchette-Mackie, E. J., and Smith, L. C., Transport of lipid across capillary endothelium, *Fed. Proc.*, 39, 2610, 1980.
65. Scow, R. O., Desnuelle, P., and Verger, R., Lipolysis and lipid movement in a membrane model. Action of lipoprotein lipase, *J. Biol. Chem.*, 254, 6456, 1979.

66. Smith, L. C. and Scow, R. O., Chylomicrons — mechanism of transfer of lipolytic products to cells, *Prog. Biochem. Pharmacol.*, 15, 199, 1979.

67. Wilson, F. A., Sallee, V. L., and Dietschy, J. M., Unstirred water layer in the intestine: rate determinant for fatty acid absorption from micellar solutions, *Science*, 174, 1031, 1971.

68. Westergaard, H. and Dietschy, J. M., Delineation of the dimensions and permeability characteristics of the two major diffusion barriers to passive mucosal uptake in the rabbit intestine, *J. Clin. Invest.*, 174, 718, 1974.

69. Ong, D. E., A novel retinol-binding protein from rat. Purification and partial characterization, *J. Biol. Chem.*, 259, 1476, 1984.

70. Takahashi, K., Odani, S., and Ono, T., A close structural relationship of rat liver Z-protein to cellular retinoid binding proteins and peripheral nerve myelin P2 protein, *Biochem. Biophys. Res. Commun.*, 106, 1099, 1982.

71. DeLuca, L. M., Glover, J., Heller, J., Olson, J. A., and Underwood, B., Guidelines for the eradication of vitamin A deficiency and xerophthalmia. VI. Recent advances in the metabolism and function of vitamin A and their relationship to applied nutrition, in *Report of the International Vitamin A Consultative Group*, Nutrition Foundation, New York, 1979, 10.

72. Holick, M. F. and DeLuca, H. F., Metabolism of vitamin D, in *Vitamin D*, Lawson, D. E. M., Ed., Academic Press, London, 1978, 51.

73. Losowsky, M. S., Kelleher, J., Walker, B. E., Davies, T., and Smith, C. L., Intake and absorption of tocopherol, *Ann. N.Y. Acad. Sci.*, 203, 212, 1972.

74. Huang, H. S. and Goodman, D. S., Vitamin A and carotenoids. I. Intestinal absorption and metabolism of ^{14}C-labeled vitamin A alcohol and beta-carotene in the rat, *J. Biol. Chem.*, 240, 2839, 1965.

75. Forsgren, L., Studies on the intestinal absorption of labelled fat-soluble vitamins (A, D, E and K) via the thoracic-duct lymph in the absence of bile in man, *Acta Chir. Scand.*, Suppl. 399, 1969.

76. Avioli, L. V., Current concepts of vitamin D-3 metabolism in man, in *The Fat-Soluble Vitamins*, DeLuca, F. H. and Suttie, J. W., Eds., University of Wisconsin Press, Madison, 1970, 159.

77. Gallo, L., Cheriathundam, E., Vahouny, G. V., and Treadwell, C. R., Immunological comparison of cholesterol esterases, *Arch. Biochem. Biophys.*, 191, 42, 1978.

78. Gallo, L., Chiang, Y., Vahouny, G. V., and Treadwell, C. R., Localization and origin of rat intestinal cholesterol esterase determined by immunocytochemistry, *J. Lipid Res.*, 21, 537, 1980.

79. Borgstrom, B., Randomization of glyceride fatty acids during absorption from the small intestine of the rat, *J. Biol. Chem.*, 214, 671, 1955.

80. Fabisch, W., Fermentative ester synthese emulsionen, *Biochem. Z.*, 259, 420, 1933.

81. Kuksis, A. and Manganaro, F., Purification and biochemical characterization of intestinal acylglycerol acyltransferases, in *Fat Absorption*, Vol. 1, Kuksis, A., Ed., CRC Press, Boca Raton, Fla., 1987.

82. Gangl, A. and Ockner, R. K., Intestinal metabolism of lipids and lipoproteins, *Gastroenterology*, 68, 167, 1975.

83. Smith, J. E., Muto, Y., and Goodman, D. S., Tissue distribution and subcellular localization of retinol-binding protein in normal and vitamin A-deficient rats, *J. Lipid Res.*, 16, 318, 1975.

84. Norum, K. R., Helgerud, P., and Lilljeqvist, A.-C., Enzymic esterification of cholesterol in rat intestinal mucosa catalyzed by acyl-CoA:cholesterol acyltransferase, *Scand. J. Gastroenterol.*, 16, 401, 1981.

85. Goodman, D. S. and Blaner, W. S., Biosynthesis, absorption and hepatic metabolism of retinol, in *The Retinoids*, Vol. 2, Sporn, M. B., Roberts, A. B., and Goodman, D. S., Eds., Academic Press, New York, 1984.

86. Goodman, D. S., Blomstrand, R., Werner, B., Huang, H. S., and Shiratori, T., The intestinal absorption of retinol and metabolism of vitamin A and beta-carotene in man, *J. Clin. Invest.*, 45, 1615, 1966.

87. Bashor, M. M. and Chytil, F., Cellular retinol-binding protein, *Biochim. Biophys. Acta*, 411, 87, 1975.

88. Berman, E. R., Horowitz, J., Segal, N., Fisher, S., and Feeney-Burns, L., Enzymatic esterification of vitamin A in the pigment epithelium of bovine retina, *Biochim. Biophys. Acta*, 360, 36, 1980.

89. Hollander, D., Wang, H. P., Chu, C. Y. T., and Badawi, M. A., Preliminary characterization of a small intestinal binding component for retinol and fatty acids in the rat, *Life Sci.*, 23, 1101, 1978.

90. Behrens, W. A. and Madere, R., Occurrence of a rat liver alpha-tocopherol binding protein *in vivo*, *Fed. Proc.*, 40, 894, 1981.

91. Gangl, A., Kornauth, W., Mlczoch, J., Sulm, O., and Klose, B., Different metabolism of saturated and unsaturated long chain plasma free fatty acids by intestinal mucosa of rats, *Lipids*, 15, 75, 1980.

92. Blomstrand, R. and Forsgren, L., Intestinal absorption and esterification of vitamin D3-1,2-3H in man, *Acta Chem. Scand.*, 21, 1662, 1967.

93. Blomstrand, R. and Forsgren, L., Vitamin K_1-^{3}H in man, its intestinal absorption and transport in the thoracic duct lymph, *Int. J. Vitam. Res.*, 38, 45, 1968.

94. Blomstrand, R. and Forsgren, L., Labelled tocopherols in man, intestinal absorption and thoracic duct lymph transport of dl-alpha-tocopheryl-3,4-^{14}C$_2$-acetate, dl-alpha-tocopheramine-3,4-^{14}C$_2$, dl-alpha-tocopherol-(5-methyl-^{3}H) and N-(methyl-^{3}H)-dl-f-tocopheramine, *Int. J. Vitam. Res.* 38, 328, 1968.

95. Blomstrand, R. and Werner, B., Studies on the intestinal absorption of radioactive beta-carotene and vitamin A in man, conversion of beta-carotene into vitamin A, *Scand. J. Clin. Lab. Invest.*, 19, 339, 1967.

96. Shearer, M. J., McBurney, A., and Barkhan, P., Studies on the absorption and metabolism of phylloquinone (vitamin K$_1$) in man, *Vitam. Horm.*, 32, 513, 1974.

97. Blomhoff, R., Helgerud, P., Dueland, S., Berg, T., Pedersen, J. I., Norum, K. R., and Drevon, C. A., Lymphatic absorption and transport of retinol and vitamin D$_3$ from rat intestine. Evidence for different pathways, *Biochim. Biophys. Acta*, 772, 109, 1984.

98. Tucker, G., Gagnon, R. E., and Haussler, M. R., Vitamin D$_3$-25-hydroxylase: tissue occurrence and apparent lack of equilibration, *Arch. Biochem. Biophys.*, 155, 47, 1973.

99. Mann, J. D., Mann, F. D., and Bollman, J. L., Hypoprothrombinemia due to loss of intestinal lymph, *Am. J. Physiol.*, 158, 311, 1949.

100. Rietz, P., Gloor, U., and Wiss, O., Menachinone aus menschlicher Leber und Faulschlamm, *Int. Z. Vitaminforsch.*, 40, 351, 1970.

101. Wolf, G., Vitamin A, in *Human Nutrition*, Vol. 3B, Alfin-Slater, R. B. and Kritchevsky, D., Eds., Plenum Press, New York, 1980, 97.

102. Losowsky, M. S., Vitamin E in human nutrition, in *The Importance of Vitamins to Human Health*, Taylor, T. G., Ed., MTP Press, Lancaster, England, 1979, 101.

103. Handelman, G. J., Machlin, L. J., Fitch, K., Weiter, J. J., and Dratz, E. D., Oral *alpha*-tocopherol supplements decrease plasma *gamma*-tocopherol levels in humans, *J. Nutr.*, 115, 807, 1985.

104. Jansson, L., Lindroth, M., and Tyopponen, J., Intestinal absorption of vitamin E in low birth weight infants, *Acta Paediatr. Scand.*, 73, 329, 1984.

105. Wiss, O. and Gloor, U., Absorption, distribution, storage and metabolites of vitamin K and related quinones, *Vitam. Horm.*, 24, 575, 1966.

106. Shearer, M. J., Barkhan, P., and Webster, C. R., Absorption and excretion of an oral dose of tritiated vitamin K$_1$ in man, *Br. J. Haematol.*, 18, 297, 1970.

107. Lefevere, M. F., DeLeenheer, A. P., and Claeys, A. E., High performance liquid chromatographic assay of vitamin K in human serum, *J. Chromatogr.*, 186, 749, 1979.

108. Shearer, M. J., Rahim, S., Barkhan, P., and Stimmler, L., Plasma vitamin K$_1$ in mothers and their newborn babies, *Lancet*, 2, 460, 1982.

109. Bjornsson, T. D., Meffin, P. J., Swezey, S. E., and Blaschke, T. F., Effects of clofibrate and warfarin alone and in combination on the disposition of vitamin K1, *J. Pharmacol. Exp. Ther.*, 210, 322, 1979.

110. Gallo-Torres, H. E., Absorption, in *Vitamin E, A Comprehensive Treatise*, Machlin, L. J., Ed., Marcel Dekker, New York, 1980, 170.

111. Peake, S. R., Windmueller, H. G., and Bieri, J. G., A comparison of the intestinal absorption, lymph and plasma transport, and tissue uptake of alpha- and gamma-tocopherol in the rat, *Biochem. Biophys. Acta*, 260, 679, 1972.

112. Desai, I. D., Parekh, C. K., and Scott, M. L., Absorption of D- and L-alpha-tocopheryl acetates in normal and dystrophic chicks, *Biochim. Biophys. Acta*, 100, 280, 1965.

113. Kasper, H. and Hotzel, D., Untersuchungen zur Frage der Vitaminresorption in Dickdarm, *Z. Ernahrungswiss.*, 4, 34, 1963.

114. Schmandke, H. and Schmidt, G., Untersuchungen uber die Resorption des alpha-Tocopherols aus oliger und wassriger Losung, *Int. Z. Vitaminforsch.*, 35, 128, 1965.

115. Kelleher, J., Davies, T., Smith, C. L., Walker, B. E., and Losowsky, M. S., The absorption of alpha-tocopherol in the rat. I. The effect of different carriers and different dose levels, *Int. J. Vitam. Nutr. Res.*, 42, 394, 1972.

116. Kelleher, J. and Losowsky, M. S., The absorption of alpha-tocopherol in man, *Br. J. Nutr.*, 24, 1033, 1970.

117. Kasper, H. and Kuhn, H. A., Der Einfluss von Triglyceriden mittekettiger Fettsauren auf die Vitamin A-Resorption beim Menschen, *Klin. Wochenschr.*, 46, 1227, 1968.

118. Williams, M. L. and Oski, F. A., Vitamin E status of infants fed formula containing medium-chain triglycerides, *J. Pediatr.*, 96, 70, 1980.

119. Nishigaki, R., Awazu, S., Hanano, M., and Fuwa, T., The effect of dosage form on absorption of vitamin A into lymph, *Chem. Pharm. Bull.*, 24, 3207, 1976.

120. Weber, F., Weiser, H., and Wiss, O., Bedarf an Vitamin E in Abhangingheit von der Zufuhr an Linolsaure, *Z. Ernahrungswiss.*, 4, 245, 1964.

121. Weber, F. and Wiss, O., Wechselwirkung zwischen Vitamin E und anderen Nahrungsbestandteilen, in *Nutritio et Dieta*, Vol. 8, S. Karger, Basel, 1966, 54.

122. Thomson, A. R. B., Keelan, N., Clandinin, M. T., Walker, K., McIntyre, Y., McLeod, J., Tavernini, M., Terrell, J., and Poland, L., Dietary fat selectively alters transport properties of rat jejunum, *J. Clin. Invest.*, 77, 279, 1986.

123. Weber, F., Gloor, U., Wursch, J., and Wiss, O., Synergism of d- and l-alpha-tocopherol during absorption, *Biochem. Biophys. Res. Commun.*, 14, 186, 1964.

124. Weber, F., Gloor, U., Wursch, J., and Wiss, O., Studies on the absorption, distribution and metabolism of l-alpha-tocopherol in the rat, *Biochem. Biophys. Res. Commun.*, 14, 189, 1964.

125. Behrens, W. A. and Madera, A. R., Interrelationship and competition of alpha- and gamma-tocopherol at the level of intestinal absorption, plasma transport and liver uptake, *Nutr. Res.*, 3, 891, 1983.

126. Brubacher, G., Scharer, K., Studer, A., and Wiss, O., Ueber die gegenseitige Beeinflussung von Vitamin E, Vitamin A und Carotinoiden, *Z. Ernahrungswiss.*, 5, 190, 1965.

127. Combs, G. F., Jr. and Scott, M. L., Antioxidant effects on selenium and vitamin E function in the chick, *J. Nutr.*, 104, 1297, 1974.

128. Doisy, E. A., Jr. and Matschiner, J. T., Biochemistry of vitamin K, in *Fat-Soluble Vitamins*, Morton, R. A., Ed., Pergamon Press, Oxford, 1970, 293.

129. Arnrich, L. and Arthur, V. A., Interactions of fat-soluble vitamins in hypervitaminoses, *Ann. N.Y. Acad. Sci.*, 355, 109, 1980.

130. Matschiner, J. T., Ameloitti, J. M., and Doisy, E. A., Jr., Mechanism of the effect of retinoic acid and squalene on vitamin K deficiency in the rat, *J. Nutr.*, 91, 303, 1967.

131. Bieri, J. G., Wu, A.-L., and Tolliver, T. J., Reduced intestinal absorption of vitamin E by low dietary levels of retinoic acid in rats, *J. Nutr.*, 111, 458, 1981.

132. Prystowsky, J. H., Smith, J. E., and Goodman, D. S., Retinyl palmitate hydrolase activity in normal rat liver, *J. Biol. Chem.*, 256, 4498, 1981.

133. Sklan, D., Vitamin A absorption and metabolism in the chick: response to high dietary intake of tocopherol, *Br. J. Nutr.*, 50, 401, 1983.

134. Matschiner, J. T., Amelotti, J. M., and Doisy, E. A., Jr., Mechanism of the effect of retinoic acid and squalene on vitamin K deficiency in the rat, *J. Nutr.*, 91, 303, 1967.

135. Olson, R. E. and Jones, J. P., Inhibition of vitamin K action by dietary vitamin E, *Proc. Soc. Exp. Biol. Med.*, submitted.

136. Bettger, W. J. and Olson, R. E., Effect of alpha-tocopherol and alpha-tocopherol-quinone on vitamin K-dependent carboxylation in the rat, *Fed. Proc.*, 41, 344, 1982.

137. Mahadevan, S., Malathi, P., and Ganguly, J., Influence of proteins on absorption and metabolism of vitamin A, *World Rev. Nutr. Diet*, 5, 209, 1965.

138. Rajaram, O. V., Fatterpaker, P., and Sreenivasan, A., Effect of protein deficiency on absorption, transport and distribution of alpha-tocopherol in the rat, *Br. J. Nutr.*, 37, 157, 1977.

139. Green, P. H. R. and Tall, A. R., Drugs, alcohol and malabsorption, *Am. J. Med.*, 67, 1066, 1979.

140. Silvakumar, B. and Reddy, V., Studies in vitamin A absorption in children, *World Rev. Nutr. Diet*, 31, 125, 1978.

Chapter 4

DIGESTION AND ABSORPTION OF FAT-SOLUBLE XENOBIOTICS

A. Kuksis

TABLE OF CONTENTS

I. INTRODUCTION

The fat-soluble xenobiotics may be defined as both noxious and relatively harmless substances derived from natural[1] and synthetic[2] sources. Like the fat-soluble vitamins, they have provided important clues to the principles of intestinal digestion and absorption of dietary fats in the past and are likely to do so in the future. The great variety of chemical structures and wide range of their molecular weights allow an assessment of the critical factors of micellarization and membrane uptake, which is not possible with the more limited forms of the dietary lipids. In addition, the study of the intestinal digestion and absorption of the fat-soluble xenobiotics is of interest, because many of these substances are toxic to animal bodies, and their increasing presence in the environment and in the food supplies constitutes a growing danger to the well-being of all living systems. Some of the topics considered here have been dealt with previously in an abbreviated form.[3]

II. STRUCTURE, NOMENCLATURE, AND ORIGIN

The molecular structures of the major classes of fat-soluble xenobiotics are given in Figures 1 to 5. For the present purposes these materials have been grouped according to their general physicochemical properties, with the specific origin and metabolic basis of any toxicity mentioned only in passing. The free alcohol and acid functions may be esterified with fatty acids and fatty alcohols or with glycerol and other polyols, respectively; these may be removed completely or partially during the enzymatic hydrolysis in the gastrointestinal tract. The partial esterification products represent highly surface-active molecules, as discussed below.

Paraffins (Figure 1) are largely harmless compounds widely distributed in nature and extensively employed in various commercial products, including food packaging materials. The hydrocarbon components of the surface waxes are usually *n*-paraffins with odd numbers between 21 and 35 carbon atoms. These molecules are believed to be derived from long-chain fatty acids with an even number of carbon atoms by decarboxylation, with the C_{29} species predominating.[4] The various hydrocarbon fractions derived from petroleum residues contain both odd and even carbon numbers as well as branched chain molecules of various molecular weights (light and heavy mineral oil). A common C_{30} unsaturated hydrocarbon, squalene, is made up of isoprene units.

Aromatic polycyclic hydrocarbons (Figure 2) in the environment have received much attention, ever since benzo(*a*)pyrene was first demonstrated to be carcinogenic. About 100 polycyclic hydrocarbons have been identified in the environment and in foods.[5] It is believed that they are formed during pyrolysis of organic substances by a series of reactions involving the combination of free radicals,[6] although some are biosynthesized or are geochemical in origin.

Polychlorinated (PCB) and polybrominated (PBB) biphenyls and triphenyls developed in the 1930s have been used for many years as heat-transfer agents, dielectric fluids in electrical transformers, and as sealants, adhesives, and paint additives in hundreds of other products. Animals exposed to them suffer from cancer, birth defects, and nervous disorders. Among humans the effects are less clear. These polycyclics have been found to contaminate cereals, milk fat, poultry products, and fish products and have been detected in human adipose tissue and milk. Safe[6] has prepared an extensive review of the biochemistry, toxicology, and mechanism of action of the polyhalogenated biphenyls. The most toxic halogenated biphenyls are the 3,3′,4,4′-tetra-, 3,3′,4,4′,5-penta-, and 3,3′,4,4′,5,5′-hexahalo biphenyls, which are approximate isostereomers of 2,3,7,8-tetra chlorodibenzo-*p*-dioxin (Figure 3). Analytical studies, however, indicate that these compounds are present only in trace amounts in commer-

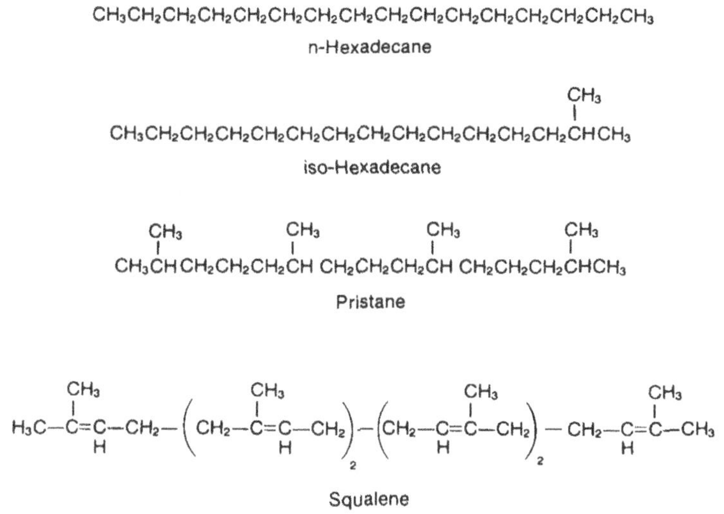

CH₃CH₂CH₂CH₂CH₂CH₂CH₂CH₂CH₂CH₂CH₂CH₂CH₂CH₂CH₂CH₃

n-Hexadecane

CH₃CH₂CH₂CH₂CH₂CH₂CH₂CH₂CH₂CH₂CH₂CH₂CH₂CHCH₃ (with CH₃ branch)

iso-Hexadecane

CH₃CHCH₂CH₂CH₂CH CH₂CH₂CH₂CH CH₂CH₂CH₂CHCH₃ (with four CH₃ branches)

Pristane

Squalene

FIGURE 1. Structures of some aliphatic hydrocarbons: *n*-hexadecane; iso-hexadecane; pristane; and squalene.

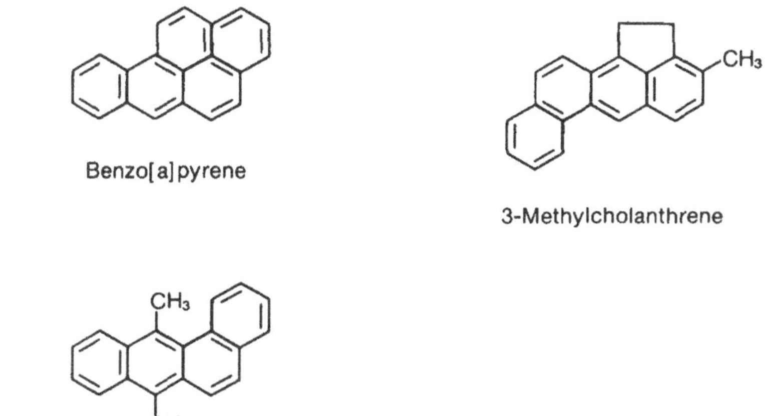

Benzo[a]pyrene

3-Methylcholanthrene

7,12-Dimethylbenz[a]anthracene

FIGURE 2. Structures of some polycyclic hydrocarbons: benzo(a)pyrene or 3,4-benzopyrene (occurs in coal tar); 3-methylcholanthrene or 20-methylcholanthrene (pyrolytic degradation product of cholesterol derivatives); 7,12-dimethylbenz(a)anthracene or 9,10-dimethyl-1,2-benzanthracene (synthetic compound).

cial polychlorinated and polybrominated biphenyls and cannot account for the toxicity of these mixtures.

Fat-soluble organic pesticides (Figure 4), such as dichlorodiphenyl trichloroethane (DDT) and dichlorodiphenyl dichloroethylene (DDE), are found in small quantities in body fat of normal human subjects following their use for agricultural and farm purposes.[7] Food products of animal origin usually contain larger quantities of DDT and DDE than vegetables and fruit. These pesticides readily penetrate the intestinal surface of mammals.

Long-chain acids and esters include both naturally occurring and synthetic compounds. The aliphatic carboxylic acids occur naturally as glycerol and wax esters, while

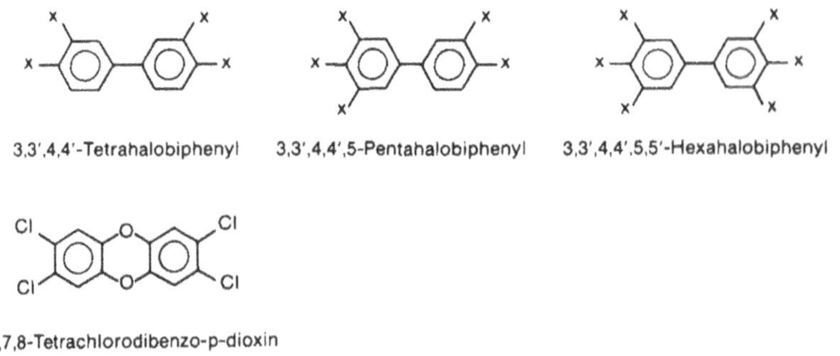

3,3′,4,4′-Tetrahalobiphenyl   3,3′,4,4′,5-Pentahalobiphenyl   3,3′,4,4′,5,5′-Hexahalobiphenyl

2,3,7,8-Tetrachlorodibenzo-p-dioxin

FIGURE 3.   Structures of selected biphenyls: 3,3′, 4,4′-tetrahalobiphenyl; 3,3′, 4,4′, 5-pentahalobiphenyl; 3,3′, 4,4′, 5,5′-hexahalobiphenyl; 2,3,7,8-tetrachlorodibenzo-*p*-dioxin.

DDD

DDE

DDT

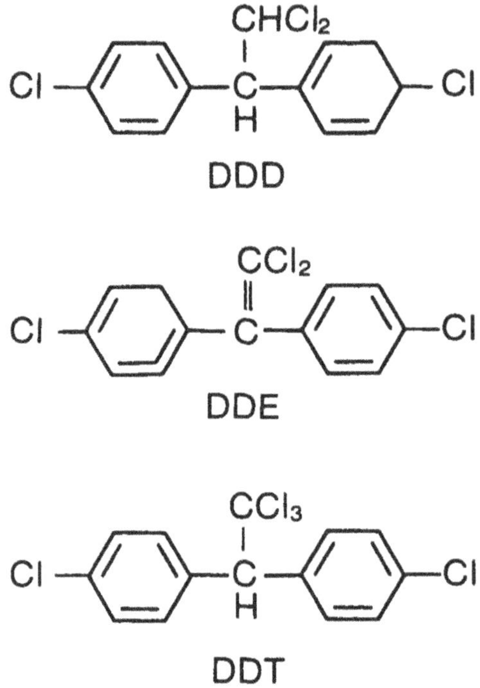

FIGURE 4.   Structures of some fat-soluble organic pesticides: DDD, 1,1-dichloro-2,2-bis(*p*-chlorophenyl)ethane; DDE, 1,1-dichloro-2,2-bis-(*p*-chlorophenyl)ethylene; DDT, 1,1,1-trichlor-2,2-bis-(*p*-chlorophenyl)ethane or dichlorophenyltrichloroethane.

the aromatic carboxylic acids are produced industrially and may find their way into the food supplies in the form of short-chain alcohol esters[8] (Figure 5). The alkyl and aryl sulfonates are present mainly as soaps.

All of these compounds are widely encountered in the environment, but their physiological effects, with a few notable exceptions, are relatively mild.[2] In addition, various fat-soluble drugs are absorbed via the lymphatics and studies on the mechanism of their solubilization and absorption have contributed to the understanding of fat absorption. Their absorption will be considered along with appropriate classes of other fat-soluble xenobiotics.

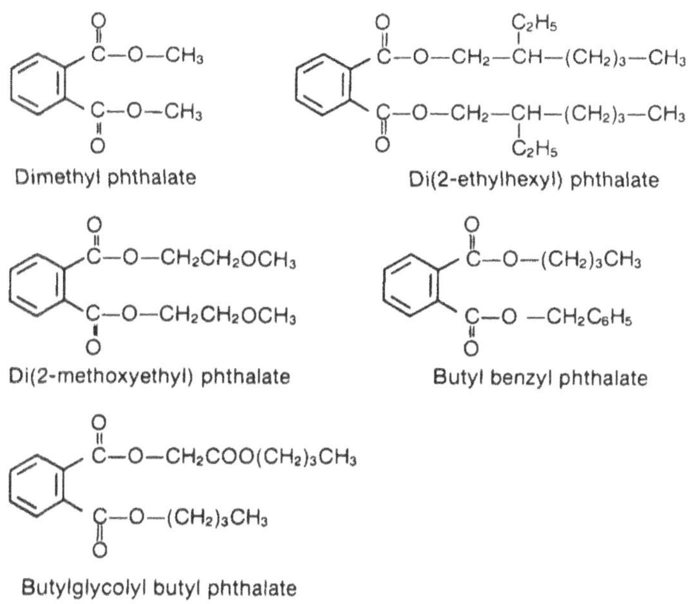

FIGURE 5. Structures of some phthalic acid esters: dimethyl phthalate; di(2-ethylhexyl)phthalate; di(2-methoxyethyl)phthalate; butylbenzylphthalate; butylglycolylbutyl phthalate.

III. METHODS OF ASSESSMENT OF XENOBIOTIC ABSORPTION

Evidence of absorption of xenobiotics may be obtained indirectly by appropriate balance studies based on mass or radioactivity of labeled tracers in the intact gastrointestinal system, or directly by means of segments of intestine, isolated absorptive cells, and vesicles derived from purified brush border membranes, which are perfused or incubated with isotope-labeled tracers. In most instances the assessment of the absorption of xenobiotics has involved the extraction and recovery of the absorbed compounds by some chromatographic method followed by a specific quantitative measurement, occasionally including mass spectrometry (MS).[9-11]

A. Animal and Tissue Preparations

The methods may be divided into those which assess absorption of xenobiotics indirectly by measuring the level of the substance in the blood or in excretory products or its distribution in the body, and those which assess it directly in the intestine.

1. Indirect Methods

The classical method of estimating absorption is by balance studies. The total output of a substance (e.g., plant sterols) is subtracted from the total intake over a period of time.[12] This study is limited in scope, because some dietary components are altered during passage through the gut, so that any difference noted may be a result of both absorption and intestinal destruction. The use of radiolabeled tracers does not overcome this difficulty, although it may simplify the measurement of the recovery of any unabsorbed material.[13]

Estimation of the levels of substances in the blood is used mainly for ease of access to the human body. The amounts of any xenobiotics detected in blood provide an incomplete reflection of the actual absorption because of continued clearance by the various organs. In addition, the substances measured in blood may have become subject to metabolic transformation.

2. Direct Methods

The use of rings, everted sacs,[3] and isolated loops[14] of the intestine and, more recently, isolated villus cells[15] and brush border vesicles[15,16] has permitted a direct demonstration of the mucosal uptake of various xenobiotics. These methods do not take into account the natural motility of the gut and the imperfect methods of isolation and preservation of the gut preparations. In the case of the villus cells, isolation results in an exposure of the basolateral surface to the penetration by the xenobiotics, which normally would be confined to the brush-border side.

In several instances the absorption of xenobiotics has been assessed by perfusion of the intestine following intubation with special sampling tubes. Either double or triple lumen tubes may be used. The rate of absorption in this open-ended perfusion depends to some extent on the speed of the infusion and tends to fall off at a flow greater than 20 mℓ/min.[17] It is open to question whether the equilibrated segment, after 30 min of perfusion, is in a physiological state for absorption or not.

B. Solvent Extraction

In general, the neutral aliphatic and aromatic hydrocarbons are readily extracted from tissues and biological fluids with the common organic solvents, such as hexane, diethyl ether, and benzene or toluene. The more polar aliphatic and aromatic alcohols and phenols may require the use of chloroform, while still more complex polar lipids are recovered only by chloroform-methanol 2:1 and other solvent mixtures of comparable polarity, as already noted for dietary fats and other lipids.[3] The efficiency of the extraction solvent and the overall procedure must be determined initially by fortifying control samples with the xenobiotic in question. This should be done over a range to include the levels expected in real samples. Plotting recovery data vs. fortification level on a log-log scale provides a good picture of the extraction efficiency.[18] Selecting fortification levels with even spacing for data points allows coverage of a wide range with a minimum of samples. Quadruplicate runs at each level provide sufficient data for calculating the mean and standard deviation.

A very important consideration in developing an effective method for measurement of trace components is the vast difference between recovery of compounds added to the sample in the laboratory and recovery of biologically incorporated compounds.[19] Usually the endogenous residues are much more difficult to extract and chemical or enzymatic hydrolysis may be necessary to free bound or conjugated residues. Exhaustive extraction of a sample matrix using different solvents can give an indication of how vigorous the extraction must be to remove all the residue present. However, the more polar the extraction solvent the more extensive the cleanup and the separation procedures that may be necessary for eventual isolation and quantitation of the compound of interest. High purity extracts of both relatively polar and nonpolar lipid-soluble compounds may be obtained by the use of reverse-phase octadecylsilane-bonded silica cartridges.[20,21]

C. Chromatography and Mass Spectrometry

The choice of the chromatographic step to be used in the identification and quantitation of the xenobiotic is dictated by the nature of the compound. The nonpolar and the chlorinated hydrocarbons are routinely analyzed by gas-liquid chromatography (GLC) with flame ionization or electron capture detection.[22] Polar compounds can be converted to nonpolar derivatives prior to gas-liquid chromatography. Chemical derivatization may also serve to stabilize labile compounds for chromatographic examination at elevated temperatures. Many compounds which cannot be analyzed by GLC may be analyzed by high-pressure liquid chromatography (HPLC), which may also serve as a method of isolation of the xenobiotic.[23] Since ambient or minimally elevated temperatures are employed, heat-labile xenobiotics present no difficulties. A wide variety of column materials are available as well as reagents for the preparation of UV absorbing derivatives of xenobiotics which do not possess chromophores of their own.

Reliance on chromatographic peaks generated by a single analytical system may not be sufficient and the results should be confirmed by some other independent physicochemical means. Thin-layer chromatography (TLC) and HPLC are effective methods to use in conjunction with GLC.[24] MS analysis in combination with GLC or liquid-liquid chromatography provides solid evidence for confirmation of the identity of any unknowns.[25] Furthermore, combined chromatographic and MS techniques can provide definitive quantitative estimates of any xenobiotics, when using stable isotope-labeled internal standards.[26]

IV. LIPID DIGESTION AND ABSORPTION

Dietary fat serves as a vehicle for the entry of most fat-soluble xenobiotics and the fatty digestion products facilitate absorption of all xenobiotics by the mucosal cells. At present there are few data for their digestion and absorption by the intestinal mucosa. It is believed, however, that many of the fat-soluble xenobiotics are absorbed by mechanisms similar to those responsible for the uptake of the fat-soluble micronutrients, such as vitamins A, D, E, and K. Of special interest is the micellarization of the dietary fats during the course of normal digestion, and the passive nature of the absorption process, which involves a membrane uptake of partially degraded lipid esters and free alcohols in the form of surface-active materials.

A. Enzymatic Hydrolyses

Some of the fat-soluble xenobiotics are alcohols and enter the body in the form of fatty acid esters, while others are acids and may be present in the food in the form of fatty alcohol esters. These esters are believed to be extensively hydrolyzed during their passage from the mouth to the jejunum. Several lipases of markedly different specificity, pH optima, and co-factor requirements are believed to be involved, although only limited observations have been made with specific xenobiotics.

1. Lingual Lipase and Milk Lipase

The dietary fats are first attacked by the lingual lipase of the mouth.[27] This enzyme hydrolyzes both long- and short-chain triacylglycerols, and there is evidence that it attacks preferentially the *sn*-3-position, but it also hydrolyzes the primary ester bonds in alkyldiacylglycerols. Thus, when a racemic mixture of alkyldiacylglycerols consisting of equimolar amounts of 1-*O*-octadecyl-2,3-dioctadecenoyl-*sn*-glycerol (labeled with either ^{3}H in the alkyl group or with ^{14}C in both acyl groups) and 3-*O*-octadecyl-1,2-dioctadecenoyl-*sn*-glycerol (labeled with either ^{14}C in the alkyl group or with ^{3}H in the acyl groups) was subjected to rat lingual lipase hydrolysis, the *sn*-3-position was hydrolyzed twice as fast as the *sn*-1-position.[28] The lingual lipase of a human infant was shown to hydrolyze the *sn*-3-position of synthetic triacylglycerols about four times faster than the *sn*-1-position.[27] This enzyme has a broad pH range (2.2 to 6.0), which permits the digestion of dietary fat immediately after food enters the mouth and continues the digestion in the stomach as well as after leaving it. The ester hydrolase originally thought to be secreted by the gastric mucosa is now believed to be either lingual lipase or a milk lipase, which also is not denatured at the acid pH of the stomach. The latter enzyme requires an activation by bile salts for the hydrolysis of long-chain fatty acid esters. Wang et al.[29] have purified the lipase from human milk and have determined that the enzyme had no significant stereospecificity, but that it attacked the short-chain fatty acids faster than the long-chain acids, regardless in which position they were found. The fatty acids from the secondary alcohol position appeared to be released at a slower rate than those from the primary positions. The bile salt-activated milk lipase also hydrolyzes other esters, such as p-nitrophenylacetate and other simple esters.[30]

2. Pancreatic Lipases

In the small intestine the glycerol esters of fatty acids become subject to the action of pancreatic lipase, which acts on triacylglycerols at the oil-water interface of the lipid particles. Stereospecific analyses have shown that pancreatic lipase attacks the acids at the sn-1- and sn-3-positions to yield the sn-1,2- and sn-2,3-diacylglycerols as intermediates in equal proportions and does not attack the secondary position directly.[31,32] However, pancreatic lipase also attacks O-alkyldiacyl glycerols and S-alkyldiacylglycerols. Paltauf et al.[28] showed that pancreatic lipase hydrolyzes the sn-3-ester bond in 1-alkyl-2,3-diacyl-sn-glycerol and the sn-1-ester bond in 1,2-diacyl-3-alkyl-sn-glycerol at equal rates. Similar observations were made by Åkesson et al.[33] Surprisingly, the enzyme attacked the 2-acyl-3-alkyl-sn-glycerol faster than the 1-alkyl-2-acyl-sn-glycerol.[28] The latter study also showed that pancreatic lipase hydrolyzed the 1-acyl-3-alkyl-sn-glycerol faster than its enantiomer. Åkesson and Michelsen[34] demonstrated that the triacylglycerol analog, X-1,2-diacyl-3-S-alkyl 3-thioglycerol-S-oxide, also is deacylated in the intestinal tract. By means of optical dispersion studies it was possible to show that the 2,3-diacyl-1-S-tetradecyl-1-thio-sn-glycerol-S-oxide was recovered in significantly higher amounts than the 1,2-diacyl-3-S-tetradecyl-3-thio-sn-glycerol-S-oxide. Although this difference in absorption of the two enantiomers could have been due to differences in the mucosal lipolysis of these compounds, the preferential reacylation of the 1-S-tetradecyl-1-thio-sn-glycerol-S-oxide in the intestinal cell remains a possibility.[35] Paltauf[36] had found that 1-O-alkyl-sn-glycerol, but not 3-O-alkyl-sn-glycerol, is acylated in the intestinal mucosa. Noda et al.[37] reported that 2,3-dioleoyloxybutane and 1-hexadecyloxy-3-oleoyloxybutane were degraded more slowly by pancreatic lipase than the corresponding acylglycerols and that 1-hexyloxy-2-octanoyloxy propane was not attacked at all. The pancreatic lipase also catalyzes the hydrolysis of 2-monoacyl-glycerols, but this apparently happens following a prior isomerization to the sn-1- and sn-3-monoacylglycerols.[38] Pancreatic lipase is activated by bile salts and by colipase, as discussed elsewhere in this book.[27]

The pancreatic juice also contains a nonspecific carboxylic ester hydrolase. This enzyme is believed to be responsible for the hydrolysis of the cholesteryl esters and the esters of the fat-soluble vitamins, which cannot be absorbed in the ester form.[39] In addition, this enzyme has been shown to hydrolyze a variety of other esters, such as p-nitrophenylacetate, β-naphthol acetate and laurate, the long-chain monoacylglycerols,[40,41] and the long-chain fatty esters of other hydroxy fatty alcohols.[42]

The pancreatic lipases also include a phospholipase A_2 type of enzyme, which converts the glycerophospholipids to the corresponding lysophospholipids.[43] Unlike the snake venom phospholipase A_2 the pancreatic phospholipase shows a marked preference for anionic phospholipids, such as phosphatidic acid, cardiolipin, and phosphatidylglycerol. The enzyme has a requirement for calcium and bile salts. Mansbach et al.[44] have isolated a novel phospholipase from the intestine with a narrow range of substrate specificity. It was active with phosphatidylglycerol as substrate in the surface barostat assay technique, but not with phosphatidylcholine. In the gut, activity was observed with both glycerol and choline phosphatides. Since the enzyme activity increased with pancreatic disease, it was concluded that this was not due to pancreatic phospholipase. The sphingomyelins are attacked by two other phospholipases active in the intestinal lumen; one releases the ceramide moiety of the sphingomyelin and the other degrades the ceramide to free nitrogenous base and free fatty acid.[45,46] Only limited studies have been performed with these enzymes, and it is not known if they are active with any other substrates.

B. Formation of Mixed Micelles

According to the original hypothesis of Hofmann and Borgstrom,[47] the lipolysis products of dietary fats, along with any other fat-soluble substances such as free sterols and fat-soluble vitamins, interact with bile salts to form mixed aggregates or micelles. The lipolytic products are dispersed as if there were two phases in the intestinal lumen. One of these is an emulsion phase of lipid particles in the 2000- to 50,000-A range suspended in an aqueous medium; the other is a micellar phase made up of particles 20 to 30 A, which is formed in the presence of bile salts. The fat absorption is thought to take place from the micellar phase, although intact micelles are not absorbed. The fat-soluble xenobiotics are similar in physical characteristics to cholesterol, free fatty acids, monoacylglycerols, and the fat-soluble vitamins and, therefore, would be expected to form mixed micelles by interaction with the biological detergents: bile acids, phospholipids, and the lipolysis products of neutral fats. There is good reason to believe that the neutral, fat-soluble xenobiotics become dissolved within the nonpolar interior of the micellar particles, while the more polar fat-soluble xenobiotics may interdigitate with the surfactant molecules of the micelles.[48]

C. Mechanism of Uptake

Until recently, it seemed that the mechanism of the absorption of the lipid molecules involved the penetration by the mixed micelles of an unstirred water layer covering the brush border membrane and the formation of a monomeric phase of the lipid molecules, which partitioned into the villus cell membranes.[49] This assumption, however, now appears to require a reexamination, especially since the findings from studies with pure micellar solutions, while physicochemically sound, may have little physiological relevance.

Several laboratories have suggested that the Hofmann-Borgstrom hypothesis may be an oversimplification. Patton and Carey[50] have reexamined the digestion of stomach emulsions and pure olive oil and have observed the sequential appearance of a birefringent calcium soap phase and a 1:1 monoglyceride-fatty acid phase. Stafford and Carey[51] have shown that upon ultracentrifugation of a turbid triglyceride-free aqueous emulsion, a bile salt micellar phase, saturated with the products of lipolysis, and a liquid-crystalline phase, saturated with bile salts, are obtained. These workers have demonstrated that the liquid crystalline phase and the micelles coexist in the aqueous phase. The mixed micelles had hydrodynamic radii of about 200 A, while the nonmicellar particles (liquid crystals) had hydrodynamic radii of 400 to 600 A. The dispersed lipids are of a bilayer nature and presumably very soluble in the lipid bilayers of cell membranes. Carey et al.[52] have suggested that owing to a high intraluminal concentration such lipids could move passively and efficiently down their concentration gradients from the lumen via apical membranes of enterocytes to the cytosol. This mechanism is related to the hydrocarbon continuum suggested by Patton[53] and would differ from the absorption via the lipid monomers in the interparticle aqueous phase suggested by Westergaard and Dietschy,[54] Thompson and Dietschy,[49] and subsequent workers.[55] Carey et al.[52] have attempted to reconcile these two views of fat absorption by speculating that during absorption a series of diffusional barriers must be overcome in order for micellar or liposomal particles to make contact with enterocyte membranes and that during this diffusion the luminal micelles may become gradually diluted and transformed into unilamellar liposomes or converted into monomers. It remains to be established how the molecules of dispersed lipid products in the intestinal lumen gain entrance into the absorptive enterocytes.

There is considerable evidence that the mixed micelles do not remain intact during the uptake by the mucosal cell. When the jejunum is perfused with a mixed micellar solution, some of the micellar solutes, such as fatty acids and monoacylglycerols, are taken up very rapidly, while cholesterol is taken up more slowly and the conjugated bile salts are excluded.[56] Uptake of different solutes at independent rates has also been demonstrated with everted sacs.[57] Comparisons of uptake of micellar oleic acid and of a monooleoylglycerol, which is a nonhydrolyzable analog of monooleoylglycerol, have shown[56] that the ratio of lipid uptake was different from the ratio in the micelles. The uptake ratio was predictable from the independent uptake ratios and from the micellar concentrations of the co-solutes. Since the micelle does not penetrate the cell membrane, it must be concluded that the lipid is taken up as simple molecules, at least from micellar perfusion media.

D. Efficiency of Uptake

According to current theories, the maximum rate of uptake of a solute would occur when the monomer solution at the interface is saturated. The micellar solubilization would increase the driving force for transport across the unstirred water layer and indirectly for penetration of the cell membrane.[56] The efficiency of uptake is also influenced by factors other than the micellar solubilization.

Thompson[58] has demonstrated that the age of the animal affects the rate of uptake of a homologous series of saturated fatty acids by the mucosa in vitro. At each rate of stirring of the solution bathing the mucosal segments, the unstirred water layer was lowest in the tissue derived from the suckling and the mature than from the old animals. From these values the increment for energy change associated with the addition of each methylene group to the fatty acid chain was 50% higher in the suckling and mature animals than in the old animals. The results suggested that discrepancies in uptake of fatty acids in animals of different ages can be explained by differences in the passive permeability properties and functional surface area of the membrane and by differences in the overlaying unstirred water layer. In mature animals the short-chain acetic acid showed a higher rate of uptake than expected from a linear extrapolation of the relation from the longer-chain fatty acids. A similar finding was made for *n*-butyric and *n*-hexanoic acids. This effect has been attributed in the past to a carrier-mediated difference.[59]

By assessing differences in potential across the small bowel, Hollander and Dadufalza[60] have demonstrated a gradual decrease in the thickness of the unstirred water layer as the rats aged. On the other hand, using linoleic acid as a probe, a fourfold increase in the surface area of the layer was found both in the jejunum and ileum and, subsequently,[61] their maximal capacity to absorb linoleic acid increased fivefold in both the jejunum and ileum. The decrease observed in the resistance of the unstirred water layer to linoleic acid with aging was not explained.

There have not been any parallel studies with lipid-soluble drugs or xenobiotics, but the above-noted changes in the efficiency of absorption of the saturated and unsaturated fatty acids with aging should be considered in relation to the fat-soluble xenobiotics, regardless of the actual nature of the underlying physiological mechanism.

V. ABSORPTION OF FAT-SOLUBLE XENOBIOTICS

Fat-soluble xenobiotics have become a source of much concern for public health because of their widespread distribution, high resistance to biodegradation, and potential toxic or carcinogenic properties. As a result, considerable effort has been expended in determining the extent of absorption of various fat-soluble compounds in vivo and in vitro. In most instances, the in vivo studies have involved the recovery of the xenobiotics from blood, liver, or other tissues, with very few measurements of the lymphatic uptake and transport of these compounds. The following section of the chapter provides short notes on the recovery of dietary fat-soluble xenobiotics from animal tissues and in vitro preparations. For this purpose the various fat-soluble xenobiotics are considered in the order listed under Structure and Nomenclature.

A. Aliphatic and Polycyclic Hydrocarbons

The intestinal absorption of the neutral hydrocarbons has challenged the imagination of lipid physiologists and biochemists for many decades and has served as proof of either absorbability or nonabsorbability of given lipid forms. Studies with isotopic tracers have demonstrated that significant amounts of aliphatic hydrocarbons may be absorbed and metabolized by the intestinal cells, while the polycyclic hydrocarbons are subject to more limited uptake and metabolic transformation.

1. Paraffins

Channon and Collinson[62] first demonstrated the absorption of liquid paraffins from the alimentary tract of rats and pigs. Their findings were confirmed and extended by later workers.[63,64] Stetten[63] fed deuterated hexadecane to rats and found that it was taken up by the intestinal tract to a considerable degree and converted to fatty acids. More systematic studies of the absorption of deuterium-labeled normal C_8 to C_{18} hydrocarbons were reported by Bernhard.[65] When a 20% solution of hydrocarbons was fed in olive oil, about 20% was recovered in the lymph lipids. The chain length did not seem to have any effect on the extent of absorption. In contrast, Albro and Fishbein[66] found that aliphatic hydrocarbons with carbon numbers from C_{14} to C_{32} fed intragastrically to rats are absorbed with an efficiency that correlated inversely with their carbon number.

McWeeny[67] found that 13 to 16% of a 2% solution of n-[1-^{14}C]hexadecane in olive oil administered intragastrically to rats was absorbed over a period of 15 hr, with 20 to 30% of the absorbed radioactivity appearing in the lymph lipids as fatty acids. Kollattukudy and Hankin[68] reported that nonadecane was absorbed by the rat and converted to the corresponding fatty acid. Although a great variety of bacteria are known to oxidize alkanes to fatty acids, a rigid exclusion of intestinal bacteria was not always achieved in the above experiments.

Vost and Maclean[69] have reexamined the lymphatic transport of [^{14}C]hexadecane and [^{14}C]octadecane in relation to [^{3}H]glycerol-labeled triacylglycerols in the rat. These studies indicate that the gut-derived hydrocarbons are transported in chylomicrons as triacylglycerol solutes. High-density lipoprotein is the plasma acceptor of these hydrocarbons, but the rate of efflux from the chylomicrons is regulated by the hydrolysis of their triacylglycerols. Chylomicrons were obtained at high and low concentrations of hydrocarbons. High concentrations of these solutes (0.5 to 3.0% of triacylglycerol weight) were achieved by feeding safflower oil with 2 to 8% hydrocarbon. The potential quantitative importance of this pathway was apparent, since, after high dietary loads, octadecane was recovered at 1 to 3% concentrations in lymph chylomicron triacylglycerols.

Consumption of diets rich in squalene has been shown to result in increased concentration in the lymph[70] and serum.[71] Absorption of the saturated analog, squalane, however, was insignificant.[70] Direct feeding of squalene has increased squalene and sterol concentration in serum and liver of the rat and enhanced fecal excretion of bile acids,[71] which suggested that squalene absorbed from the diet participated in the overall cholesterol synthesis. Tilvis and Miettinen[72] have recently demonstrated that [^{3}H]squalene is, like [4-^{14}C]cholesterol, absorbed through the lymphatic route and that approximately 20% of absorbed [^{3}H]squalene is cyclized to sterols during the transit through the intestinal wall. An increase in dietary squalene load from 8 to 48 mg decreased the absorption percentage of [^{3}H]squalene from 46 to 26%, but did not affect the absorption of [^{14}C]cholesterol (47%). The more rapid appearance of dietary [^{3}H]squalene than [^{14}C]cholesterol in chyle and serum was rationalized by the fairly high difference in squalene concentration between the test mixture and intestinal mucosal cells, which was not the case for the dietary [^{14}C]cholesterol.

2. Polycyclics

The absorption of polycyclic hydrocarbons in experimental animals has been studied extensively.[73] In most instances, however, the test compounds were administered by i.p. injection, and the results may not have much relevance to the evaluation of orally ingested polycyclics. The data obtained from the oral administration of several radioactive polycyclics indicate that they are rapidly absorbed and eliminated through biliary excretion or in the feces and urine. Some of the absorbed material is retained in the adrenals, the ovaries, and in body fat, where it can be detected after 8 days. Extensive testing[74,75] indicates that only a few of the known carcinogenic polycyclic aromatic hydrocarbons are capable of inducing cancer in animals when administered orally.

Pocock and Vost[76] have reported high efficiency of intestinal absorption and transport into intestinal lymph for [^{14}C]1,1,1-trichloro-2,2-bis (p-chlorophenol)ethane (DDT), with 60% recovered from lymph within 12 hr of gastric feeding. Vost and Maclean[69] have investigated the lymphatic absorption of [^{3}H]benzo(a)pyrene in comparison to that of [^{14}C]DDT and the normal chain C_{16} and C_{18} hydrocarbons. All hydrocarbons were transported in the triacylglycerol oil phase of chylomicrons. The quantitative association of all four hydrocarbons with triacylglycerol supports the suggestion that core lipid provides an oil phase for nonspecific transport of hydrocarbons of low polarity.[76] High-density lipoprotein was the major plasma acceptor of all labeled hydrocarbons. It selectively concentrated hydrocarbons from their chylomicron triacylglycerol phase both during passive transfer of DDT from unhydrolyzed chylomicrons in vitro and with other solutes in vivo. Plasma chemical fluxes were measured for octadecane and DDT and both showed net fluxes from chylomicrons to high-density lipoprotein. [^{3}H]Benzo(a)pyrene in vivo did not transfer rapidly to other lipoproteins either from chylomicrons or from injected high-density lipoprotein. This differs from a report in which chylomicrons, labeled with [^{3}H]benzo(a)pyrene by incubation, were injected into rats and [^{3}H]benzo(a)pyrene was recovered predominantly in very low-density lipoproteins and the ultracentrifuged residual protein fraction with highest specific radioactivity in the lung.[77] These discrepancies may have reflected damage from incubation to chylomicrons and from ultracentrifugation shear to very low-density lipoproteins and chylomicrons that generate lipoprotein material in the residual fraction.[78] Results with hydrocarbons introduced into lipoproteins in vitro indicate subsequent passive diffusion of benzo(a)pyrene and its hydroxylated metabolites among major human serum lipoproteins[79] and similar diffusion of chlorinated hydrocarbons.[80] Simple passive diffusion of hydrocarbons introduced biologically in lipoproteins has been reported only for chylomicron DDT.[76]

Laher et al.[81] have assessed the effect of concomitant lipid absorption on the bioavailability and lymphatic transport in rats of benzo(a)pyrene, a carcinogenic polycyclic aromatic hydrocarbon. Conscious animals equipped with biliary and mesenteric lymphatic catheters received intraduodenally [^{3}H]-labeled benzo(a)pyrene completely dissolved in either 50 or 500 μmol of olive oil. The tenfold variation in the mass of the carrier vehicle did not significantly affect the disposition of the hydrocarbon, and portal, not lymphatic transport, was the major route postabsorptively. The results showed that the rat enterocyte quickly adapts to dietary fat contaminated with polycyclic aromatic hydrocarbons, even during the assimilation of a single dose of fat. It was concluded that during the postabsorptive synthesis of chylomicrons, benzo(a)pyrene is metabolized and removed from the triglyceride oil droplets. The site where benzo(a)pyrene is separated from dietary fat in the rat intestine is uncertain. In reporting recent studies, Vetter et al.[82] have pointed out that in the killfish, benzo(a)pyrene follows dietary fat through the processes of digestion, dispersion, absorption, and resynthesis to reappear inside the cell once again within triglyceride droplets. Benzo(a)pyrene is then metabolized, separated from the fat, and finally appears in the gall bladder of the fish. If benzo(a)pyrene is also coassimilated with dietary fat in the rat, then it must be separated from the intracellular fat droplets during their migration from the apical region of the cell, through the endoplasmic reticulum to the Golgi complex, where they become complete chylomicrons.[83] During this lipid processing in the rat enterocyte, the fat droplets are enveloped by smooth endoplasmic reticulum membranes, and separation of benzo(a)pyrene from fats probably occurs via the benzo(a)pyrene-metabolizing enzymes of the smooth endoplasmic reticulum. These observations indicate that although initial lymphatic transport of polycyclic aromatic hydrocarbons may require chylomicrons, a large flux is not needed and increased lipid transport in lymph does not promote increased benzo(a)pyrene transport.

In an earlier study, Laher et al.[84] had shown that in the rat, the presence of bile salts and long-chain partial glycerides in the lumen maximizes the uptake of 7,12-dimethylbenz(a)anthracene. This study was undertaken to identify certain dietary and physiological factors in the mammalian gastrointestinal lumen which might influence the uptake of this representative polycyclic aromatic hydrocarbon. Luminal bile significantly enhanced biliary recovery of radiolabel following instillation in both long-chain triglyceride vehicles, but did not affect the recovery from medium-chain oil. This study demonstrates that while bile is not absolutely necessary for the uptake of this orally ingested hydrocarbon carcinogen, it significantly facilitates absorption from a long-chain triglyceride vehicle. The appearance of this xenobiotic or its metabolites in the bile, along with the evidence for extensive biliary recycling, calls attention to the fact that the intestinal epithelium and the liver can be exposed for relatively long periods to ingested carcinogens. The rapid metabolism and excretion into the bile of polycyclic aromatic hydrocarbons following a systemic entry in rats has long been recognized[85] and the biliary excretion of radiolabel provides an index of relative [³H]DMBA absorption from various lipid vehicles following duodenal administration.

Borm et al.[15] have investigated in rats the metabolism of 7-ethoxycoumarin and 7-ethoxycoumarin following pretreatment with 3-methylcholanthrene. Twenty-four hours after 3-methylcholanthrene pretreatment (20 mg/kg) induced monooxygenase activity varied from 2.5 to 6-fold, depending on the method of cell preparation. The formation of glucuronides in cells was significantly lowered by the 3-methylcholanthrene pretreatment, while sulfation was unaffected.

3. Biphenyls and Triphenyls

Although the use of many of the biphenyl and triphenyl preparations has recently been reduced or entirely banned, they have contaminated the environment widely, because they are relatively resistant to biodegradation. The extent of food contamination is variable and may be considerable at certain times in specific geographic areas, although generally the amounts detectable in food substances are low.[86,87] Wasserman et al.[88] and Lo and Sandi[5] have prepared extensive reviews on polycyclic residues in the biologic environment and in foods. The polychlorinated biphenyls are excreted in breast milk[89] and residues have been detected in human adipose tissue.[90] The effects of PBBs on humans have been determined by examining two highly exposed groups, the Michigan Chemical Co. workers and the farm residents who consumed the PBB-contaminated meat, milk, cheese, butter, and related products. Clinical studies did not show any significant differences in the exposed and nonexposed (control) groups.[91]

However, the interpretation of some of the results on the PBB effects has been disputed.[92] Allen and Norback[75] have shown that these agents, when administered orally, induce gastric hyperplasia and dysplasia in rhesus monkeys. Although histopathological changes resembling precarcinomatous lesions have been shown in the gastric submucosa of animals receiving these agents, their effect on the human alimentary tract has remained uncertain.

Since the polyhalogenated biphenyls and triphenyls are readily solubilized by the digestion products of dietary fats, it is obvious that they will participate in the same digestive and dispersive processes as trace nutrient lipids in the intestinal lumen. Thus, Pocock and Vost[76] have demonstrated high efficiency of intestinal absorption and transport into intestinal lymph, with 60% of administered [14C]DDT being recovered from lymph within 12 hr of gastric feeding. Pocock and Vost[76] and Maliwal and Guthrie[80] report simple passive diffusion of DDT between chylomicrons and other lipoproteins. More recently, Vost and Maclean[69] have shown that the gut transport of the pesticide is comparable to that of other polycyclic hydrocarbons and normal chain alkanes. Plasma chemical fluxes were measured for octadecane and DDT and both showed net fluxes from chylomicrons to high-density lipoproteins. Earlier, Sieber[93,94] had studied the absorption of [14C]labeled compounds structurally related to *p,p'-*DDT in thoracic duct cannulated rats and identified the parent DDT compounds and their metabolites in the lymph. The DDT compounds varied in their lipid solubility and in the extent of their lymphatic absorption, but a strict correlation between lipid solubility and lymphatic absorption was not established, possibly because of other factors such as differences in rate and routes of excretion of each compound.

B. Long-Chain Alcohols, Acids, and Esters

The absorption of long-chain alcohols, acids, and their esters has been studied more extensively than that of other xenobiotics, but clear-cut results have not been obtained. The wax esters and long-chain alcohol ethers have served as model compounds for the absorption of both glyceryl esters and fat-soluble vitamins. Their physiological effects, with some exceptions, are relatively mild.

1. Long-Chain Alcohols

Although the long-chain alcohols are widely distributed in nature in the form of wax esters and these esters are extensively employed in personal care products, there appear to have been no systematic studies of the intestinal absorption of these fat-soluble xenobiotics.[95] Baxter et al.[96] have reported on the lymphatic transport of the esters of phytol and cetyl alcohols.

Among the long-chain alcohols may be included the alkyl ethers of glycerol, which are absorbed intact and also become subject to acylation by the acyltransferases of the intestinal mucosa. Sherr and Treadwell[97] and Gallo et al.[98] have proposed the use of the 1- and 2-octadecylglyceryl ethers as model compounds for study of triacylglycerol resynthesis in the intestinal mucosa. Sherr and Treadwell,[97] using everted sacs of rat intestine, have demonstrated the conversion of 2-octadecylglyceryl ether into alkoxymono- and alkoxydiacylglycerols. Gallo et al.[98] showed that the microsomes of rat intestinal mucosa convert both the 1- and 2-octadecyl glyceryl ethers into alkoxymonoacylglycerols. The 2-octadecylglyceryl ether is converted into X-1-acyl-2-alkylglyceryl ether followed by a further conversion into the alkoxydiacyl glycerol, while the 1-octadecylglyceryl ether is converted mainly to X-1-alkyl-3-acylglyceryl ether, with small proportions of the X-1-alkyl-2-acyl glyceryl ether. The small amount of the X-1,2-diradylglycerol formed in the latter instance may account for the conversion of the X-1-octadecylglyceryl ether to the alkoxydiacylglycerol. Subsequently, Paltauf and Johnston[99] have shown that the *sn*-3-octadecylglycerol is not acylated by the acyltransferases of the intestinal mucosa. An attempt to utilize the 2-monoalkylglycerol as a substrate for probing the stereochemical course of acylation via the monoacylglycerol pathway has failed[100] because of the loss of the stereospecificity of both phospholipase C and D for the *sn*-1-acyl-2-alkylglycerophosphocholine employed as intermediate in the assay.

2. Long-Chain Aliphatic Acids

This group of xenobiotics comprises the naturally occurring long-chain saturated and monosaturated fatty acids, the positional and geometric isomerization products of natural unsaturated fatty acids arising from industrial hydrogenation of natural fats and oils, some naturally occurring hydroxy fatty acids, and the brominated fatty acids used in various foods, drinks, and industrial products.

Several laboratories have reported that the consumption by weanling rats of triacylglycerols containing the C_{22} monosaturated acids, erucate and cetoleate, results in a marked accumulation of lipids in the heart.[101] However, Mattson and Streck[102] did not observe this for fat containing behenic acid, nor when the dietary fat was 2-behenoyl-dilinoleoylglycerol. Following feeding of this compound, only 24% of the behenate moiety was found in the lymph. Tomarelli et al.[103] and Filer et al.[104] have shown that a saturated fatty acid attains maximum absorbability when it is esterified at the *sn*-2-position of a triacylglycerol that contains unsaturated fatty acids in the *sn*-1- and *sn*-3-positions. However, behenic acid was still poorly absorbed. Szlam and Sgoutas[105] have studied the absorption in rats of 20:1 and 22:1 acids in the form of rapeseed oil and showed that the accumulation of dietary 20:1 and 22:1 acids in serum, lipoproteins, and their lipid subfractions was highly dependent on the alimentary condition of the animal. Unfasted rats incorporated these acids in all lipoproteins and their lipid subclasses, whereas starved rats only in high-density lipoprotein triacylglycerols. Kritchevsky et al.[106] have reported that native peanut oils are more atherogenic than randomized oils of the same fatty acid composition. Manganaro et al.[107] have discussed the possible effect on the absorption of natural peanut oils of the presence of long-chain fatty acids, largely in the *sn*-3-position. Ammon et al.[108] have reported adverse effects of long-chain fatty acids on solute absorption by the intestinal mucosa of man during perfusion of the jejunum.

Fatty acids with *trans*-double bonds occur in some natural oils from leaves and seeds of plants and ruminant fats, but the main source in our diets is from partial hydrogenation of vegetable and marine oils. During this process a wide variety of geometrical and positional isomers of fatty acids arise, depending on the original composition of the oils.[109] Emken[110] has reviewed evidence indicating that the *trans* and positional fatty acid isomers are absorbed at rates comparable to those noted for the natural monoenoic and dienoic fatty acids. Blomstrand and Svensson[111] have studied the incorporation of dietary isomeric fatty acids derived from different partially hydrogenated marine oils into individual phospholipids of mitochondrial membranes of the rat heart. Dietary *cis*-isomers of 22:1 seemed to have a specific ability to interfere with cardiac ATP synthesis and also to alter the fatty acid composition of cardiolipin of rat heart.

Bergstrom et al.[112] have investigated the effect of steric hindrance at the carboxyl group on the absorption and metabolism in the rat of 2,2-dimethylstearic acid. The acid was well absorbed when dissolved in olive oil and administered to rats by stomach tube. It was then transported via the lymphatics and incorporated into triacylglycerols and glycerophospholipids. Dimethylstearic acid did not take part in the formation of ester bonds during hydrolysis of olive oil with pancreatic lipase. When the lymph acylglycerols containing this acid were treated with pancreatic juice, this ester bond appeared virtually resistant to the action of lipase. Other lipases responsible for the splitting of these bonds in the tissues apparently were not affected, as evidenced by the appearance of a [carboxy-^{14}C]-2:2-dimethyladipic acid in the urine.

Brominated vegetable oils have been widely used in the soft drink industry as dispersing agents for flavoring citrus oils.[113] Recently, these oils have been used as flame retardants[114] providing another source of potential contamination of food supplies. Although the brominated oils were originally classified as safe for human consumption, subsequent studies in several laboratories[115-118] have indicated toxicity when fed to rats. Conacher et al.[119] have shown that brominated olive, sesame, corn, and cottonseed oils were hydrolyzed by pancreatic lipase with the same specificity as nonbrominated oils, but with a decrease in the activity of the enzyme. Under conditions that gave 50% hydrolysis of corn oil, the brominated olive oil, sesame oil, corn, and cottonseed oils gave approximately 20, 16, 12, and 12% hydrolysis, respectively. As the content of tetrabromostearate increased, resulting in a higher melting point of the substrate, the activity of the lipase decreased. Since the brominated oils were degraded in the same manner as the common vegetable oils, they were assumed to be absorbed and deposited in the tissues in a similar fashion. The 9,10,12,13-tetrabromo stearates were detected in the livers and hearts of rats after administration of brominated cottonseed oils.

Lipid-bound bromine has been demonstrated by several investigators[115,116,120] in rat tissues following consumption of brominated vegetable oils by the animals. Rats fed brominated corn oil had higher lipid bromine concentration than those fed either the dibromo- or tetrabromostearates.[115] It was speculated that the brominated monoacylglycerols are better absorbed than the fatty acids. Whether the increase in tissue lipids involves increased levels of triacylglycerols or simply an increase in lipid mass from the substitution of brominated fatty acids for nonbrominated has not been determined.[116] The tetrabromostearate appeared to be more active than other bromoacids in producing adverse changes, particularly serious intracellular fatty degeneration.[120] In comparison to the monoacylglycerol of tetrabromomonostearate, monoacylglycerol of bromostearate, or a mixture of the two monoacylglycerols, which provided proportions of brominated acids comparable to that found in the brominated corn oil, the most pronounced effects were observed with corn oil. It was speculated that this was due to the presence of small amounts of other brominated acids, e.g., hexabromostearate derived from linolenic acid, or brominated sterols. Another consideration was the position of the brominated acid in the glycerol molecule. In corn oil, most of the linoleic acid is found in the *sn*-2-position and therefore most of the tetrabrominated stearate would be present in the form of 2-monoacylglycerol following pancreatic hydrolysis, whereas the synthetic monoacylglycerols would be largely of the *sn*-1- and the *sn*-3-types. Mohamed et al.[121] have reported that 9,10-dibromo palmitate is not oxidized by the β-oxidation system of mitochondria. James and Kestell[122] identified 5,6-dibromosebacic acid as an excretion product following dosing of rats with 9,10-dibromostearic acid. This could have been formed by a combination of β- and ω-oxidation followed by β-oxidation. Earlier work has shown that α-halogenated acids inhibit fatty acid oxidation.[123,124]

Castor oil contains ricinoleic acid as a major component. This acid has a hydroxyl group at carbon 12. The intestinal release of ricinoleic acid leads to a purgative action by inducing vigorous peristalsis due to irritation of the bowel mucosa. Watson and Gordon[125] have reported that this effect may be due to the poor absorbability of ricinoleic acid, which would then accumulate in the intestine and form soaps or soap solution. Maenz and Forsyth[126] have shown that in brush border vesicles, the secretory effects caused by ricinoleate are expressed at concentrations significantly below those associated with detergent effects or altered epithelial morphology. It was concluded that ricinoleate is calcium ionophore in the jejunal brush border vesicles and could have a significant intestinal secretory activity due to this Ca^{2+} ionophore property.

Gaginella et al.[127] have reported cytotoxic effects of ricinoleic acid and other surfactants on isolated intestinal epithelial cells.

Although castor oil is easily hydrolyzed in the small bowel, the activation and absorption of the free acid is not efficient. The result is a rapid accumulation of ricinoleic acid and its mineral salts. Nevertheless, some 5 to 10% of the fatty acids in the carcass triacylglycerols can be shown to be ricinoleic acid, when it or triricinoleoylglycerol is fed to rats.[125,128,129] Despite the rather heavy accumulation in triacylglycerols, no ricinoleic acid was incorporated into tissue phospholipids. On the basis of these results, Watson and Murray[128] suggest that the incorporation of ricinoleate into triacylglycerols may proceed via the acylation of monoacylglycerols rather than via the *de novo* formation of phosphatidic acid. Barber et al.[130] have shown that ricinoleoyl-CoA is essentially inactive with 1-acyl or 2-acyl-*sn*-glycerol-3-phosphorylcholine as acceptors. However, it can serve as acyl donor when glycerol-3-phosphate or 1-acyl-*sn*-glycerol-3-phosphate is available to yield di- and monoricinoleoyl-3-glycerophosphates, respectively. However, the possibility remains that ricinoleic acid could have become esterified to the *sn*-1,2-diacylglycerols to yield *sn*-3-ricinoleoyl-1,2-diacylglycerols. Therefore, the acylation at the *sn*-3-position may be the principal site of entry of ricinoleate into the triacylglycerols in vivo. This mechanism of synthesis of ricinoylglycerols does not appear to have received experimental attention.

3. Aromatic Acids

Sodium benzoate is widely used as a preservative in foods. Conacher et al.[131] have reported that all commercial drinks analyzed contained sodium benzoate as a preservative. The extensive studies of Hogben[132] on the absorption of organic acids from the intestine in vivo failed to reveal any specificity with respect to chemical structure other than a dependence on pKa and the lipid solubility of the nonionized form. Jackson et al.[133] have demonstrated competitive interactions between benzoic and phenylacetic acids, and benzoic and pentanoic acids, but the interpretation of these interactions is complex. Transport of benzoic acid was observed only in rat jejunum and not in rat ileum. Spencer and Brody[134] have shown that the hamster intestine accumulated a number of aromatic acids, including phenylacetic acid, and Spencer et al.[135] have shown that several benzoic acid derivatives were accumulated by the intestine of the mouse.

During the course of a detailed biochemical investigation into the mechanism of action of the hypolipidemic agent, 4-benzyloxybenzoate (BRL 14280), Fears et al.[136] indicated that this acid behaves as a fatty acid and participates in glycerolipid metabolism. With the aromatic acid in the diet of rats, similar compounds accumulated in the adipose tissue. Chemical characterization of the material synthesized in vivo showed that the metabolite was a triacylglycerol in which one of the fatty acid moieties was substituted by the 4-benzyloxybenzoate. Participation of an unnatural acid in the formation of triacylglycerols presumably necessitates prior formation of the acyl-CoA derivative. It is, therefore, relevant to note that benzoyl-CoA and phenylacetyl-CoA can be produced in vivo[137] and medium-chain fatty acid-CoA ligase is active with a variety of aliphatic and aromatic carboxylic acids. It may also be recalled that the principles of the β-oxidation of fatty acids were derived by Knoop[138] from work with phenyl-substituted and odd-number carbon fatty acids.

The widespread use of organic plasticizers and stabilizers has increased the hazard of environmental contamination by various aromatic acid esters, with a variety of metabolic effects. Nikinorow et al.[139] have studied the oral toxicity of several of these compounds, such as di-*n*-butyl phthalate, di(2-ethyl-hexyl) phthalate, di(*n*-octyl) tin *S*,*S*'-bis (isooctylmercaptoacetate), and dibenzyl tin *S*,*S*'-bis(isooctylmercaptoacetate), and have shown in the rat that these agents can induce congestion of the small intestine and mucosal sloughing in the stomach and intestines. Di(2-ethylhexyl)phthalate

(DEHP) is the most widely used commercial plasticizer. DEHP contaminates virtually all ecosystems[140] and has been found in human tissue,[141] in the food supply of man,[142] and in the most common material for medical devices such as flexible bags and tubing.[143] In recent years there has been rising concern over the safety of phthalate esters, which have been implicated in mutagenic, teratogenic, and cytotoxic effects, and in inhibition of enzymatic activities.[143] In certain applications, DEHP has been replaced by di(2-ethylhexyl)adipate, also known as DOA, which is also an inhibitor of enzymes. DOA is metabolized in the gut and other tissues to yield the monoester, adipic acid,[144] and presumably, 2-ethylhexanol.[145]

4. Sulfonic Acids and Other Detergents

The alkyl and aryl sulfates and sulfonates and other detergents used as surface-active agents for cleaning and disinfecting in bottling plants, canneries, dairies, and restaurants may also become ingested and absorbed by the intestinal mucosa. It is well established that decyl sulfate, as well as natural detergents such as bile and sodium taurocholate, alter the gastric mucosal barrier with consequent changes in permeability, resulting in increased influx of hydrogen ions and increased efflux of sodium and potassium ions.[146] Feldman et al.[147] and Feldman and Reinhard[148] demonstrated that sodium taurocholate accelerates the release of total phosphorus, lipid phosphorus, and protein from the everted rat small intestine at concentrations above critical micellar concentration (CMC). Feldman and Reinhard[148] examined the effects of a series of anionic detergents (surfactants) and showed a loss of protein from the everted rat intestine at surfactant concentrations above CMC. Kirkpatrick et al.[149] have examined the differential solubilization of protein, phospholipid, and cholesterol of erythrocyte membrane by detergents.

Early studies by Freeman et al.[150] demonstrated that oral ingestion (4 months, 100 mg/day) of a mixture of alkyl aryl sodium sulfonates by human subjects did not significantly alter digestion and absorption of other foodstuffs, as determined by body weight changes and analysis of fecal fat and nitrogen. However, larger doses given to dogs irritated the gastrointestinal tract and resulted in loose stools and increased fecal mucus. Chronic toxicity studies incorporating oral feeding regimens of alkyl dimethylbenzyl ammonium chloride, sodium dioctylsulfosuccinate, sodium lauryl sulfate, polyethyleneglycol monoisooctyl phenol ether, and sulfoethylmethyl oleoylamide in rats have shown that all of the detergents except polyethyleneglycol monoisooctyl phenyl ether induced diarrhea, gastrointestinal mucosal irritation, intestinal bloating, and small hemorrhages of the gastric mucosa.[151] More recently, Weaver and Griffith[152] tested dogs with a variety of detergents, including sodium alkylbenzene sulfonate, which is currently the most widely used surfactant in detergent formulations. It is readily absorbed in the intestinal tract of rats, enters the portal venous blood,[153] and has been observed to induce histopathology of the rat gastrointestinal tract after long-term feeding (0.02 to 0.5% of diet).

C. Polyglycerol and Diol Esters

The polyglycerol esters of fatty acids have been used for more than 50 years as food emulsifiers and are known to be well digested and absorbed in the rat. Recent investigations by King et al.[154] have demonstrated that 2.5 to 10% polyglycerol ester (decaglycerol dodecaoleate) added to the diet of rats is relatively nontoxic, but nevertheless interferes with the absorption of dietary fat. This effect is mediated through an undefined biochemical mechanism.

The acylation of mono- and diacylglycerols with adipic acid produces a series of viscous compounds with a number of potentially useful properties. Thus, polymers of saturated fatty acids, adipic acid, and glycerol tend to be relatively low melting, possess a resistance to oxidation, and have been considered for use in the food industry. Shull et al.[155] determined the digestibility, absorption, and in vivo oxidation of the two types of adipic acid esters of acylglycerols: an *sn*-1,3-diacyl-2-adipate and polymers of fatty acids, adipic acid, and glycerol. In female rats these products were absorbed as readily as ordinary fats. The digestibility coefficient of 93% correlated well with the degree of absorption of the [14C]stearic acid moiety of both diacylglycerol adipate and the polymeric fat. However, the absorbed [14C]material that was expired during the 8-hr experimental period differed markedly for the two fats. Four times as much radioactivity from stearic acid appeared in the expired CO_2 when the diacylglycerol adipate was fed than when the polymeric fat was given.

Industrial glycols, such as ethanediol or 1,3-butanediol and their fatty acid esters, are commonly encountered in the environment and are frequently ingested in small amounts along with the diet. These esters are subject to complete or partial hydrolysis by pancreatic lipase, in which case they may be absorbed along with the dietary fat. Parthasarathy et al.[156] have demonstrated that the ethanediol monoesters are hydrolyzed by the endogenous monoacylglycerol lipase associated with rat liver microsomal fraction. Since the diol-derived choline phospholipids are highly membrane-active and enzyme-active agents, efficient hydrolysis of these esters was claimed to be a significant step in maintaining physiologically tolerable levels of acyldiolphosphocholines in the system. Valisek et al.[157] have identified diesters of 3-chloro-1,2-propanediol as components of protein hydrolysates of vegetable meals and flours that contain triacylglycerols, while Gardner et al.[158] have isolated and identified the C_{16} and C_{18} fatty acid esters of chloropropanediol in adulterated Spanish cooking oils. In addition, Cerbulis et al.[159] have isolated 3-chloropropanediol esters of fatty acids as minor components of goat's milk and in a study[160] have compared the fatty acid composition of the chloropropanediol diesters to that of the glycerol esters of goat's milk. Only long-chain fatty acids were present in the chloropropanediol esters, while the glycerol esters contained both short- and long-chain fatty acids. Myher et al.[161] have shown that the chlorine is distributed about equally between the *sn*-1- and *sn*-3-positions of rac-1,2-diacyl-chloropropanediol, which appears to exclude their biosynthetic origin. It is not known to what extent these types of compounds may have been involved in the toxic oil syndrome in Spain. However, the presence of chloropropanediol esters in oils exposed to hydrochloric acid during the refining process warrants an investigation of their contribution to any toxicity of such dietary fat.

VI. MODULATION OF XENOBIOTIC ABSORPTION

In most instances the potential absorbability of the fat-soluble xenobiotics has been implied from studies of their partition between organic solvents (octanol) and water, which has been demonstrated to be closely related to the actual appearance of the chemical in lymph, blood, and tissues.[164,165] There is evidence that the lipophilic xenobiotics are absorbed in a similar fashion to the trace nutrient lipids and fat-soluble vitamins, the uptake of which is intimately dependent on the normal processes of fat digestion and absorption.[2,3] Therefore, factors that affect the normal digestion and absorption of dietary fat and fat-soluble vitamins would be anticipated to also affect the absorption of fat-soluble xenobiotics. However, other factors related to the specific xenobiotic must also be considered.

A. Micellar Solubilization

The use of natural fats or triacylglycerol oils for carrying medicinal drugs is an ancient art, but the physicochemical basis of promoting the absorption of fat-soluble drugs and xenobiotics by means of digestible oil vehicles is not fully understood. In fact, there have been very few studies of the solubility of drugs and xenobiotics even in triacylglycerol oils, and the fat solubility of a particular component has been assumed on the basis of its solubility in organic solvents. Recently, Patton et al.[164] have determined the solubilities of some long-chain fatty acids, alcohols, alkanes, and triacylglycerols, and of some aromatic, chlorinated aromatic, and chlorinated aliphatic hydrocarbons in trioleoylglycerol. It was concluded that above their melting temperature, all test compounds are miscible with liquid fat. Below their melting temperature the solubility of all test compounds could be estimated by the equation: log (mole fraction solubility) = $-\delta$ Sf(Tm $-$ T)/2.303 RT, where δ Sf, the entropy of fusion, can be estimated from chemical structure according to Yalkowsky and Valvani,[165] and the melting point (Tm) is either known or experimentally determined. It was observed that for long-chain compounds, solubility in triacylglycerol dropped precipitously with an increase in melting point. For aromatic and chlorinated compounds, the drop was more gradual. Since the entropy of fusion of rigid aromatic compounds is approximately 13.5 e.u. at room temperature, their solubility in triacylglycerol is a linear function of melting point. Direct measurements of solubility in the luminal contents during fat digestion, however, have not been made for either fat-soluble vitamins, drugs, or xenobiotics.

The need for the simultaneous administration of triacylglycerols for the absorption of hydrocarbons appears to have been firmly established. Elbert et al.[166] reported that administration of pure mineral oil to rats resulted, at most, in 3% absorption, whereas feeding of equimolar amounts of octadecane and trioleoylglycerol led to the absorption of about 50% of the octadecane, which was recovered in the lymph partly as fatty acid. Krabisch and Borgstrom[167] found the solubility of octadecane in 8 mM sodium taurocholate to be less than 0.2 μmol/mℓ, but if 7.2 μmol of 1-monooleoylglycerol was included in the solution, the solubility increased more than tenfold. The ready absorption of hydrocarbons in the presence of the lipolytic products of triacylglycerols favors the idea that the absorption of the nonpolar lipids takes place via a mixed micellar phase, which may be taken up intact by the mucosal membrane or carried into the cell membrane via the hydrocarbon continuum proposed by Patton.[53]

Laher and Barrowman[168] have investigated the physicochemical behavior of xenobiotic hydrocarbons in simulated intestinal contents. The simulated luminal contents contained concentrations of fatty acids, monoacylglycerols, triacylglycerol, phospholipid, and sodium ion likely to be found postprandially in the small intestine. Laher and Barrowman[168] estimated that approximately six molecules of bile salt solubilized one molecule of hydrocarbon when allowed to form mixed fatty acid-monoacylglycerol-bile salt micelles. It was evident that depletion of the triacylglycerol content at fixed fatty acid and monoacylglycerol concentration significantly promoted the partition of the polycyclic aromatic hydrocarbons (7,12-dimethylbenzanthracene, 3-methylcholanthrene, and benzo(a)pyrene) from an oil phase into an aqueous micellar phase. Triacylglycerol hydrolysis was, therefore, necessary for an effective partition of hydrocarbon from an oil phase into a water-dispersible bile salt solution. As hydrolysis proceeded, increasing quantities of dimethylbenzanthracene partitioned into the aqueous phase, reaching a maximum of 13% of the hydrocarbon in the aqueous phase at 40% hydrolysis.

Laher and Barrowman[168] have also examined the partition of a polychlorinated biphenyl mixture (Arochlor® 1242), containing 42% chlorine, from an oil phase into an aqueous micellar phase and have found it to be promoted by free fatty acids and monoacylglycerols. As for the transfer of polycyclic aromatic hydrocarbons, hydrolysis of triacylglycerol was necessary for an appreciable partition from the oil phase into the aqueous bile salt solution.

B. Effect of Dietary Fat

Both micellar solubilization and the eventual recovery in lymph of the various xenobiotics are affected, not only by quantity, but also by the nature of the dietary fat. According to Laher and Barrowman,[168] the partition of dimethylbenzanthracene and methylcholanthrene into micelles was optimized by long-chain monounsaturated (oleic) acid and monooleoylglycerol as compared with their octanoic and linoleic acid counterparts. This was consistent with the studies of Borgstrom[169] who found that cholesterol partitioned more in favor of an aqueous phase when the fatty acid species was oleic rather than linoleic. In contrast, the polychlorinated biphenyls appeared to favor the aqueous phase when the diunsaturated rather than the monounsaturated fatty acid was present. Laher and Barrowman[168] have suggested that these contradictory observations, in spite of otherwise similar behavior in the simulated intestinal system, could be explained by a stereochemical fit with the lipid cores of the mixed micelles, if a true micellar fit was occurring. There was speculation that the various polycyclic hydrocarbons studied might have a better fit when the oleic acid lipids were incorporated into the micelles rather than the linoleic acid and monolinoleoylglycerol, while the reverse might be true for the polychlorinated biphenyls.

Laher et al.[81] have compared the intestinal absorption of an orally ingested radioactive carcinogen, [^{3}H(G)]-7,12-dimethylbenz(a)anthracene dissolved in olive oil, safflower oil, or medium chain triacylglycerols. Luminal bile significantly enhanced biliary recovery of radiolabel following instillation in both long-chain triacylglycerol vehicles, but did not affect the recovery from the medium-chain oil. In the presence of luminal bile, plasma levels and biliary recovery of radiolabel using medium-chain triacylglycerols were significantly less than with either long-chain triacylglycerol. It was concluded that under physiological conditions, a long-chain triacylglycerol vehicle will provide greater systemic availability for the hydrocarbon than a medium-chain vehicle. The differences observed in the recovery of the dimethylbenz(a)anthracene dissolved in olive oil or medium-chain triacylglycerols are in agreement with studies by Dao,[170] who found olive and sesame oils superior to trioctanoylglycerol in promoting the absorption of orally ingested 3-methylcholanthrene in mice. These findings can be explained by the relative abilities of the long-chain and medium-chain lipids to create the hydrocarbon continuum necessary for the absorption of the xenobiotic. Laher et al.[84] point out that the ability of the dietary components to participate in the hydrocarbon continuum will not only increase the exposure of the enterocyte to a parent xenobiotic, but will determine repeated exposure to metabolites as a result of the initial absorption.

In a subsequent study, Laher et al.[81] have demonstrated that varying the mass of the carrier triacylglycerol oil by one order of magnitude does not significantly alter the disposition of benzo(a)pyrene in the rat. Although the chylomicrons produced from both fat doses were initially contaminated with benzo(a)pyrene, within 1 to 1.5 hr the radioactivity in lymph began to drop, such that by 3 hr in the animals fed high fat, the chylomicrons were essentially free of the carcinogen. Since polycyclic aromatic hydrocarbons, like benzopyrene, are sequestered in dietary fats,[168] Laher et al.[81] point out that a high fat diet may be a high carcinogen diet, and that the lymphatic export of fat-soluble polycyclic aromatic hydrocarbons can be minimized by maintaining an induced monooxygenase system in the intestine. However, not all polycyclic aromatic hydrocarbons may be subject to the same disposal as benzo(a)pyrene. In those instances the gastrointestinal lymphatics are the major route for transport of trace lipid-soluble compounds such as DDT.[76]

Aside from facilitating the absorption, the nature of dietary fat may also affect the metabolism of the fat-soluble xenobiotics. Early studies indicated that rats on diets deficient in essential fatty acids had decreased rates of drug metabolism, as measured in vitro.[171] High unsaturated fatty acid levels increased mixed-function oxidase and

UDP-glucuronyl transferase activity in some studies[171] and decreased activity in others.[172] Unsaturated fatty acids were necessary for the induction of cytochrome P-450 and UDP-glucuronyl transferase by phenobarbital.[171,173] However, recent work has shown[174] that more than 70% of the phospholipid in rat liver microsomes can be replaced with highly saturated egg yolk lecithin without affecting activities linked to cytochrome P-450.[175] Furthermore, changes in drug metabolism were not compared in humans fed unsaturated and saturated fat diets. Other nutritional factors that affect metabolic processes involving bioactivation and detoxification of fat-soluble xenobiotics are discussed in a recent review by Guengerich.[176]

Since the fat-soluble xenobiotics are sequestered in dietary fats,[168] it follows that their absorption would be impaired if the fat was not hydrolyzed and absorbed. Sucrose polyester is made up of a mixture of the hexa-, hepta-, and octa-esters that are formed by esterification of sucrose with long-chain fatty acids.[177] It has the appearance and physical properties of triacylglycerols, but differs from them in not being digested or absorbed.[178-180] This polyester has been shown to lower plasma cholesterol concentrations when incorporated into the diet of rats,[181] human volunteers,[182,183] and African green monkeys.[184] It is anticipated that a similar impairment in absorption would be observed for lipid-soluble xenobiotics, which are also distributed between the micellar phase and the unhydrolyzed oil phase of the sucrose polyester.

C. Absorptive Interactions of Fat-Soluble Xenobiotics

The luminal and tissue interactions among the fat-soluble xenobiotics are very well known in some instances and not so well known in others. Surfactants is one of the most important groups of adjuvants in pharmaceutical preparations and their effects on the absorption of various drugs have been studied extensively.[93,94] These can influence overall critical determinants of the rate and extent of drug absorption, e.g., drug solubility and dissolution rate, gastric emptying, and membrane permeation. Various surfactants are known to inhibit the intestinal absorption of inorganic electrolytes and water-soluble nutrients, which are considered to be transported by active and carrier-mediated processes. In an extensive series of experiments, Yasuhara et al.[185] have determined the effect of surfactants on the absorption of p-aminobenzoic acid from the rat small intestine. The order of magnitude of the absorption-enhancing effect was as follows: Triton® X-100 > sodium dodecyl sulfate > cetyltrimethylammonium chloride > Tween® 80. Further quantitative investigations must be awaited to propose exact mechanisms for the action of surfactants in altering the membrane permeability of lipid-soluble drugs.

In those instances where one of the fat-soluble xenobiotics is present in large excess, the absorptive interference with any other fat-soluble xenobiotic may be similar to that of certain components of dietary fat, e.g., nondigestible polyglycerol esters and related food additives. Thus, the excretion of lipophilic xenobiotics such as hexachlorobenzene, which are metabolized very slowly, is stimulated by nonabsorbable lipids, such as squalene, paraffin, and sucrose polyester.[186] The nonabsorbable lipid is believed to act as a sink for the highly lipophilic hexachlorobenzene by creating a gradient between blood and the aqueous phase of gut lumen. The interference of plant sterols with the absorption of cholesterol and other sterols is also well documented, although the mechanism of this effect is not well understood.[15]

Amphiphilic molecules that participate in formation of mixed micelles are capable of increasing the neutral sterol output in feces. Rodgers et al.[187] have shown that the nondigestible dialkyl-phosphatidylcholine inhibits absorption of cholesterol, without showing any appreciable effect on fatty acid absorption, and have suggested that this agent has therapeutic potential. Occasionally, the nondigestible components have been observed to present unexpected effects. The glycerol triethers proposed[188] as triacylglycerol markers for indirect measurement of fat absorption in man have been found to separate from an aqueous phase at a faster rate than the triacylglycerol.[189] The dissociation apparently occurred in the stomach. The triethers also tend to separate from triacylglycerols in the small intestine, where the triacylglycerol digestion products are transferred to the aqueous micellar phase from the oily phase which contains the triether.

VII. SUMMARY AND CONCLUSIONS

Over the last decade the study of the digestion and absorption of fat-soluble xenobiotics has progressed from an idle curiosity to a genuine concern. This is due mainly to the increasing levels of the fat-soluble pollutants in the environment and in the food supply, and to the direct or indirect effects of their ingestion and absorption upon the well-being of man and other living creatures. Since the intestinal mucosal surface constitutes the main contact area with these substances, their physiological and pathological fate at the intestinal level has become a major area of investigation in man and experimental animals.

This review of recent studies on uptake and lymphatic transport, in general, supports the current belief that fat-soluble xenobiotics are digested and absorbed along with dietary fat. However, there is evidence to indicate that fat solubility alone does not guarantee lymphatic transport of any xenobiotic. Detailed examination of the fate of various ingested fat-soluble xenobiotics indicates that elaborate mechanisms exist for differentiation between these and normal components of dietary fat at the various levels of absorption. This difference is manifested first of all by the fact that a true solubilization in the dietary fat may not exist, although it is usually assumed to be so. Further, the fatty esters of many of the xenobiotic alcohols are not readily hydrolyzed either by the lipases or esterases present in the mouth, stomach, and intestinal lumen. Next, there is evidence that the solubility of the original xenobiotic molecule or its hydrolysis products decreases progressively as the lipolysis of the dietary fat proceeds. This leads to a differential micellarization of the lipolytic products of dietary fats and of the fat-soluble xenobiotics. This differentiation is further manifested during the breakdown and absorption of the micellar or liposomal aggregates at the microvillus membrane. Finally, there is an effective discrimination among the lipid components that become reesterified and dissolved in the lipid cores of the chylomicrons. There is further hydrolysis by endogenous lipases of foreign lipid esters that have escaped luminal hydrolysis, and oxidation of polycyclic hydrocarbons and phenols by microsomal oxidases, which effectively segregates them from the lipid molecules packaged into the chylomicrons.

Nevertheless, dietary fats frequently facilitate the absorption of fat-soluble xenobiotics, as evidenced by the uptake of hydrocarbons, which does not occur in the absence of lumenal fat. The absorptive influences among the different fat-soluble xenobiotics have been less extensively investigated, except for studies on the promotion of absorption of fat-soluble drugs by the presence of actual or potential amphiphiles.

In conclusion, detailed studies of the digestion and absorption of fat-soluble xenobiotics have clearly succeeded in generating new concepts and new data. These will help to formulate new experimental approaches to fill in the many uncertain areas remaining.

ACKNOWLEDGMENTS

The studies of the author and collaborators referred to in this review were supported by grants from the Medical Research Council of Canada and the Ontario Heart Foundation, Toronto, Canada.

REFERENCES

1. Dhopeshwarkar, G. A., Naturally occurring food toxicants: toxic lipids, *Prog. Lipid Res.*, 19, 107, 1981.
2. Pfeiffer, C. J., Gastroenterologic response to environmental agents — absorption and interactions, in *Handbook of Physiology, Section 9: Reactions to Environmental Agents*, Lee, D. H. K., Ed., American Physiological Society, Bethesda, Md., 1977, 349.
3. Kuksis, A., Intestinal digestion and absorption of fat-soluble environmental agents, in *Intestinal Toxicology*, Schiller, C. M., Ed., Raven Press, New York, 1984, 69.
4. Kolattukudy, P. E., Introduction to natural waxes, in *Chemistry and Biochemistry of Natural Waxes*, Kolattukudy, P. E., Ed., Elsevier, Amsterdam, 1976, 1.
5. Lo, M. T. and Sandi, E., Polycyclic aromatic hydrocarbons (polynuclears) in foods, *Residue Rev.*, 69, 35, 1978.
6. Safe, S., Polychlorinated biphenyls (PCBs) and polybrominated biphenyls (PBBs): biochemistry, toxicology, and mechanism of action, *CRC Crit. Rev. Toxicol.*, 13, 319, 1984.
7. Bristol, D. W., MacLeod, K. E., and Lewis, R. G., Direct and indirect chemical methods for exposure assessment, in *Determination and Assessment of Pesticide Exposure*, Siewierski, M., Ed., Elsevier, Amsterdam, 1984, 79.
8. Thomas, J. A. and Thomas, M. J., Biological effects of di-(2-ethylhexyl)phthalate and other phthalic acid esters, *CRC Crit. Rev. Toxicol.*, 13, 282, 1984.
9. Cairns, T., Siegmund, E. G., Jacobson, R. A., Barry, T., Petzinger, G., Morris, W., and Heikes, D., Application of mass spectrometry in the regulatory analysis of pesticides and industrial chemicals in food and feed commodities, *Biomed. Mass Spectrom.*, 10, 301, 1983.
10. Gilbert, J., Confirmation and quantification of contaminants by MS SIM, in *Analysis of Food Contaminants*, Gilbert, J., Ed., Elsevier, London, 1984, 266.
11. Kuksis, A. and Myher, J. J., Lipids, in *Mass Spectrometry: Application in Clinical Biochemistry*, Lawson, A. M., Ed., Walter de Gruyter, Berlin, 1986.
12. Samuel, P., Crouse, J. R., and Ahrens, E. H., Jr., Evaluation of an isotope ratio method for measurement of cholesterol absorption in man, *J. Lipid Res.*, 19, 82, 1978.
13. Hassan, A. S. and Rampone, A. J., Intestinal absorption and lymphatic transport of cholesterol and *beta*-sitostanol in the rat, *J. Lipid Res.*, 20, 646, 1979.
14. Richter, E. and Strugala, G. J., All-glass perfusator for investigation on the intestinal transport and metabolism of foreign compounds, *in vitro*, *J. Pharmacol. Methods*, 14, 297, 1985.
15. Borm, P. J. A., Koster, A. S. J., Frankhuijzen-Siereevogel, J. C., and Noordhoek, J., A comparison of two-cell isolation procedures to study in vitro intestinal wall biotransformation in control and 3-methylcholanthrene pretreated, *Cell Biochem. Function*, 1, 161, 1983.
16. Sallee, V. L., Fatty acid and alcohol partitioning with intestinal brush border and erythrocyte membranes, *J. Membrane Biol.*, 43, 187, 1978.
17. Duthie, H. L., Methods for studies of intestinal absorption in man, *Br. Med. Bull.*, 23, 213, 1967.
18. Moseman, R. F. and Oswald, E. O., Development of analytical methodology for assessment of human exposure to pesticides, in *Human Exposure to Pesticides*, American Chemical Society, Washington, D.C., 1980, 251.
19. Edgerton, T. R. and Moseman, R. F., Determination of pentachlorophenol in urine: the importance of hydrolysis, *J. Agric. Food Chem.*, 27, 197, 1979.
20. Setchell, K. D. R. and Worthington, J., A rapid method for the quantitative extraction of bile acids and their conjugates from serum using commercially available reverse-phase octadecyl silane bonded cartridges, *Clin. Chim. Acta*, 125, 135, 1982.
21. Powell, W. S., Rapid extraction of arachidonic acid metabolites from biological samples using octadecylsilyl silica, *Methods Enzymol.*, 86, 467, 1982.
22. Safe, S., Halogenated hydrocarbons and aryl hydrocarbons identified in human tissues, *Toxicol. Environ. Chem.*, 5, 153, 1982.
23. Duane, W. C., Behrens, J. C., Kelly, S. G., and Levine, A. S., A method for measurement of nanogram quantities of 3-methylcholanthrene in stool samples, *J. Lipid Res.*, 25, 523, 1983.

24. Watts, R. R., Ed., Manual of analytical methods for the analysis of pesticide residues in human and environmental media, in *Technical Report, EPA-600/8-80-038, U.S.E.P.A.,* Health Effects Research Laboratory, Research Triangle Park, N.C., 1980.

25. Self, R., Recent developments in the estimation of trace toxic substances in food, *Biomed. Mass Spectrom.,* 6, 361, 1979.

26. Tindall, G. W. and Winninger, P. E., Gas chromatography-mass spectrometry method for identifying and determining polychlorinated biphenyls, *J. Chromatogr.,* 196, 109, 1980.

27. Wang, C.-S., Hydrolysis of dietary glycerides and phosphoglycerides: fatty acid and positional specificity of lipases and phospholipases, in *Fat Absorption,* Vol. 1, Kuksis, A., Ed., CRC Press, Boca Raton, Fla., 1987.

28. Paltauf, F., Esfandi, F., and Holasek, A., Stereospecificity of lipases. Enzyme hydrolysis of enantiomeric alkyldiacylglycerols by lipoprotein lipases, lingual lipase and pancreatic lipase, *FEBS Lett.,* 40, 119, 1974.

29. Wang, C.-S., Kuksis, A., Manganaro, F., Myher, J. J., Downs, D., and Bass, H. G., Studies on substrate specificity of purified human milk bile salt-activated lipase, *J. Biol. Chem.,* 258, 9197, 1983.

30. Wang, C.-S., Human milk bile salt-activated lipase. Further characterization and kinetic studies, *J. Biol. Chem.,* 256, 10198, 1981.

31. Morley, N. H., Kuksis, A., and Buchnea, D., Hydrolysis of synthetic triacylglycerols by pancreatic and lipoprotein lipase, *Lipids,* 9, 481, 1974.

32. Brockerhoff, H., Substrate specificity of pancreatic lipase. Influence of the structure of fatty acids on the reactivity of the esters, *Biochim. Biophys. Acta,* 212, 92, 1970.

33. Akesson, B., Gronowitz, S., Herslof, B., Michelsen, P., and Olivecrona, T., Stereospecificity of different lipases, *Lipids,* 18, 313, 1983.

34. Akesson, B. and Michelsen, P., Digestion and absorption of a sulphoxide analogue of triacylglycerol in the rat, *Chem. Phys. Lipids,* 29, 341, 1981.

35. Akesson, B., Digestion and absorption of sulphoxide analogs of triacylglycerols, in *Fat Absorption,* Vol. 2, Kuksis, A., Ed., CRC Press, Boca Raton, Fla., 1987.

36. Paltauf, F., Metabolism of the enantiomeric 1-O-alkylglycerol ethers in the rat intestinal mucosa *in vivo:* incorporation into 1-O-alkyl and 1-O-alk-1-enyl glycerol lipids, *Biochim. Biophys. Acta,* 239, 38, 1971.

37. Noda, M., Tsukahara, H., and Ogafa, M., Enzymic hydrolysis of diol lipids by pancreatic lipase, *Biochim. Biophys. Acta,* 529, 270, 1978.

38. Borgstrom, B., Influence of bile salt, pH and time on the action of pancreatic lipase; physiological implications, *J. Lipid Res.,* 5, 522, 1964.

39. Kuksis, A., Digestion and absorption of fat-soluble vitamins, in *Fat Absorption,* Kuksis, A., Ed., CRC Press, Boca Raton, Fla., 1987.

40. Mattson, F. H. and Volpenhein, R. A., Hydrolysis of primary and secondary esters of glycerol by pancreatic juice, *J. Lipid Res.,* 9, 79, 1968.

41. Morgan, R. G. H., Barrowman, J., Filipek-Wender, H., and Borgstrom, B., The lipolytic enzymes of rat pancreatic juice, *Biochim. Biophys. Acta,* 167, 355, 1968.

42. Mattson, R. H. and Volpenhein, R. A., Hydrolysis of fully esterified alcohols containing from one to eight hydroxyl groups by the lipolytic enzymes of rat pancreatic juice, *J. Lipid Res.,* 13, 325, 1972.

43. Scow, R. O., Stein, Y., and Stein, O., Incorporation of dietary lecithin and lysolecithin into lymph chylomicrons in the rat, *J. Biol. Chem.,* 242, 4919, 1967.

44. Mansbach, C. M., II, Peroni, G., and Verger, R., Intestinal phospholipase. A novel enzyme, *J. Clin. Invest.,* 69, 368, 1968.

45. Nilsson, A., Metabolism of sphingomyelin in the intestinal tract of the rat, *Biochim. Biophys. Acta,* 164, 575, 1968.

46. Nilsson, A., The presence of sphingomyelin and ceramide-cleaving enzymes in the small intestinal tract, *Biochim. Biophys. Acta,* 176, 339, 1969.

47. Hofmann, A. F. and Borgstrom, B., The intraluminal phase of fat digestion in man: the lipid content of the micellar and oil phases of intestinal content obtained during fat digestion and absorption, *J. Clin. Invest.,* 43, 247, 1964.

48. Kuksis, A., Intestinal digestion and absorption of fat-soluble environmental agents, in *Intestinal Toxicology,* Schiller, C. M., Ed., Raven Press, New York, 1984, 69.

49. Thompson, A. B. R. and Dietschy, J. M., Intestinal lipid absorption: major extracellular and intracellular events, in *Physiology of the Gastrointestinal Tract,* Vol. 2, Johnson, L. R., Ed., Raven Press, New York, 1981, 1147.

50. Patton, J. S. and Carey, M. C., Watching fat digestion, *Science,* 204, 145, 1979.

51. Stafford, R. J. and Carey, M. C., Physical-chemical nature of the aqueous lipids in intestinal content after a fatty meal: revision of the Hofmann-Borgstrom hypothesis, *Clin. Res.,* 28(Abstr.), 511A, 1981.

52. Carey, M. C., Small, D. M., and Bliss, C. M., Lipid digestion and absorption, *Ann. Rev. Physiol.*, 45, 651, 1983.
53. Patton, J. S., Gastrointestinal lipid absorption, in *Physiology of the Gastrointestinal Tract*, Vol. 2, Johnson, L. R., Ed., Raven Press, New York, 1981, 1123.
54. Westergaard, H. and Dietschy, J. M., Delineation of the dimensions and permeability characteristics of the two major diffusion barriers to passive mucosal uptake in the rabbit intestine, *J. Clin. Invest.*, 174, 718, 1974.
55. Sundqvist, T., Magnusson, K. E., Sjodahl, R., Stjernstrom, I., and Tagesson, C., Passage of molecules through the wall of the gastrointestinal tract, *Gut*, 21, 208, 1980.
56. Simmonds, W. J., Uptake of fatty acid and monoglyceride, in *Lipid Absorption: Biochemical and Clinical Aspects*, Rommel, K., Goebell, H., and Bohmer, R., Eds., MTP Press, Lancaster, England, 1976, 51.
57. Westergaard, H. and Dietschy, J. M., The mechanism whereby bile acid micelles increase the rate of fatty acid and cholesterol uptake into the intestinal mucosal cell, *J. Clin. Invest.*, 58, 97, 1976.
58. Thompson, A. B. R., Effect of age on uptake of homologous series of saturated fatty acids into rabbit jejunum, *Am. J. Physiol.*, 239, G363, 1980.
59. Mishkin, S., Yalowsky, M., and Kessler, J. I., Stages of uptake and incorporation of micellar palmitic acid by hamster proximal intestinal mucosa, *J. Lipid Res.*, 13, 155, 1972.
60. Hollander, D. and Dadufalza, V. D., Aging: its influence on the intestinal unstirred water layer thickness, surface area, and resistance in the unanaesthetized rat, *Can. J. Physiol. Pharmacol.*, 61, 1501, 1983.
61. Hollander, D., Dadufalza, V. D., and Sletten, E. G., Does essential fatty acid absorption change with aging?, *J. Lipid Res.*, 25, 129, 1984.
62. Channon, H. J. and Collinson, G. A., LXXVII. The unsaponifiable fraction of liver oils. V. The absorption of liquid paraffin from the alimentary tract of the rat and pig, *Biochem. J.*, 23, 676, 1929.
63. Stetten, D. W., Jr., Metabolism of paraffin, *J. Biol. Chem.*, 147, 327, 1943.
64. Frazer, A. C., The absorption of triglyceride fat from the intestine, *Physiol. Rev.*, 26, 103, 1946.
65. Bernhard, K., Absorption of aliphatic hydrocarbons, carotene and vitamin A in the rat, *Fette Seifen Anstrichmittel*, 55, 160, 1953.
66. Albro, P. W. and Fishbein, L., Absorption of aliphatic hydrocarbons by rats, *Biochim. Biophys. Acta*, 219, 437, 1970.
67. McWeeny, D. J., Ph.D. thesis, University of Birmingham, 1957; cited in *Mitchell, M. P. and Hubscher, G.*, Oxidation of n-hexadacane by subcellular preparations of guinea pig small intestine, *Eur. J. Biochem.*, 7, 90, 1968.
68. Kolattukudy, P. E. and Hankin, L., Metabolism of plant wax paraffin (*n*-nonacosane) in the rat, *J. Nutr.*, 90, 167, 1966.
69. Vost, A. and Maclean, N., Hydrocarbon transport in chylomicrons and high density lipoproteins in rat, *Lipids*, 19, 423, 1984.
70. Schoen, H. and Fahsold, W., Untersuchungen zu Frage der Resorption von Squalen, *Klin. Wochenschr.*, 38, 177, 1960.
71. Liu, G. C. K., Ahrens, E. H., Jr., Schreibman, P. H., and Crouse, J. R., Measurement of squalene in human tissues and plasma: validation and application, *J. Lipid Res.*, 17, 38, 1976.
72. Tilvis, R. S. and Miettinen, T. A., Absorption and metabolic fate of dietary [³H]squalene in the rat, *Lipids*, 18, 233, 1983.
73. U.S. Environmental Protection Agency Scientific and Technical Assessment Report on Particulate Polycyclic Organic Matter, Washington, D.C., 1972, 1975.
74. Campbell, A. D., Korwitz, W., Burke, J. A., Jelinek, C. F., Rodricks, J. V., and Shibko, S. I., *Handbook of Physiology, Section 9, Reactions to Environmental Agents*, Lee, D. H. K., Ed., American Physiological Society, Bethesda, Md., 1977, 167.
75. Allen, J. R. and Norback, D. H., Polychlorinated biphenyl- and triphenyl-induced gastric mucosal hyperplasia in primates, *Science*, 179, 498, 1973.
76. Pocock, D. M. and Vost, A., DDT absorption and chylomicron transport in rat, *Lipids*, 9, 374, 1974.
77. Vauhkonen, M., Kuusi, T., and Kinnunen, P. K. J., Serum and tissue distribution of benzo(a)pyrene from intravenously injected chylomicrons in a rat, *in vivo*, *Cancer Lett.*, 11, 113, 1980.
78. Herbert, P. N., Forte, T., Schulman, R., Lapiana, M., Gond, E., Levy, R., Frederickson, D. S., and Nichols, A. V., Structural and compositional changes attending the ultracentrifugation of very low density lipoproteins, *Prep. Biochem.*, 5, 93, 1975.
79. Shu, H. P. and Nichols, A. V., Uptake of lipophilic carcinogens by plasma lipoproteins. Structure-activity studies, *Biochim. Biophys. Acta*, 665, 376, 1981.
80. Maliwal, B. P. and Guthrie, F. E., *In vitro* uptake and transfer of chlorinated hydrocarbons among human lipoproteins, *J. Lipid Res.*, 23, 474, 1982.

81. Laher, J. M., Rigler, M. W., Vetter, R. D., Barrowman, J. A., and Patton, J. S., Similar bioavailability and lymphatic transport of benzo(a)pyrene when administered to rats in different amounts of dietary fat, *J. Lipid Res.,* 25, 1337, 1984.

82. Vetter, R. D., Carey, M. C., and Patton, J. S., Coassimilation of dietary fat and benzo(a)pyrene in the small intestine: as absorption model using the killfish, *J. Lipid Res.,* 26, 428, 1985.

83. Sabesin, S. M. and Frase, S., Electron microscopic studies of the assembly, intracellular transport and secretion of chylomicrons by rat intestine, *J. Lipid Res.,* 18, 496, 1977.

84. Laher, J. M., Chernenko, G. A., and Barrowman, J. A., Studies of the absorption and enterohepatic circulation of 7,12-dimethylbenz(a)anthracene in the rat, *Can. J. Physiol. Pharmacol.,* 61, 1368, 1983.

85. Daniel, P. M., Pratt, O. E., and Prichard, M. M. L., Metabolism of labelled carcinogen hydrocarbons in rats, *Nature (London),* 215, 1142, 1967.

86. Fujiwara, K., Environmental and food contamination with PCBs in Japan, *Sci. Total Environ.,* 4, 219, 1975.

87. Kay, K., Polybrominated biphenyls (PBB) — environmental contamination in Michigan, 1973—1976, *Environ. Res.,* 13, 79, 1977.

88. Wasserman, M., Wasserman, D., Cucos, S., and Miller, H. J., World PCBS map: storage and effects in man and his biologic environment in the 1970s, *Ann. N.Y. Acad. Sci.,* 320, 69, 1979.

89. Wickizer, T. and Brilliant, L. B., Testing for polychlorinated biphenyls in human milk, *Pediatrics,* 68, 411, 1981.

90. Mes, J., Davies, D. J., and Turton, D., Polychlorinated biphenyls and other chlorinated hydrocarbon residues in adipose tissues of Canadians, *Bull. Environ. Contam. Toxicol.,* 28, 97, 1982.

91. Anderson, H. A., Wolff, M. S., Lilis, R., Holstein, E. C., Valcuikas, J. A., Anderson, K. E., Petrocci, M., Sarkozi, L., and Selikoff, I. J., Symptoms and clinical abnormalities following ingestion of polybrominated biphenyl-contaminated food products, *Ann. N.Y. Acad. Sci.,* 320, 684, 1979.

92. Seagull, E. A. W., Developmental abilities of children exposed to polybrominated biphenyls, *Am. J. Public Health,* 73, 281, 1983.

93. Sieber, S. M., The entry of foreign compounds into the thoracic duct lymph of the rat, *Xenobiotica,* 4, 265, 1974.

94. Sieber, S. M., The lymphatic absorption of p,p'-DDT and some structurally-related compounds in the rat, *Pharmacology,* 14, 443, 1976.

95. Mahadevan, V., Fatty alcohols: chemistry and metabolism, *Prog. Fats Other Lipids,* 15, 255, 1978.

96. Baxter, J. H., Steinberg, D., Mize, C. E., and Avigan, J., Absorption and metabolism of uniformly [^{14}C]-labeled phytol and phytanic acid by the intestine of the rat studied with thoracic duct cannulation, *Biochim. Biophys. Acta,* 137, 277, 1967.

97. Sherr, S. I. and Treadwell, C. R., Triglyceride biosynthesis from monoglycerides in isolated segments of intestinal mucosa. Utilization of an ether analogue of 2-monostearin, *Biochim. Biophys. Acta,* 98, 539, 1965.

98. Gallo, L., Vahouny, G. V., and Treadwell, C. R., The 1- and 2-octadecyl glyceryl ethers as model compounds for study of triglyceride resynthesis in cell fractions of intestinal mucosa, *Proc. Soc. Exp. Biol. Med.,* 127, 156, 1968.

99. Paltauf, F. and Johnston, J. M., The metabolism in vitro of enantiomeric 1-O-alkylglycerols and 1,2- and 1,3-alkylacylglycerols in the intestinal mucosa, *Biochim. Biophys. Acta,* 239, 47, 1971.

100. Bugaut, M., Kuksis, A., and Myher, J. J., Loss of stereospecificity of phospholipases C and D upon introduction of an sn-2-alkyl group into rac-1,2-diacyl-sn-glyceryl-3-phosphorylcholine, *Biochim. Biophys. Acta,* 835, 304, 1985.

101. Beare-Rogers, J. L., Trans and positional isomers of common fatty acids, *Adv. Nutr. Res.,* 5, 172, 1983.

102. Mattson, F. H. and Streck, J. A., Effect of the composition of glycerides containing behenic acid on the lipid content of the heart of weanling rats, *J. Nutr.,* 104, 483, 1974.

103. Tomarelli, R. M., Meyer, B. J., Weaber, J. R., and Bernhart, F. W., Effect of positional distribution of the fatty acids of human milk and infant formulas, *J. Nutr.,* 95, 583, 1968.

104. Filer, L. J., Jr., Mattson, F. H., and Fomon, S. J., Triglyceride composition and fat absorption by the human infant, *J. Nutr.,* 99, 293, 1969.

105. Szlam, F. and Sgoutas, D. S., Eicosenoic and docosenoic acid incorporation in serum lipoproteins in rats fed rapeseed oil, *Lipids,* 13, 121, 1978.

106. Kritchevsky, D., Tepper, S. A., Vasselinovitch, D., and Wissler, R. W., Cholesterol vehicle in experimental atherosclerosis. XIII. Randomized peanut oil, *Atherosclerosis,* 17, 225, 1973.

107. Manganaro, F., Myher, J. J., Kuksis, A., and Kritchevsky, D., Acylglycerol structure of genetic varieties of peanut oils of varying atherogenic potential, *Lipids,* 16, 508, 1981.

108. Ammon, H. V., Thomas, P. J., and Phillips, S. F., Effect of long-chain fatty acids on solute absorption: perfusion studies in the human jejunum, *Gut,* 18, 805, 1977.

109. Dutton, H. J., Hydrogenation of fats and its significance, in *Geometrical and Positional Fatty Acid Isomers*, Emken, E. A. and Dutton, H. J., Eds., American Oil Chemists' Society, Champaign, Ill., 1979, 1.

110. Emken, E. A., Nutrition and biochemistry of *trans* and positional fatty acid isomers in hydrogenated oils, *Ann. Rev. Nutr.*, 4, 339, 1984.

111. Blomstrand, R. and Svensson, L., The effect of partially hydrogenated marine oils on the mitochondrial function and membrane phospholipid fatty acids in rat heart, *Lipids*, 18, 151, 1983.

112. Bergstrom, S., Borgstrom, B., Tryding, N., and Westoo, G., Intestinal absorption and metabolism of 2:2-dimethylstearic acid in the rat, *Biochem. J.*, 58, 604, 1954.

113. Lehman, A. J., Brominated olive oil and carbonated beverages, *Assoc. Food Drug Off. U.S. Bull.*, 20, 71, 1958.

114. Kuryla, W. C., Available flame retardants, *Flame Retardancy Polymer Materials*, 1, 1, 1973.

115. Gaunt, I. F., Grosso, P., and Gangoli, S. D., Brominated maize oil. I. Short-term toxicity and bromine storage studies in rats fed brominated maize oil, *Fed. Cosmet. Toxicol.*, 9, 13, 1971.

116. Jones, B. A., Tinsley, I. J., Wilson, G., and Lowrey, R. R., Toxicology of brominated fatty acids: metabolite concentration and heart and liver changes, *Lipids*, 18, 327, 1983.

117. Munro, I. C., Middleton, E. J., and Grice, H. C., Biochemical and pathological changes in rats receiving brominated cottonseed oil for 80 days, *Fed. Cosmet. Toxicol.*, 7, 25, 1969.

118. Munro, I. C., Hand, B., Middleton, E. J., Heggveit, H. A., and Grice, H. C., Toxic effects of brominated vegetable oils in rats, *Toxicol. Appl. Pharmacol.*, 2, 432, 1972.

119. Conacher, H. B. S., Hartman, D. K. J., and Chadka, R. K., Pancreatic lipolysis of some brominated vegetable oils, *Lipids*, 5, 497, 1970.

120. Jones, B. A., Tinsley, I. J., and Lowrey, R. R., Bromine levels in tissue lipids of rats fed brominated fatty acids, *Lipids*, 18, 319, 1983.

121. Mohamed, H. F., Andreone, T. L., and Dryer, R. L., Mitochondrial metabolism of D,L-threo-9,10-dibromopalmitic acid, *Lipids*, 15, 255, 1980.

122. James, S. P. and Kestell, P., The fate of brominated soya oil in the animal body, *Xenobiotica*, 8, 557, 1978.

123. Burgess, R. A., Butt, W. D., and Baggaley, A., Some effects of alpha-bromo-palmitate, an inhibitor of fatty acid oxidation, on carbohydrate metabolism in the rat, *Biochem. J.*, 109, 38P, 1968.

124. Pande, S. V., Siddiqui, A. W., and Gatereau, A., Inhibition of long chain fatty acid activation by *alpha*-bromopalmitate and phytanate, *Biochim. Biophys. Acta*, 248, 156, 1971.

125. Watson, W. C. and Gordon, R. S., Jr., Studies on the digestion, absorption and metabolism of castor oil, *Biochem. Pharmacol.*, 11, 229, 1962.

126. Maenz, D. D. and Forsyth, G. W., Ricinoleate and deoxycholate are calcium ionophores in jejunal brush border vesicles, *J. Membrane Biol.*, 70, 125, 1982.

127. Gaginella, T. S., Phillips, S. F., Dozois, R. R., and Go, V. L. M., Stimulation of adenylate cyclase in homogenates of isolated intestinal epithelial cells from hamsters, *Gastroenterology*, 74, 11, 1978.

128. Watson, W. C. and Murray, E. S., Triricinolein synthesis *in vivo*, *Biochim. Biophys. Acta*, 106, 311, 1965.

129. Perkins, E. G., Endres, J. G., and Kummerow, F. A., The metabolism of fats. Effect of dietary hydroxy acids and their triglycerides on growth, carcass, and fecal fat composition in the rat, *J. Nutr.*, 73, 291, 1961.

130. Barber, E. D., Smith, W. L., and Lands, W. E. M., Incorporation of ricinoleic acid into glycerolipids, *Biochim. Biophys. Acta*, 248, 171, 1971.

131. Conacher, H. B. S., Chadha, R. K., and Sahasrabudhe, M. R., Determination of brominated vegetable oils in soft drinks by gas-liquid chromatography, *J. Am. Oil Chem. Soc.*, 46, 558, 1969.

132. Hogben, C. A. M., The alimentary tract, *Ann. Rev. Physiol.*, 22, 381, 1960.

133. Jackson, M. J., Shiau, Y.-E., Bane, S., and Fox, M., Intestinal transport of weak electrolytes. Evidence in favor of a three compartment system, *J. Gen. Physiol.*, 63, 187, 1974.

134. Spencer, R. P. and Brody, K. R., Biotin transport by small intestine of rat, hamster and other species, *Am. J. Physiol.*, 206, 653, 1964.

135. Spencer, R. P., Brody, K. R., and Vishno, F., Species differences in the intestinal transport of p-aminobenzoic acid, *Comp. Biochem. Physiol.*, 17, 883, 1966.

136. Fears, R., Baggaby, K. H., Alexander, R., Morgan, B., and Hindley, R. M., The participation of ethyl 4-benzyloxy benzoate (BRL 10894) and other aryl-substituted acids in glycerolipid metabolism, *J. Lipid Res.*, 19, 3, 1978.

137. Londesborough, J. C. and Webster, L. T., Jr., Fatty acyl CoA synthesis, *Enzymes*, 10, 469, 1974.

138. Knoop, F., Der Abbau aromatischer, Fettsauren im Tier Korper, *Beitr. Chem. Physiol. Pathol.*, 6, 150, 1905.

139. Nikinorow, M., Mazyr, H., and Piekaca, H., Effect of orally administered plasticizers and polyvinyl chloride stabilizers in the rat, *Toxicol. Pharmacol.*, 26, 253, 1973.

140. Peakall, D. B., Phthalate esters: occurrence and biological effects, *Residue Rev.,* 54, 1, 1975.
141. Overturf, M. L., Druilhet, R. E., Liehr, J. G., Kirkendahl, W. M., and Caprioli, R. M., Phthalate ester in normal and pathological human kidneys, *Bull. Environ. Contam. Toxicol.,* 22, 536, 1979.
142. Tomita, I., Nakamura, Y., and Yagi, Y., Phthalic acid esters in various foodstuffs and biological materials, *Exotoxicol. Environ. Safety,* 1, 275, 1977.
143. Shintani, A., Determination of phthalic acid, mono-(2-ethyl-hexyl)phthalate and di-(2-ethyl-hexyl)phthalate in human plasma and in blood products, *J. Chromatogr.,* 337, 279, 1985.
144. Takahashi, T., Tanaka, A., and Yamaha, T., Elimination, distribution and metabolism of di(2-ethylhexyl)adipate (DEHA) in rats, *Toxicology,* 22, 223, 1981.
145. Albro, D. W., The metabolism of 2-ethylhexanol in rats, *Xenobiotica,* 5, 625, 1975.
146. Davenport, H. W., Destruction of the gastric mucosal barrier by detergents and urea, *Gastroenterology,* 54, 175, 1968.
147. Feldman, S., Reinhard, M., and Wilson, C., Effect of sodium taurodeoxycholate on biological membranes. Release of phosphorus, phospholipid, and protein from everted rat small intestine, *J. Pharm. Sci.,* 62, 1961, 1973.
148. Feldman, S. and Reinhard, M., Interaction of sodium alkyl sulfates with everted rat small intestinal membrane, *J. Pharm. Sci.,* 65, 1460, 1976.
149. Kirkpatrick, F. H., Gordesky, S. E., and Marinetti, G. V., Differential solubilization of proteins, phospholipids and cholesterol of erythrocyte membranes by detergents, *Biochim. Biophys. Acta,* 345, 154, 1974.
150. Freeman, S., Burrill, M. W., Li, T. W., and Ivy, A. C., The enzyme inhibitory action of an alkyl aryl sulfonate and studies of its toxicity when ingested by rats, dogs and humans, *Gastroenterology,* 4, 332, 1945.
151. Fitzhugh, O. G. and Nelson, A. A., Chronic toxicities of surface active agents, *J. Am. Pharm. Assoc.,* 37, 29, 1948.
152. Weaver, J. E. and Griffith, J. F., Induction of emesis by detergent ingredients and formulations, *Toxicol. Appl. Pharmacol.,* 14, 214, 1969.
153. Michael, W. R., Metabolism of linear alkylate sulfonate and alkyl benzene sulfonate in albino rats, *Toxicol. Appl. Pharmacol.,* 12, 473, 1968.
154. King, W. R., Michael, W. R., and Coots, R. H., Subacute oral toxicity of polyglycerol ester, *Toxicol. Appl. Pharmacol.,* 20, 327, 1971.
155. Shull, R. L., Gayle, L. A., Coleman, R. D., and Alfin-Slater, R. B., Metabolic studies of glyceride esters and adipic acid, *J. Am. Oil Chem. Soc.,* 38, 84, 1961.
156. Parthasarathy, S., Lin, J.-T., and Baumann, W. J., Ethanediol monoester hydrolysis by monoacylglycerol lipase of rat liver microsomes, *Biochim. Biophys. Acta,* 573, 107, 1979.
157. Valisek, J., Davidek, J., Kubelka, V., Janicek, G., Svobodova, Z., and Simicova, Z., New chlorine-containing organic compounds in protein hydrolysates, *J. Agric. Food Chem.,* 28, 1142, 1980.
158. Gardner, A. M., Yurawecz, M. P., Cunningham, W. C., Diachenko, G. W., Mazzola, E. P., and Brumley, W. C., Isolation and identification of C_{16} and C_{18} fatty acid esters of chloropropanediol in adulterated Spanish cooking oils, *Bull. Environ. Contam. Toxicol.,* 31, 625, 1983.
159. Cerbulis, J., Parks, O. W., Liu, R. H., Piotrowski, E. G., and Farell, H. M., Jr., Occurrence of diesters of 3-chloro-1,2-propanediol in the neutral lipid fraction of goat's milk, *J. Agric. Food Chem.,* 32, 474, 1984.
160. Kuksis, A., Marai, L., Myher, J. J., Cerbulis, J., and Farrell, H. M., Jr., Comparative study of the molecular species of chloropropanediol diester and triacylglycerols in milk fats, *Lipids,* 21, 1986.
161. Myher, J. J., Kuksis, A., Marai, L., and Cerbulis, J., Stereospecific analysis of fatty acid esters of chloropropanediol isolated from fresh goat milk, *Lipids,* 21, 1986.
162. Lien, E. J. and Wang, P. H., Lipophilicity, molecular weight, and drug action: reexamination of parabolic and bilinear models, *J. Pharm. Sci.,* 69, 648, 1980.
163. Chiou, C. T., Schmedding, D. W., and Block, J. H., Correlation of water solubility with octanol-water partition coefficient, *J. Pharm. Sci.,* 70, 1176, 1981.
164. Patton, J. S., Stone, B., Papa, C., Abramowitz, R., and Yalkowsky, S. H., Solubility of fatty acids and other hydrophobic molecules in liquid trioleoylglycerol, *J. Lipid Res.,* 25, 189, 1984.
165. Yalkowsky, S. H. and Valvani, S. C., Solubility and partitioning. I. Solubility of non-electrolytes in water, *J. Pharm. Sci.,* 69, 912, 1980.
166. Elbert, A. G., Schleifer, C. R., and Hees, S. M., Absorption, digestion and excretion of [³H]-mineral oil in rats, *J. Pharm. Sci.,* 55, 923, 1966.
167. Borgstrom, B., Fat digestion and absorption, in *Biomembranes 4B, Intestinal Absorption,* Smyth, D. H., Ed., Plenum Press, New York, 1974, 55.
168. Laher, J. M. and Barrowman, J. A., Polycyclic hydrocarbon and polychlorinated biphenyl solubilization in aqueous solutions of mixed micelles, *Lipids,* 18, 216, 1983.
169. Borgstrom, B., Partition of lipids between emulsified oil and micellar phases of glyceride-bile salt dispersions, *J. Lipid Res.,* 8, 598, 1967.

170. Dao, T. L. and Chan, P. C., Effect of duration of high fat intake on enhancement of mammary carcinogenesis in rats, *J. Natl. Cancer Inst.,* 71, 201, 1983.
171. Norred, W. P. and Wade, A. E., Effect of dietary lipid ingestion on the induction of drug-metabolizing enzymes by phenobarbital, *Biochem. Pharmacol.,* 22, 432, 1973.
172. Hietanen, E., Hanninen, O., Laitenin, M., and Lang, M., Regulation of hepatic drug metabolism by elaidic and linoleic acids in rats, *Enzyme,* 23, 127, 1978.
173. Marshall, W. J. and McLean, A. E. M., A requirement for dietary lipids for induction of cytochrome P-450 by phenobarbitone in rat liver microsomal fraction, *Biochem. J.,* 122, 569, 1971.
174. Muller-Enoch, D., Churchill, P., Fleischer, S., and Guengerich, F. P., Interaction of microsomal cytochrome P-450 and NADPH-cytochrome P-450 reductase in the presence and absence of lipids, *J. Biol. Chem.,* 259, 8174, 1984.
175. Anderson, K. E., Conney, A. H., and Kappas, A., Nutritional influences on chemical biotransformations in humans, *Nutr. Rev.,* 40, 161, 1982.
176. Guengerich, F. P., Effects of nutritive factors on metabolic processes involving bioactivation and detoxification of chemicals, *Ann. Rev. Nutr.,* 4, 207, 1984.
177. Mattson, F. H. and Nolen, G. A., Absorbability by rats of compounds containing from one to eight ester groups, *J. Nutr.,* 102, 1171, 1972.
178. Mattson, F. H. and Volpenhein, R. A., Rate and extent of absorption of the fatty acids of fully esterified glycerol, erythritol, xylitol and sucrose as recovered in thoracic duct cannulated rats, *J. Nutr.,* 102, 1177, 1972.
179. Mattson, F. H. and Volpenhein, R. A., Hydrolysis of fully esterified alcohols containing from one to eight hydroxyl groups by the lipolytic enzymes of rat pancreatic juice, *J. Lipid Res.,* 13, 325, 1972.
180. Rizzi, G. P. and Taylor, H. M., A solvent-free synthesis of sucrose polyesters, *J. Am. Oil Chem. Soc.,* 55, 398, 1978.
181. Mattson, F. H., Jandrack, R. J., and Webb, M. R., The effect of a nonabsorbable lipid, sucrose polyester, on the absorption of dietary cholesterol by the rat, *J. Nutr.,* 106, 747, 1976.
182. Krause, J. R. and Grundy, S. M., Effects of sucrose polyester (SPE) on cholesterol metabolism in man, *Metabolism,* 28, 994, 1979.
183. Glueck, C. J., Jandacek, R., Hogg, E., Allen, C., Baehler, L., and Tewksbury, M., Sucrose polyester: substitution for dietary fats in hypocaloric diets in the treatment of familial hypercholesterolemia, *Am. J. Clin. Nutr.,* 37, 347, 1983.
184. Adams, M. R., McMahan, M. R., Mattson, F. H., and Clarkson, T. B., The long-term effects of dietary sucrose polyester on African green monkeys, *Proc. Soc. Exp. Biol. Med.,* 167, 346, 1981.
185. Yasuhara, M., Kobayashi, H., Kurosaki, Y., Kimura, T., Muranishi, S., and Sezaki, H., Comparative studies on the absorption mechanism of p-aminobenzoic acid and p-acetamidobenzoic acid from the rat intestine, *J. Pharm. Dyn.,* 2, 177, 1979.
186. Richter, E., Stimulation of the faecal excretion of hexachlorobenzene (HCB) — a review, in press.
187. Rodgers, J. B., Fondacaro, J. D., and Kot, J., The effect of synthetic diether phospholipid on lipid absorption in the rat, *J. Lab. Clin. Med.,* 89, 147, 1977.
188. Morgan, R. G. H. and Hofmann, A. F., Synthesis and metabolism of glycerol-^{3}H triether, a nonabsorbable oil-phase marker for lipid absorption studies, *J. Lipid Res.,* 11, 223, 1970.
189. Saunders, D. R. and O'Brien, T. K., Disappointment with triethers as markers for measuring triglyceride absorption in man, *Gut,* 13, 867, 1972.

Chapter 5

DIGESTION AND ABSORPTION OF GLYCERIDE ANALOGS

Björn Åkesson and Peter Michelsen

TABLE OF CONTENTS

I. INTRODUCTION

Most glycerides in nature contain acyl groups linked via ester bonds to the glycerol moiety as in triacylglycerols and diacyl phospholipids. Among phospholipids, plasmalogens, and alkyl analogs are common, and among neutral lipids 1-*O*-alkyl-2,3-diacylglycerol* is the most abundant analog. Acyl lipids are degraded by lipases in mammalian, microbial, and other systems, but other enzymes are involved in the degradation of ether bonds. The difference between ester and ether lipids has several implications. First, the intestinal utilization of dietary triacylglycerols and their ether analogs may differ. Second, much work has been performed to elucidate the stereospecificity, positional, and fatty acid specificity of lipases degrading triacylglycerols in model systems. Acyl migration and the rapid sequential hydrolysis of several ester bonds has made this task more difficult. To overcome this problem glyceride analogs containing lipase-resistant bonds instead of ester bonds have been used to define the substrate specificity of lipases.

II. CHEMISTRY AND NATURAL OCCURRENCE OF GLYCERIDE ANALOGS

A. Neutral Ether Lipids in Nature

1-Alkyl-2,3-diacylglycerol is the most abundant neutral ether lipid in nature, and it is especially prominent in some species of fish and in tumor tissue.[1-3] In many tissues hexadecyl, octadecyl, and octadecenyl are the dominating alkyl groups. Ether phospholipids are also ubiquitous in nature, but they will not be covered in this review. The biological function of ether lipids is undefined, although it is known that ether phospholipids act as membrane lipids. Several effects have been observed after the administration of glycerol ethers to cells or animals,[4] but whether these represent pharmacological or physiological effects is not known at present.

Also, *S*-alkyl glycerol ethers have been detected in nature in bovine[5] and human[6] heart and more recently in Ehrlich ascites cells.[7] The isolation procedures involved saponification and, therefore, it is unknown whether natural glycerol thioethers occur as dihydroxy compounds or contain acyl groups esterified at the hydroxyl groups. In human heart the dominating alkyl component in glycerol thioethers was found to be hexadecyl, followed by octadecyl and octadecenyl.[6] The biosynthesis of glycerol thioethers can take place from the corresponding mercaptan and dihydroxyacetone phosphate, analogous to the biosynthesis of *O*-alkyl glycerol ethers.[8] On the other hand, glycerol thioethers do not seem to be degraded in vivo in the same way as *O*-alkyl glycerol ether analogs.[9] As already noted for *O*-alkyl glycerol ethers, no biological function of *S*-alkyl glycerol ethers has been defined.

B. Chiroptical Properties of Glyceride Analogs

Glycerol by itself is not optically active, but if groups attached to the primary hydroxyls of glycerol differ from each other structurally, the compound can be resolved into two enantiomeric forms.

Optical activity is associated with electronic transitions of a chromophore occurring in a chiral environment. Since the optical activity varies as a function of wavelength, it can be measured by spectroscopic techniques, such as optical rotatory dispersion (ORD) or circular dichroism (CD).[10,11]

Relatively few lipid classes have been studied with these techniques. At this institute, several enantiomeric acylglycerols and their derivatives have been studied with ORD

---

* Unless otherwise indicated *sn* (stereospecific numbering) nomenclature is used.

and CD by Gronowitz and collaborators.[10,11] The results indicated that it should be possible to use the techniques for stereochemical analysis of a wider range of lipids, and more recent data are summarized in the following.

In one study different 1,2-diacyl-3-*O*-alkylglycerols were synthesized, and their chiroptical properties were investigated with ORD and CD.[12] It was found that when the acyl chain length of 1,2-diacyl-3-*O*-alkyl was varied, the optical activity remained almost unchanged (1,2-diacetyl- and 1,2-dilauroyl-3-*O*-hexylglycerol, Figure 1). If the alkyl chain was increased from hexyl to hexadecyl (1,2-dilauroyl-3-*O*-hexylglycerol and 1,2-dimyristoyl-3-*O*-hexadecylglycerol) only a small effect on the optical activity was observed (Figure 1). When the alkyl was methyl, the activity decreased, and when the alkyl group was replaced by hydrogen, i.e., 1,2-dimyristoylglycerol, the optical activity was lower. This is probably due to the fact that the number of possible conformations is larger, when the alkyl chain is very short at position 3. Only when the alkyl chain contained a strong chromophore such as in 1,2-dioleoyl-3-*S*-tetradecyl-3-thioglycerol, stronger optical activity resulted.

The CD curves of all 1,2-diacyl-3-*O*-alkylglycerols investigated had the same shape (Figure 1). They all exhibited negative Cotton effects around 215 nm which were assigned to the n → π* transitions of the ester chromophores. This was also observed for mono- and diacylglycerol.[13]

As shown in Figure 1, a sulfur-containing glyceride analog had higher optical activity than the corresponding oxygen analog. Since we found that this could be useful as an analytical method for the determination of stereoconfiguration of lipids in feeding experiments, a wide range of sulfur-containing glycerides were prepared and analyzed for their chiroptical properties (Table 1, Figure 2). The analog containing a glycerol thioether moiety had higher optical activity than the corresponding *O*-alkyl derivative, but the difference was not as pronounced as for triacylglycerols compared to triacylthioglycerols. The data may be useful in future studies of the stereochemical configuration of naturally occurring sulfur-containing glycerides. The methods and compounds have also been used for the analysis of chyle lipids after feeding the analogs to rats (see Section IV).

III. DEGRADATION OF GLYCERIDE ANALOGS BY LIPASES

The first step in the degradation of triacylglycerol by lipases generally involves splitting of the ester bonds at one or both of the primary positions of glycerol. Initial studies with pancreatic lipase indicated that the lipase did not discriminate between positions 1 and 3.[17] This has later been confirmed using both triacylglycerols and alkyldiacylglycerols as substrates.[18-20] When triacylglycerols were used, the demonstration of a stereospecificity was experimentally more difficult, since acyl migration could influence the chemical structure of the products. A more straightforward method was to use enantiomeric alkyldiacylglycerols, where the acyl moieties in one enantiomer were labeled with ³H and in the other enantiomer with ¹⁴C (Figure 3). A preferential attack by a lipase on either position 1 or 3 could then be demonstrated if the ¹⁴C/³H ratio in liberated fatty acids deviated from that in the substrate.

Using this approach we found that lipolysis of alkyldiacylglycerols by liver homogenates resulted in different ¹⁴C/³H ratios, depending on the pH of the incubation (Figure 4). At pH 5 both enantiomers were attacked at the same rate, indicating that the lysosomal lipase had no stereospecificity.[21] At pH 8, 1,2-dioleoyl-3-alkylglycerol was more rapidly degraded than 1-alkyl-2,3-dioleoylglycerol, indicating a preferential attack at position 1. In the next series of experiments the heparin-releasable lipase from rat liver was shown to exhibit the same type of stereospecificity, indicating that this enzyme accounted for the stereospecificity observed in the liver homogenate.[21] The

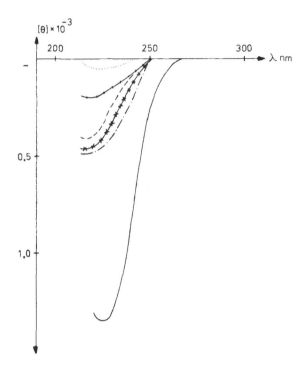

FIGURE 1. CD spectra (ethanol) of 1,2-dimyristoylglycerol (dots), 1,2-dimyristoyl-3-*O*-methylglycerol (dots with solid line), 1,2-diacetyl-3-*O*-hexylglycerol (dashes), 1,2-dilauroyl-3-*O*-hexylglycerol (x with solid line), 1,2-dimyristoyl-3-*O*-hexadecylglycerol (dots, dashes), and 1,2-dioleoyl-3-*S*-tetradecyl-3-thioglycerol (solid line) (the last compound in hexane). (From Michelsen, P. and Herslöf, B., *Chem. Phys. Lipids,* 32, 27, 1983. With permission.)

same stereospecificity was also observed when the substrate was dispersed in different agents, indicating that the specificity was not very sensitive to the physical form of the substrate.

As pointed out in several studies,[18,21] the stereospecificities observed in these types of experiments may be exerted at the alkylacylglycerol level rather than the alkyldiacylglycerol level. This possibility was investigated in experiments where the lipolysis of enantiomeric pairs of alkylacylglycerols was measured. The heparin-releasable lipase from rat liver preferentially degraded 2-acyl-3-alkylglycerol compared to 1-alkyl-2-acylglycerol, but it did not discriminate between 1,3-alkylacyl enantiomers.[21] The stereospecificity was also less pronounced than that observed for alkyldiacylglycerols, indicating the existence of a true stereospecificity vs. the latter substrate.

The preferential attack at position 1 of alkyldiacylglycerol by the heparin-releasable lipase[21] is the same type of stereospecificity as demonstrated for lipoprotein lipase from bovine milk by Paltauf and collaborators.[18] They also found a preferential lipolysis by lipoprotein lipase of 2-acyl-3-alkylglycerol compared to 1-alkyl-2-acylglycerol, and in addition, of 1-acyl-3-alkylglycerol compared to 1-alkyl-3-acylglycerol. The latter specificity was not demonstrated for the heparin-releasable lipase.[21] In addition, Paltauf et al.[18] found that rat lingual lipase had an opposite stereospecificity with a preferential attack at position 3 of alkyldiacylglycerol.

In a later study we extended these studies to another lipase in milk, the bile salt-stimulated lipase, and compared its specificity to that of pancreatic lipase, a fungal

Table 1
SYNTHESIZED OPTICALLY ACTIVE
DERIVATIVES OF 1-THIOGLYCEROL[14]

Derivative

1,2-*O*-Isopropylidene-3-thio-*sn*-glycerol
1,2-*O*-Isopropylidene-3-*S*-acetyl-3-thio-*sn*-glycerol
1,2-*O*-Isopropylidene-3-*S*-oleoyl-3-thio-*sn*-glycerol
Bis-1,2-*O*-isopropylidene-3-thio-*sn*-glycerol
3-Thio-*sn*-glycerol
Bis-3-thio-*sn*-glycerol
3-*S*-Acetyl-3-thio-*sn*-glycerol
3-*S*-Oleoyl-3-thio-*sn*-glycerol
1,2-*O*-Distearoyl-3-*S*-acetyl-3-thio-*sn*-glycerol
1,2-*O*-Diacetyl-3-*S*-acetyl-3-thio-*sn*-glycerol
1,2-*O*-Dioleoyl-3-*S*-oleoyl-3-thio-*sn*-glycerol
1,2-*O*-Isopropylidene-3-*S*-tetradecyl-3-thio-*sn*-glycerol
3-*S*-Tetradecyl-3-thio-*sn*-glycerol
1,2-*O*-Dioleoyl-3-*S*-tetradecyl-3-thio-*sn*-glycerol
1,2-*O*-Dioleoyl-3-*S*-tetradecyl-3-thio-*sn*-glycerol-*S*-oxide R,R
1,2-*O*-Dioleoyl-3-*S*-tetradecyl-3-thio-*sn*-glycerol-*S*-oxide R,S
1,2-*O*-Dioleoyl-3-*S*-tetradecyl-3-thio-*sn*-glycerol-*S*-dioxide

lipase *(Rhizopus arrhizus)*, and a bacterial lipase *(Pseudomonas fluorescens)*.[20] When the latter enzyme was incubated with equal amounts of 1,2-di(^{3}H)oleoyl-3-tetradecylglycerol and 1-tetradecyl-2,3-di(^{14}C)oleoylglycerol, both isomers were degraded at the same rate as evidenced by a constant ^{14}C/^{3}H ratio in the remaining substrate at all degrees of lipolysis (Figure 5). The ^{14}C/^{3}H ratio in alkylacylglycerol deviated markedly from unity especially at extensive lipolysis, indicating that 2-oleoyl-3-tetradecylglycerol was split to oleic acid and tetradecylglycerol to a larger degree than 1-tetradecyl-2-oleoylglycerol. This indicated a stereospecificity vs. alkylacylglycerols, although no specificity vs. enantiomeric alkyldiacylglycerols could be demonstrated.

When the substrate specificity vs. enantiomers of *rac*-1-alkyl-2-acylglycerol was investigated, the same preferential degradation of the 2-acyl-3-alkyl isomer was observed. Also, the lipolysis of X-1-alkyl-3-acylglycerol isomers was studied. Lipase from *P. fluorescens* degraded 1-acyl-3-alkylglycerol faster than 1-alkyl-3-acylglycerol, but the stereospecificity was not as pronounced as that observed with X-1-alkyl-2-acylglycerol isomers.[20] This indicates that the degradation of the latter compounds did not proceed via acyl migration to X-1-alkyl-3-acylglycerol and subsequent ester cleavage. In addition, model experiments showed that acyl migration under the conditions used was negligible.

The stereospecificity vs. the three pairs of alkylacylglycerol enantiomers mentioned was also studied for pancreatic lipase, *R. arrhizus* lipase, and bile salt-stimulated lipase.[20] As found for *P. fluorescens* the lipases attacked alkyldiacylglycerols without any stereospecificity, but they degraded 2-oleoyl-3-tetradecylglycerol faster than 1-tetradecyl-2-oleoylglycerol (Figure 6), although this specificity was less marked for the bile salt-stimulated lipase. *R. arrhizus* lipase (Figure 7) and pancreatic lipase in addition degraded 1-acyl-3-alkylglycerol faster than 1-alkyl-3-acylglycerol, which was also found for *P. fluorescens* lipase.

This study indicated that the four lipases studied lacked stereospecificity vs. enantiomeric alkyldiacylglycerols. Instead, another type of stereospecificity was detected, namely, that a lipase which degrades alkyldiacylglycerol without stereospecificity will attack 2-acyl-3-alkylglycerol preferentially to 1-alkyl-2-acylglycerol. An indication for

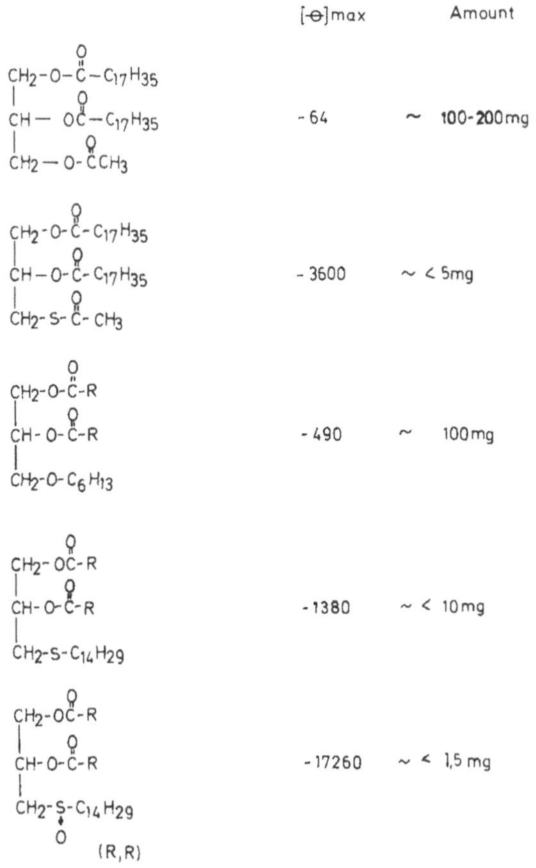

FIGURE 2. The maximum optical activity by CD and the amount of sample required for recording spectra of some optically active glycerol derivatives. (From Michelsen, P., Synthesis and Chiroptical Properties of Some Glycerolipids, Thesis, University of Lund, Sweden, 1981. With permission.)

this type of specificity for pancreatic lipase was found previously,[18] although alkyl-monoacylglycerols were not tested as substrates.

The mechanism behind the stereospecificity vs. alkylmonoacylglycerol still remains unknown. The enzymes must have attacked the ester bond of the secondary alcohol directly, since the hydrolysis of 1,3-diacylglycerol analogs, which are the possible products after acyl migration, was less stereospecific than that of X-1,2-diacylglycerol analogs. Noda et al.[22] found that 2,3-dioleoyloxybutane and 1-hexadecyloxy-3-oleoyloxybutane were degraded slowly by pancreatic lipase, and that 1-hexyloxy-2-octanoyloxypropane resisted hydrolysis. Under our conditions, X-1-alkyl-3-acylglycerol was hydrolyzed at least 20-fold faster than X-1-alkyl-2-acylglycerol by pancreatic lipase and R. arrhizus lipase, approximately tenfold faster by P. fluorescens lipase, but less than twofold faster by bile salt-stimulated lipase.[20]

It was also demonstrated that 1-acyl-3-alkylglycerol was degraded faster than its enantiomer by pancreatic lipase, and by lipases from P. fluorescens and R. arrhizus. A similar specificity was noted for bovine milk lipoprotein lipase,[18] but no clear stereospecificity was observed in the degradation of 1,3-diacylglycerols by pancreatic lipase.[23]

It has been found that pancreatic lipase is more easily bound to a triacylglycerol surface than to a trialkylglycerol surface,[24] indicating the importance of the ester car-

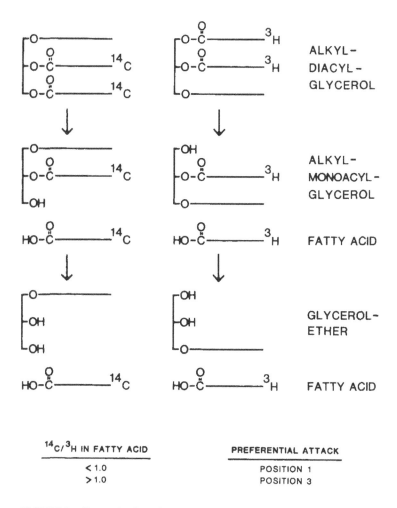

$^{14}C/^{3}H$ IN FATTY ACID	PREFERENTIAL ATTACK
< 1.0	POSITION 1
> 1.0	POSITION 3

FIGURE 3. Determination of stereospecificity of lipases by using enantiomeric alkyldiacylglycerols as substrate.

bonyl groups of the substrate in this process. The carbonyl groups attached at the secondary carbon of glycerol is probably important for the interaction of substrate with lipoprotein lipase, since no stereospecificity for this enzyme was observed when it was incubated with *rac*-1,2-dialkyl-3-acyl-glycerol.[25] Another example which shows the role of the substituent at the secondary carbon is the fact that *R. arrhizus* lipase does not hydrolyze 2-phosphatidylcholine, but shows activity toward the dimethylester of 1,3-diacylglycerol-2-phosphoric acid.[26] In spite of this, it is obvious that a 2-acyl group in the substrate is not always necessary for stereospecificity, since X-1-alkyl-3-acylglycerols can be degraded with specificity by some lipases. It can thus be concluded that different lipases degrade alkyldiacylglycerols and alkylmonoacylglycerols with varying degrees of stereospecificity. Whether the stereospecificities observed are due to interactions of the substrate with the active center of the enzyme remains to be studied. A summary of the stereospecificities of different lipases is given in Table 2.

A review on other aspects of the substrate specificity of lipases has appeared.[27] Cambou and Klibanov recently described interesting new applications where the stereospecificity of lipases is utilized.[28] Using yeast lipase they prepared optically active alcohols and butyric ester from the corresponding racemates.

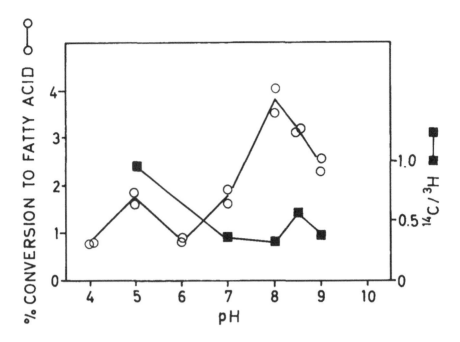

FIGURE 4.   Hydrolysis of enantiomeric alkyldiacylglycerols in a homogenate from rat liver at different pH. (From Åkesson, B., Gronowitz, S., and Herslöf, B., *FEBS Lett.*, 71, 241, 1976. With permission.)

IV. STEREOSPECIFICITY IN THE INTESTINAL ABSORPTION OF GLYCERIDE ANALOGS

A. *O*-Alkyl Glycerides

In the intestinal digestion and absorption, specificity vs. enantiomeric glycerides may be exerted at different stages of the process, such as lipolysis, transport through the intestinal wall, and synthesis of intestinal lipoproteins. Fat absorption has been extensively studied by feeding animals with different fats and measuring the appearance of fat in chyle and feces. Most often the fatty acid composition of the dietary fat used is given, but in only a few studies has the dietary fat been characterized stereochemically.

Previous studies showed that chimylalcohol and its dioleate was absorbed in rat and man,[29,30] but stereoisomers were not compared in those studies. Paltauf[31] fed 1-*O*-octadecylglycerol and 3-*O*-octadecylglycerol to rats and recovered both isomers in the intestinal mucosa, although the former isomer was preferentially utilized in biosynthetic reactions. To study the metabolic fate of the fatty acids of alkyldiacylglycerol isomers we fed a 50:50 mixture of 1-tetradecyl-2,3-di(^{14}C)oleoylglycerol and 1,2-di(^{3}H)oleoyl-3-tetradecylglycerol to rats with a thoracic duct cannula.[32] Chyle and feces were collected for 30 hr, and then the intestine and its contents also were sampled. In all analyses, the ratio of ^{3}H to ^{14}C never deviated significantly from that in the original mixture, indicating that oleic acid-containing compounds were metabolized in the same way, irrespective of whether the fatty acid originated from positions 1 plus 2 or positions 2 plus 3. Most of the lipid radioactivity in chyle was located in triacylglycerol, but significant amounts were also found in alkyldiacylglycerol. In chyle, 55 to 79% of the total tetradecylglycerol was recovered in alkyldiacylglycerol, and the rest was mainly found in the alkylacylglycerol fraction. Also, when 1,3-dioctadecenoyl-2-hexadecylglycerol was fed to rats, 1,3-diacyl-2-alkylglycerol was recovered in chyle, and

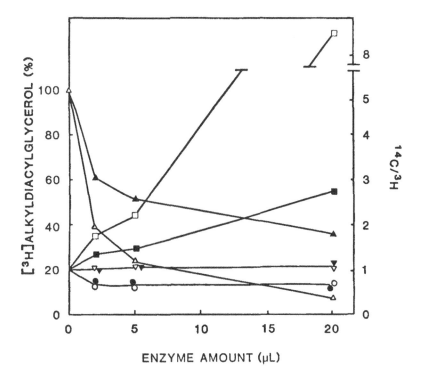

FIGURE 5. Lipolysis of racemic alkyldiacylglycerol by lipase from *P. fluorescens;* 25 μg (open symbols) or 100 μg (closed symbols) of 1,2-di(³H)oleoyl-3-tetradecylglycerol plus 1-tetradecyl-2,3-di(¹⁴C)oleoylglycerol was incubated with different amounts of lipase. Symbols: amount of remaining (³H)alkyldiacylglycerol substrate (%) (open and closed triangles); ¹⁴C/³H ratio in alkyldiacylglycerol (inverted open and closed triangles); alkylmonoacylglycerol (open and closed squares); and fatty acid (open and closed circles). The ¹⁴C/³H ratio in the substrate was set at unity.

the ratio of the glycerolether to triacylglycerol varied in lipoproteins of different size.[33] The stereochemistry of chyle glycerides was not investigated in this experiment, but other studies showed that several 2-monoacylglycerols are preferentially acylated at position 1 in the intestinal mucosa.[34] Also, the preferential acylation of 1-*O*-alkylglycerol compared to 3-*O*-alkylglycerol has been verified from in vitro experiments.[35] It is possible that alkylglyceride isomers are also preferentially utilized in other biosynthetic reactions. In rat liver 1-alkyl-2-acylglycerol is preferred compared to 2-acyl-3-alkylglycerol by diacylglycerol kinase,[36] but so far this has not been established in the intestine.

B. Sulfur-Containing Glycerides

As mentioned in Section II, *S*-alkyl glycerols have been demonstrated and characterized in small amounts in some mammalian tissues, but very little is known about their metabolism. In addition, thioester-containing glycerides and phospholipids have been used as substrates in assays of lysophospholipase,[37] and more recently phospholipase C[38] and phospholipase A$_1$.[39] The basis for the assays is that the free thiol formed can be conveniently determined by spectrophotometry. The studies mentioned in Section II demonstrated that enantiomers of sulfur-containing glycerides could be quantified by reasonable sensitivity using ORD and CD, in contrast to natural triacylglycerols. Therefore, they could be used as model compounds in investigations of stereospecificities in glyceride metabolism.

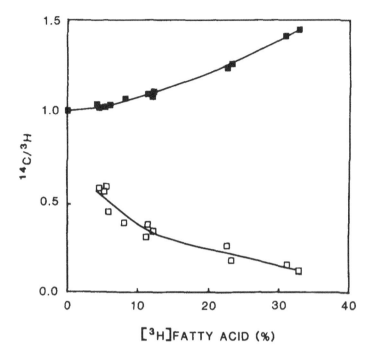

FIGURE 6.   Hydrolysis of 50 µg of 2-(^{3}H)oleoyl-3-tetradecylglycerol plus 50 µg of 1-tetradecyl-2-(^{14}C)oleoylglycerol by lipase from *R. arrhizus*. Symbols: ^{14}C/^{3}H ratio in alkylmonoacylglycerol (closed square); fatty acid (open square). The ^{14}C/^{3}H ratio in the substrate was set at unity.

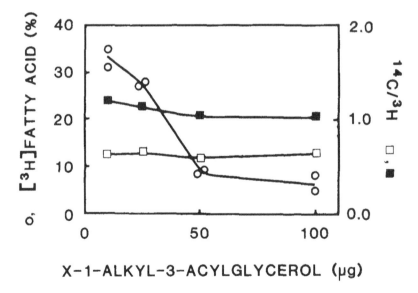

FIGURE 7.   Hydrolysis of different amounts of an equimolar mixture of 1-(^{3}H)oleoyl-3-tetradecylglycerol and 1-tetradecyl-3-(^{14}C)oleoylglycerol by lipase from *R. arrhizus*. Symbols: ^{3}H in fatty acid (open circle); ^{14}C/^{3}H ratio in alkylmonoacylglycerol (closed square); and fatty acid (open square). The ^{14}C/^{3}H ratio in the original substrate was set at unity.

Table 2
STEREOSPECIFICITY OF DIFFERENT LIPASES VS. ALKYLACYLGLYCEROLS[20]

Lipase	Preferred position in X-1-alkyl-2,3-diacylglycerol	Preferred isomer among 1-alkyl,2-acyl-*sn*-glycerol (1A2E) and 2-acyl 3-alkyl-*sn*-glycerol (2E3A)	Ref.
Lipoprotein lipase (milk, post-heparin plasma, adipose tissue)	1	2E3A	18
Heparin-releasable hepatic lipase	1	2E3A	21
Hepatic lysosomal lipase	Nonspec.	—	21
Pancreatic lipase	Nonspec.	2E3A	18, 20
Lingual lipase	3	—	18
Bile salt-stimulated milk lipase	Nonspec.	2E3A	20
R. arrhizus lipase	Nonspec.	2E3A	20
P. fluorescens lipase	Nonspec.	2E3A	20

In one study *rac*-trioleoyl-1-thioglycerol with a trace amount of (^{14}C)oleic acid was fed to rats together with either a trace or an equal amount of tri(^{3}H)oleoylglycerol.[40] When trioleoyl-thioglycerol constituted all or half of the fed lipid, triacylglycerol was the dominating chyle lipid. A smaller amount of radioactivity was found in a compound with the same R_f value as trioleoylthioglycerol on thin-layer chromatography (TLC). It constituted 4.8 to 7.1% of chyle lipid weight, and it was identified as trioleoyl-thioglycerol by UV spectrophotometry and mass spectrometry (MS). The ORD spectrum of the isolated compound (Figure 8) was the same as that obtained for synthetic trioleoyl-3-thioglycerol, except that it was inverted. The molecular rotations were, however, lower than that for the synthetic compound, indicating that the isolated samples did not contain only triacyl-1-thioglycerol, but instead a mixture of this compound and its antipode, triacyl-3-thioglycerol. Also, the CD measurements confirmed the identity of the compound. Quantitative determinations made from CD spectra showed that the proportions of 1-thio-isomer/3-thio-isomer were 63/37 and 78/22 in two experiments.

The data showed that trioleoyl-thioglycerol was degraded in the intestine and to a minor degree resynthesized in chyle formation. Both when trioleoyl-thioglycerol was fed alone or together with trioleoylglycerol, the amount of triacyl-thioglycerol in chyle lipids was only 1/10 to 1/20 of the amount of triacylglycerol. This is probably not due to a slower digestion by pancreatic lipase, but may reflect a low utilization of 2-acyl-thioglycerol for chyle lipid resynthesis. The specific radioactivity of (^{14}C)oleate, which was fed as free acid, in chyle X-triacyl-1-thioglycerol was approximately half of that in chyle triacylglycerol. This indicates that the unlabeled fatty acids in the fed trioleoyl-thioglycerol to some degree were retained in the corresponding compound in chyle. In this case the higher proportion of triacyl-1-thioglycerol in the chyle may reflect a preferential utilization of 2-acyl-1-thioglycerol compared to 2-acyl-3-thioglycerol. An alternative mechanism for the stereospecificity, a selective reutilization of one of the liberated thioglycerol isomers, cannot be excluded.

Chiroptical properties of 1,2-dioleoyl-3-*S*-tetradecyl-3-thioglycerol and the corresponding *S*-oxide and *S*-dioxide were studied previously,[16] and since the *S*-oxide exhibited high rotation in ORD and CD, the digestion and absorption of this compound was studied in rats with a thoracic duct cannula.[41] *rac*-1,2-di(^{3}H)-oleoyl-3-tetradecyl-3-thioglycerol-*S*-oxide, which is a mixture of diastereomeric compounds assigned the

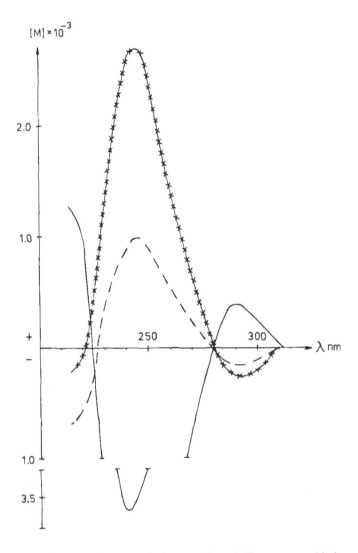

FIGURE 8.   ORD spectra of triacylthioglycerol. The spectrum with the major negative peak represents trioleoyl-3-thioglycerol. The other two spectra represent chyle X-triacyl-1-thioglycerol obtained after feeding *rac*-1,2-dioleoyl-3-*S*-oleoyl-3-thioglycerol to rats (two experiments). (From Åkesson, B., Gronowitz, S., and Michelson, P., *Chem. Phys. Lipids*, 23, 93, 1979. With permission.)

S,S-, R,R-, S,R-, and R,S-configurations (Figure 10), was fed together with a trace amount of (^{14}C)oleic acid. Chyle lipids collected up to 6 hr contained 6 to 33% of the administered ^{3}H. Of the lipid ^{3}H 75% was found in triacylglycerol and 13% in a compound identical to the administered sulfoxide as determined by MS (Figure 9). Analysis by ORD/CD shows that the isolated sulfoxide had spectra similar to those of synthetic 1,2-dioleoyl-3-*S*-tetradecyl-3-thioglycerol-*S*-oxide, except that the spectra were inverted, indicating that the 1-thioglycerol isomer was most abundant. The isolated sulfoxide was also separated by TLC into diastereomers, which were quantified. Compound A (R,R- and S,S-configuration) constituted $^2/_3$ and compound B (R,S- and S,R-configuration) $^1/_3$. Analysis by ORD/CD showed that 72 to 100% of compound A was the 1-thioglycerol isomer (i.e., S,S) and 70 to 90% of compound B was the 1-thioglycerol isomer (i.e., S,R) (Figure 10).

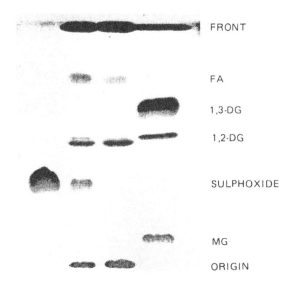

FRONT

FA

1,3-DG

1,2-DG

SULPHOXIDE

MG

ORIGIN

FIGURE 9. TLC of chyle lipid extracts obtained after feeding *rac*-1,2-dioleoyl-3-*S*-tetradecyl-3-thioglycerol-*S*-oxide to rats. The samples are from left to right: (1) administered lipid; (2) chyle collected at 0 to 6 hr; (3) *chyle collected at 6 to 18 hr*; (4) acylglycerol reference substances. (From Akesson, B. and Michelsen, P., *Chem. Phys. Lipids*, 29, 341, 1981. With permission.)

The $^{14}C/^{3}H$ ratios in the fed emulsion and in chyle triacylglycerol and sulfoxide were essentially identical, indicating that the triacylglycerol analog had been deacylated in the intestinal tract and that some of the deacylated sulfoxide had been reesterified in the intestinal mucosa. The mechanism behind the preferential incorporation of 1-thioglycerol isomers into chyle lipids is probably not a stereospecific lipolysis, since pancreatic lipase attacks alkyldiacylglycerol without stereospecificity and cleaves 2-acyl-3-alkylglycerol faster than 1-alkyl-2-acylglycerol.[18,20] If the same specificity holds for the sulfoxide, the isomer substituted at position 3 of the glycerol moiety would have been produced most rapidly. Instead, the preference for the 1-thioglycerol isomer is most probably explained by stereospecificity at the acyltransferase level, since 1-*O*-alkylglycerol but not 3-*O*-alkylglycerol is acylated in the intestinal mucosa.[35] Whether the preferential appearance in chyle of diastereomer A compared to diastereomer B also is due to selectivity among acyltransferases is still not known.

The experiments where sulfur-containing glyceride analogs were fed to rats have shown that the analogs are deacylated and to some extent absorbed and reesterified in the intestinal mucosa.[40,41] Several new examples of selective utilization of glyceride isomers have been demonstrated, using ORD and CD measurements on small amounts of lipids. Further work is necessary to explain the isomer selectivity at the enzyme level. The use of ORD and CD in the analysis of glyceride isomers may be advantageous also in studies of glyceride metabolism in organs other than the intestine.

V. CONCLUDING REMARKS

The substrate specificity of lipases includes a number of specificities,[27] and for certain substrates such as natural triacylglycerols several structural properties will affect the rate of lipolysis. To distinguish between different types of specificities, triacylglycerol analogs are valuable substrates, and data on that subject have been reviewed in Section III. Different lipases have varying degrees of stereospecificity, and the pattern

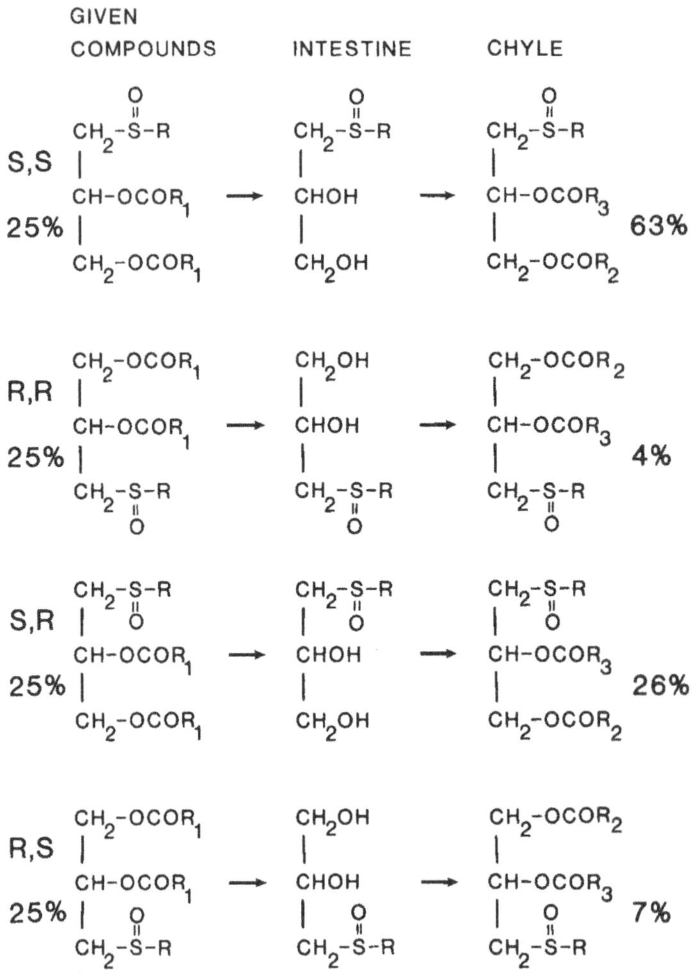

GIVEN
COMPOUNDS   INTESTINE   CHYLE

$$\text{S,S} \quad 25\%$$

$$
\begin{array}{lll}
\overset{O}{\overset{\|}{CH_2-S-R}} & \overset{O}{\overset{\|}{CH_2-S-R}} & \overset{O}{\overset{\|}{CH_2-S-R}} \\
| & | & | \\
CH-OCOR_1 \longrightarrow & CHOH \longrightarrow & CH-OCOR_3 \\
| & | & | \qquad 63\% \\
CH_2-OCOR_1 & CH_2OH & CH_2-OCOR_2
\end{array}
$$

$$\text{R,R} \quad 25\%$$

$$
\begin{array}{lll}
CH_2-OCOR_1 & CH_2OH & CH_2-OCOR_2 \\
| & | & | \\
CH-OCOR_1 \longrightarrow & CHOH \longrightarrow & CH-OCOR_3 \\
| & | & | \qquad 4\% \\
\underset{O}{\underset{\|}{CH_2-S-R}} & \underset{O}{\underset{\|}{CH_2-S-R}} & \underset{O}{\underset{\|}{CH_2-S-R}}
\end{array}
$$

$$\text{S,R} \quad 25\%$$

$$
\begin{array}{lll}
\overset{O}{\overset{\|}{CH_2-S-R}} & \overset{O}{\overset{\|}{CH_2-S-R}} & \overset{O}{\overset{\|}{CH_2-S-R}} \\
| & | & | \\
CH-OCOR_1 \longrightarrow & CHOH \longrightarrow & CH-OCOR_3 \\
| & | & | \qquad 26\% \\
CH_2-OCOR_1 & CH_2OH & CH_2-OCOR_2
\end{array}
$$

$$\text{R,S} \quad 25\%$$

$$
\begin{array}{lll}
CH_2-OCOR_1 & CH_2OH & CH_2-OCOR_2 \\
| & | & | \\
CH-OCOR_1 \longrightarrow & CHOH \longrightarrow & CH-OCOR_3 \\
| \quad \overset{O}{\overset{\|}{}} & | \quad \overset{O}{\overset{\|}{}} & | \quad \overset{O}{\overset{\|}{}} \qquad 7\% \\
CH_2-S-R & CH_2-S-R & CH_2-S-R
\end{array}
$$

FIGURE 10.   Schematic outline of the absorption of X-1,2-dioleoyl-3-*S*-tetradecyl-3-thioglycerol-*S*-oxide. To the left the configuration of the four administered isomers is shown, and to the right the proportions of the four isomers recovered in chyle are given. (From Åkesson, B. and Michelsen, P., *Chem. Phys. Lipids*, 29, 341, 1981. With permission.)

of specificity vs. different pairs of glyceride enantiomers varies. This is probably explained by the characteristics of the substrate-enzyme interaction.[42] So far, little information is available to explain the stereospecificity of lipases in relation to the structure of the enzymes, although one study indicated that the stereospecificity of lipoprotein lipase was an intrinsic property of the enzyme protein rather than of the activator, apolipoprotein CII.[43] Use of additional substrate analogs should eventually lead to more detailed understanding of the substrate specificity of lipases.

During the metabolism of glycerides in the intact organism stereospecificities both in lipolysis and in other enzymatic reactions of glycerides will influence the metabolic fate of different enantiomers.[44] The integrated influence of enzyme specificities may be studied, e.g., by comparison of the structure of a given food fat with that of the fat deposited in different tissues. Such as approach would, however, be very complicated, and the comparison would be heavily influenced by endogenous lipid synthesis and other reactions. Glyceride analogs can be used as model compounds to follow the metabolic fate of exogenous glycerides, especially if adequate techniques can be applied for this purpose. We have used ORD and CD to demonstrate stereospecificities during absorption of sulfur-containing glycerides.[40,41] The technique was sensitive enough for the analysis of milligram amounts of lipid. Other physical methods are available, especially for the characterization of optically active triacylglycerols,[45] but in general they are not applicable to analyze the small amounts of triacylglycerol available in most metabolic studies. Stereospecific analysis using enzymatic methods has been used for several purposes,[44,46] but the procedure is complex and not ideally suitable for large numbers of analyses. Therefore, a program to find new chromatographic methods for the separation of glyceride enantiomers has been initiated.[47-49]

Stereospecificities in glyceride metabolism have several metabolic implications and only a few will be discussed here. The formation of 2,3-diacylglycerol by lipases specific for position 1 may hinder the formation of phospholipids from lipolysis products, since only 1,2-diacylglycerol is utilized in phospholipid biosynthesis. 2,3-Diacylglycerol is actually the preferred isomer formed in vivo in postheparin plasma,[50] and 2,3-diacylglycerol was also found in liver.[46]

Recently, a number of physiological effects related to protein kinase C have been ascribed to diacylglycerols and their analogs.[51,52] This is a very interesting area for future research, and it may provide rationale for physiological variations in tissue diacylglycerol content[46] and for specificities in the metabolism of different glyceride isomers. Furthermore, the glyceride analogs described in this review may also be useful in studies on the relation between glyceride structure and biological effect.

ACKNOWLEDGMENTS

The authors thank Prof. S. Gronowitz, Head of the Division of Organic Chemistry, Dr. Å. Nilsson, Ms. B. Mårtensson, and Mr. J. Glans for their valuable contribution to the studies from our laboratory. The studies were supported by grants from the Swedish Medical Research Council (project 3968), the Natural Science Research Council, and the Påhlsson Foundation.

REFERENCES

1. Horrocks, L. A., Content, composition, and metabolism of mammalian and avian lipids that contain ether groups, in *Ether Lipids*, Snyder, F., Ed., Academic Press, New York, 1972, 177.
2. Snyder, F., Ether-linked lipids and fatty alcohol precursors in neoplasms, in *Ether Lipids*, Snyder, F., Ed., Academic Press, New York, 1972, 273.
3. Mangold, H. K., Ether lipids in the diet of humans and animals, in *Ether Lipids, Biochemical and Biomedical Aspects*, Mangold, H. K. and Paltauf, F., Eds., Academic Press, New York, 1983, 231.
4. Hallgren, B., Therapeutic effects of ether lipids, in *Ether Lipids, Biochemical and Biomedical Aspects*, Mangold, H. K. and Paltauf, F., Eds., Academic Press, New York, 1983, 261.
5. Ferrell, W. J. and Radloff, D. M., Glyceryl thioethers: a new naturally occurring lipid from bovine heart muscle, *Physiol. Chem. Phys.*, 2, 551, 1970.
6. Ferrell, W. J., Glycerol thioethers in human heart, *Lipids*, 8, 234, 1973.
7. Ferrell, W. J. and Garces, A., Natural occurrence of S-alkyl glycerol ethers in Ehrlich ascites cells, *Int. J. Biochem.*, 9, 33, 1978.
8. Ferrell, W. J. and Desmyter, E., Studies on the biosynthesis of S-alkyl bonds of glycerolipids, *Physiol. Chem. Phys.*, 6, 497, 1974.
9. Snyder, F., Piantadosi, C., and Wood, R., The metabolism of glyceryl thioethers, *Proc. Soc. Exp. Biol. Med.*, 130, 1170, 1969.
10. Gronowitz, S., Herslöf, B., Ohlson, R., and Töregård, B., ORD and CD studies of saturated glycerides, *Chem. Phys. Lipids*, 14, 174, 1975.
11. Gronowitz, S. and Herslöf, B., ORD and CD studies of glycerides. IV. Unsaturated glycerides, *Chem. Phys. Lipids*, 23, 101, 1979.

12. Michelsen, P. and Herslöf, B., Synthesis and chiroptical properties of neutral ether lipids, *Chem. Phys. Lipids*, 32, 27, 1983.
13. Herslöf, B., En studie över de optiska egenskaperna hos kirala glycerolderivat, thesis, University of Lund, Sweden, 1976.
14. Michelsen, P., Synthesis and Chiroptical Properties of Some Glycerolipids, thesis, University of Lund, Sweden, 1981.
15. Gronowitz, S., Herslöf, B., Michelsen, P., and Åkesson, B., Syntheses and chiroptical properties of some derivatives of 1-thioglycerol, *Chem. Phys. Lipids*, 23, 93, 1979.
16. Michelsen, P., Herslöf, B., and Åkesson, B., Chiroptical properties of S-alkyl derivatives of 1-thioglycerol, *Chem. Phys. Lipids*, 29, 177, 1981.
17. Tattrie, N. H., Bailey, R. A., and Kates, M., The action of pancreatic lipase on stereoisomeric triglycerides, *Arch. Biochem. Biophys.*, 78, 319, 1958.
18. Paltauf, F., Esfandi, F., and Holasek, A., Stereospecificity of lipases. Enzymic hydrolysis of enantiomeric alkyl diacylglycerols by lipoprotein lipase, lingual lipase and pancreatic lipase, *FEBS Lett.*, 40, 119, 1973.
19. Morley, N. H., Kuksis, A., and Buchnea, D., Hydrolysis of synthetic triacylglycerols by pancreatic and lipoprotein lipase, *Lipids*, 9, 481, 1974.
20. Åkesson, B., Gronowitz, S., Herslöf, B., Michelsen, P., and Olivecrona, T., Stereospecificity of different lipases, *Lipids*, 18, 313, 1983.
21. Åkesson, B., Gronowitz, S., and Herslöf, B., Stereospecificity of hepatic lipases, *FEBS Lett.*, 71, 241, 1976.
22. Noda, M., Tsukahara, H., and Ogata, M., Enzymic hydrolysis of diol lipids by pancreatic lipase, *Biochim. Biophys. Acta*, 529, 270, 1978.
23. Morley, N. H., Kuksis, A., Buchnea, D., and Myher, J. J., Hydrolysis of diacylglycerols by lipoprotein lipase, *J. Biol. Chem.*, 250, 3414, 1975.
24. Börgstrom, B. and Donnér, J., The polar interactions between pancreatic lipase, colipase, and the triglyceride substrate, *FEBS Lett.*, 83, 23, 1977.
25. Paltauf, F. and Wagner, E., Stereospecificity of lipases. Enzymatic hydrolysis of enantiomeric alkyldiacyl- and dialkylacylglycerols by lipoprotein lipase, *Biochim. Biophys. Acta*, 431, 359, 1976.
26. Slotboom, A. J., de Haas, G. H., Burbach-Westerhuis, G. J., and van Deenen, L. L. M., Hydrolysis of phosphoglycerides by purified lipase preparations. I. Substrate-, positional- and stereospecificity, *Chem. Phys. Lipids*, 4, 15, 1970.
27. Jensen, R. G., Dejong, F. A., and Clark, R. M., Determination of lipase specificity, *Lipids*, 18, 239, 1983.
28. Cambou, B. and Klibanov, A. M., Preparative production of optically active esters and alcohols using esterase-catalyzed stereospecific transesterification in organic media, *J. Am. Chem. Soc.*, 106, 2687, 1984.
29. Blomstrand, R., Digestion, absorption and metabolism of chimyl alcohol fed as free alcohol or as alkoxydiglyceride, *Proc. Soc. Exp. Biol. Med.*, 102, 662, 1959.
30. Blomstrand, R. and Ahrens, E. H., Absorption of chimyl alcohol in man, *Proc. Soc. Exp. Biol. Med.*, 100, 802, 1959.
31. Paltauf, F., Metabolism of the enantiomeric 1-0-alkyl glycerol ethers in the rat intestinal mucosa in vivo: incorporation into 1-0-alkyl and 1-0-alk-1'-enyl glycerol lipids, *Biochim. Biophys. Acta*, 239, 38, 1971.
32. Åkesson, B., Gronowitz, S., Herslöf, B., and Ohlson, R., Absorption of synthetic, stereochemically defined acylglycerols in the rat, *Lipids*, 13, 338, 1978.
33. Pitas, R. E., Hagerty, M. M., and Jensen, R. G., Transport of diacylalkylglycerols in chylomicrons and very low density lipoproteins of rat intestinal lymph following intragastric administration of 1,3-dioctadecenoyl-2-hexadecylglycerol, *Lipids*, 13, 844, 1978.
34. O'Doherty, P. J. A. and Kuksis, A., Microsomal synthesis of di- and triacylglycerols in rat liver and Ehrlich ascites cells, *Can. J. Biochem.*, 52, 514, 1974.
35. Paltauf, F. and Johnston, J. M., The metabolism in vitro of enantiomeric 1-0-alkyl glycerols and 1,2- and 1,3-alkyl acyl glycerols in the intestinal mucosa, *Biochim. Biophys. Acta*, 239, 47, 1971.
36. Kanoh, H. and Åkesson, B., Properties of microsomal and soluble diacylglycerol kinase in rat liver, *Eur. J. Biochem.*, 85, 225, 1978.
37. Aarsman, A. J., van Deenen, L. L. M., and van den Bosch, H., Studies on lysophospholipases, *Bioorg. Chem.*, 5, 241, 1976.
38. Cox, J. W., Snyder, W. R., and Horrocks, L. A., Synthesis of choline and ethanolamine phospholipids with thiophosphoester bonds as substrates for phospholipase C, *Chem. Phys. Lipids*, 25, 369, 1979.
39. Cox, J. W. and Horrocks, L. A., Preparation of thioester substrates and development of continuous spectrophotometric assays for phospholipase A_1 and monoacylglycerol lipase, *J. Lipid Res.*, 22, 496, 1981.

40. Åkesson, B., Gronowitz, S., and Michelsen, P., Digestion and absorption of trioleoyl-thioglycerol in the rat, *Chem. Phys. Lipids,* 23, 93, 1979.
41. Åkesson, B. and Michelsen, P., Digestion and absorption of a sulphoxide analogue of triacylglycerol in the rat, *Chem. Phys. Lipids,* 29, 341, 1981.
42. Borgström, B. and Brockman, H. L., *Lipases,* Elsevier, Amsterdam, 1984.
43. Somerharju, P., Kuusi, T., Paltauf, F., and Kinnunen, P. K. J., Stereospecificity of lipoprotein lipase is an intrinsic property of the active site of the enzyme protein, *FEBS Lett.,* 96, 170, 1978.
44. Kuksis, A., *Handbook of LIpid Research. Fatty Acids and Glycerides,* Plenum Press, New York, 1978.
45. Schlenk, W., Synthesis and analysis of optically active triglycerides, *J. Am. Oil Chem. Soc.,* 42, 945, 1965.
46. Åkesson, B., Composition of rat liver triacylglycerols and diacylglycerols, *Eur. J. Biochem.,* 9, 463, 1969.
47. Michelsen, P. and Odham, G., Highly sensitive diestereoisomeric analysis of secondary alcohols and short-chain 1,2-diacylglycerols using capillary gas chromatography/mass spectrometry, *J. Chromatogr.,* 331, 295, 1985.
48. Michelsen, P., Aronsson, E., Odham, G., and Åkesson, B., On the diastereomeric separations of natural glyceroderivatives as their 1-(1-naphthyl)ethyl carbamates by high-performance liquid chromatography, *J. Chromatogr.,* 350, 417, 1985.
49. Isaksson, R. and Michelsen, P., Analytical and preparative separation of some chiral glycerol acetonides by liquid chromatography on swollen microcrystalline triacetyl cellulose, *Chem. Scr.,* in press.
50. Morley, N., Kuksis, A., Hoffman, A. G. D., and Kakis, G., Preferential in vivo accumulation of sn-2,3-diacylglycerols in postheparin plasma of rats, *Can. J. Biochem.,* 55, 1075, 1977.
51. Berridge, M. J., Inositol triphosphate and diacylglycerol as second messengers, *Biochem. J.,* 220, 345, 1984.
52. Nishizuka, Y., The role of protein kinase C in cell surface signal transduction and tumour promotion, *Nature (London),* 308, 693, 1984.

Chapter 6

EFFECT OF DIETARY FAT ON FORMATION AND SECRETION OF CHYLOMICRONS AND OTHER LYMPH LIPOPROTEINS

A. Kuksis

TABLE OF CONTENTS

I. INTRODUCTION

It is well established that dietary fats affect plasma cholesterol and triacylglycerol levels and these are reflected in the relative proportions, absolute concentrations, and composition of various plasma lipoproteins.[1] The consumption of saturated fats leads to increases in plasma cholesterol and triacylglycerols and conversely the ingestion of unsaturated fats leads to decreases in these plasma lipoprotein components.[2] Although such alterations in the plasma would have been anticipated first in corresponding changes in lymph lipids and lipoproteins, few comparative studies have been made to demonstrate this. There have been no attempts to determine the effect of dietary fat on the subtle structural details of chylomicrons or other lipoproteins which may affect the function at the receptor and molecular levels. Biochemical and metabolic reasons for the resulting alterations in plasma lipid levels have been usually sought elsewhere.[3-5]

Since the formation and secretion of lipoproteins is an essential part of the digestion and absorption of fats, the present chapter reviews these studies and points out any differences in the structure and apoprotein composition of lymph lipoproteins resulting from the feeding of saturated and unsaturated fats in the presence and absence of cholesterol. It is shown that alterations in the lymph lipoprotein levels and in certain structural and compositional features contribute to the changes in plasma lipoproteins that are eventually observed folllowing the consumption of different dietary fats. The effect of diet on plasma lipoprotein metabolism has been recently reviewed.[6] The present discussion is opened with a brief consideration of the general features of chylomicron formation and secretion by the villus cells.

II. MECHANISM OF FORMATION AND SECRETION OF LYMPH LIPOPROTEINS

The absorbed fat is transported from the intestine in the form of special emulsion particles called chylomicrons.[7] These particles are released from the villus cells via the basolateral membranes and enter the intestinal lymphatic capillaries. They pass through the mesenteric lymphatic vessels and cysterna chyli to be drained through the thoracic duct into the bloodstream.[7,8] The formation of the chylomicrons involves the biosynthesis of the lipid and protein components, which may take place simultaneously or sequentially.[9] This process is accompanied by an increased biosynthesis of cell membranes as reflected in increased turnover of the component phospholipids, which clearly exceeds the needs of formation of the surfactant.[10,11] The load and the nature of the dietary fat affect both the composition and structure of the chylomicrons and may be reflected in different rates of formation and secretion.

A. Composition and Structure of Lymph Chylomicrons

Chylomicrons are spherical triacylglycerol-rich particles 50 to 500 nm in diameter and are formed within the intestine during triacylglycerol absorption.[12] The particle size distribution is approximately logarithmic. The particles formed during peak fat absorption are larger than those formed at other times.[13] Also, the triacylglycerol resynthesis via the phosphatidic acid pathway would appear to lead to the formation of smaller chylomicrons than a resynthesis via the monoacylglycerol pathway,[14] possibly due to differences in the overall flux of the triacylglycerol through the intestine.

The chylomicrons may be isolated as lymph lipoproteins of density less than 1.006 g/mℓ by ultracentrifugation at 3×10^6 g/min. Purification of chylomicrons from contaminating serum proteins may be accomplished by gel filtration through 2% agarose columns.[15] Recent studies suggest that low temperatures normally used in the isolation of chylomicrons may result in crystallization of the saturated core triacylglycerols,[16] increasing particle density by about 10%.[17]

1. Lipid Composition and Particle Size

The chylomicrons are composed of triacylglycerols (86 to 92%), phospholipids (6 to 8%), free sterols (1 to 2%), steryl esters (1 to 2%), and protein (1 to 2%). The fatty acid composition of the triacylglycerols reflects largely that of the dietary fat, while that of the glycerophospholipids and steryl esters may differ significantly from it.[18] The general chemical class composition of the particles may be shown by calculation[19] to be consistent with the location of the nonpolar triacylglycerols and cholesterol esters in the core and the polar phospholipids and free cholesterol in the surface monolayer, which also includes the protein (20% of the surface). The exact composition of the chylomicrons changes with particle size, because the proportion of the surface components varies with the square of particle diameter, whereas core components vary with the cube. In model systems, it has been shown that the polar surface may contain up to 3% of triacylglycerol,[20] and phase diagrams have indicated the probability that up to 3% of the cholesteryl ester may be found in the phospholipid monolayer.[21,22] Precise relationships have been calculated to describe how composition changes with lipoprotein size during metabolism.[23-25]

Fasting lymph contains triacylglycerol particles the size of very low-density lipoproteins (VLDL) (20 to 50 nm), which carry some 50% of lymph cholesterol and triacylglycerol in the fasting state. These particles may represent the endogenous intestinal and biliary lipids.[26] Mahley et al.[27] have demonstrated that rat intestinal chylomicrons and VLDL-sized particles are distinct, because they do not become mixed readily in individual Golgi vesicles.

2. Apoprotein Composition

Although a minor component, the protein associated with chylomicron consists of a complex mixture and has important roles in chylomicrons clearance and metabolism. Most of the known plasma apolipoproteins are present,[9] including apoB (10%,) apoA-IV (10%), apoE (5%), ApoA-I(15 to 35%), and apoC (45 to 50%). ApoA-II, a minor apoprotein in the rat, is present on human chylomicrons. ApoB is a high molecular weight, hydrophobic apoprotein found in the triacylglycerol-rich lipoproteins of all species studied. Early work indicated that in both rats[28] and humans[29] chylomicron apoB is a smaller protein (240,000) than the minor apo B found in low-density lipoprotein (LDL) (353,000) in plasma. More recent work[30] has demonstrated that freshly secreted rat chylomicrons contain some of the apoB in the form of a high molecular weight protein (549,000), which corresponds to the major species of apoB in LDL. The high molecular weight apoB has not yet been demonstrated in the villus cell, but the corresponding mRNA has been detected there, although in lower amounts than in liver.[31] It has been suggested[30] that all apoB of rat chylomicrons may be secreted originally in the higher molecular weight form and rapidly degraded to the lower molecular weight form during isolation. These forms share extensive immunological cross reactivity and therefore exclude organ specificity. The possibility of contamination with plasma apoB, however, would not appear to have been completely excluded. In the rat each chylomicron contains about 0.47×10^6 daltons of ApoB, which corresponds to about one molecule of the higher molecular weight apoB per particle.[14,32] Several lines of evidence indicate that in the rat apoB is newly synthesized as a chylomicron apoprotein during triacylglycerol absorption.[33,34] Likewise, apoA-I is actively synthesized during triacylglycerol absorption.[35,36] There is evidence[35] that in the rat a maximum of 5% of the apoA-I in lymph chylomicrons could have originated from plasma. Recently, isoforms of apoA-I have been described in chylomicrons.[37,38] Fat feeding also results in an increase in rat chylomicrons of the lymphatic secretion of apoA-IV[39] and in human villus cells of the apoA-IV content.[40] There is little evidence that the intestine synthesizes apoE[35,41] or apoC[39,42] in significant amounts. Both in vivo and in vitro experiments suggest that these apoproteins associate with chylomicrons after secretion.[38,43]

B. Formation and Secretion of Chylomicrons

The newly absorbed fatty acids and monoacylglycerols cross the microvillus membrane by passive diffusion and are first visible morphologically as triacylglycerol droplets within the smooth endoplasmic reticulum. Subsequently, fat droplets accumulate in the Golgi apparatus in the supranuclear portion of the villus cell. Eventually, migration of elements of the Golgi complex occurs toward the basolateral aspect of the cell and fusion of Golgi membranes with the lateral cell membrane occurs with lipoproteins being discharged into the intracellular space by reverse pinocytosis.[44] Microtubules have been implicated in the directed intracellular movement of the lipid particles through the villus cell,[45] but it is not known whether specific sites are involved in the membrane fusion.

The biochemical events involved in the biosynthesis of the triacylglycerols by the villus cells have been discussed in another chapter in this book.[46] Stremmel et al.[47] have recently isolated a fatty acid binding protein from rat jejunal microvillus membranes. The presence of this protein in the apical and lateral portions of the brush border cells of jejunum, but not on the luminal surface of esophagus or colon, suggests that it plays a role in fatty acid absorption from the gut. Previously, an active role had been sought for a cytosolic fatty acid binding protein in the triacylglycerol biosynthesis of the intestinal mucosa.[48,49]

An active synthesis of proteins is necessary for normal chylomicron secretion. Studies with inhibitors of protein synthesis have demonstrated an accumulation of triacylglycerol droplets within the intestinal cell and a decreased lymphatic secretion of chylomicrons.[50,51] The role of protein synthesis has been questioned by others,[52] who found that lipid absorption into lymph of animals with inhibited protein synthesis proceeded at almost normal levels. They attributed any decrease in absorption to delayed gastric emptying and decreased lymph flow produced by the toxic action of the drugs employed. Work with isolated villus cells,[53,54] however, has confirmed the earlier claims of decreased protein synthesis and inhibited release of chylomicrons following puromycin administration. The latter work[54] has suggested that chylomicron formation and release by the mucosal cells depend upon intact rough endoplasmic reticulum and an active protein and phospholipid biosynthesis. Chylomicrons secreted during inhibition of protein synthesis are larger and contain reduced amounts of apoA-I.[51] However, the amounts of apoA-IV and apoB on these particles are preserved, which suggests their essential requirement for chylomicron secretion.[15] Whether or not the synthesis of any other protein moiety is required is as yet unknown.

The ultrastructural localization of apoB within the rough endoplasmic reticulum and on the triacylglycerol droplets in the smooth endoplasmic reticulum suggests an addition of apoB early in chylomicron formation.[55] Detailed correlations of the rates of biosynthesis of apoB and of secretion of chylomicrons, however, have given contradictory results. Thus, work with human intestinal biopsies has shown that the rate of secretion of chylomicrons exceeds the rate of synthesis of apoB.[56] More recent studies using rat intestine have shown[57] that the cellular content of apoB does not decrease despite continued lipid stimulation. In addition, administration of protein synthesis inhibitors to rats not absorbing exogenous lipid produces a modest decrease in the apoB content of intestinal cells (20% reduction).[57] Protein synthesis inhibitors administered to rats actively absorbing triacylglycerols can deplete the intestinal mucosal apoB content by approximately 50%, but not further, suggesting a limit beyond which cellular apoB cannot be decreased.[57] Alpers et al.[58] have obtained data to show that part of the intracellular apoB is associated with LDL and high-density lipoprotein (HDL)-size particles, while in the lamina propria and lymph most of the apoB is associated with VLDL and chylomicrons. Hence, only the larger apoB-containing particles are secreted, while the "nascent" precursors remain in the cell. Work with isolated villus cells has shown[53] that puromycin added in vitro causes only a marginal reduction in chylomicron secretion. The central role of apoB in the secretion of chylomicrons is consistent with the characteristics of abetalipoproteinemia, a disease in which there is an inability to clear chylomicrons resulting in an accumulation of triacylglycerol droplets in the intestine.[50,51]

There is also evidence that fat feeding results in an increased biosynthesis and secretion of apoA-I from the intestine into lymph.[15] Green et al.[40] have shown a doubling of apoA-I content of duodenal enterocytes in man after fat feeding. Furthermore, Ghiselli et al.[59] have demonstrated a marked change in the apoA-I isoform pattern in lymph chylomicrons and VLDL following feeding with a significant increase in the relative contribution of the apoA-I-1 isoform. Current studies,[60] however, suggest that under conditions of physiological intake of dietary triacylglycerol, apoA-I synthesis in jejunal enterocytes is not regulated acutely by changes in triacylglycerol flux. Also, it has been shown[34,61] that apoA-I synthesis can proceed independently of triacylglycerol absorption, and it remains to be established whether or not apoA-I biosynthesis is essential for chylomicron secretion. The role of apoA-IV in chylomicron formation and secretion is not known either. Gordon et al.[62] have recently demonstrated that the amount of rat intestinal apoA-IV mRNA is markedly increased after fat feeding.

Strauss and Jacob[63] have shown that the externalization of chylomicrons has an absolute requirement for Ca^{2+}. The exact role of Ca^{2+} in this process is not known. Redgrave[64] has speculated that activation of some Ca^{2+}-requiring enzyme such as phospholipase might be involved in the process of fusion of adjacent Golgi and plasma membranes. The release of chylomicrons from inverted sacs of the rat intestine is inhibited by substitution of deuterium oxide for water in the incubation buffers,[63] which presumably interferes with the assembly of the microtubules.[65] Colchicine[63,66-68] also inhibits the transport of lipids by intestinal cells, which further suggests that microtubules[69] may be involved in the movement of prechylomicrons from Golgi apparatus to a site near the basolateral plasma membrane. A similar effect has been demonstrated in isolated villus cells for nocodazole and colcemid.[68] Sabesin and Frase[70] have demonstrated that chylomicrons are externalized in packets rather than as individual particles. This observation may require a redefinition of the earlier claim of reverse pinocytosis involving individual chylomicron particles.[44]

Chylomicron-like particles have been demonstrated to be released from isolated enterocytes preloaded with fat in vivo.[71,72] A differentiation between spillage resulting from cell breakage and orderly secretion was made on the basis of the known puromycin interference with lipoprotein assembly and/or secretion. However, only a pretreatment of the animals with puromycin was found to be effective.[72]

C. Origin of Other Lymph Lipoproteins

Hoffman et al.[68] have shown that the radioactive glycerolipids released into the medium by isolated villus cells were found largely in the VLDL and higher-density lipoprotein fractions, very little being recovered in the chylomicron fraction and/or in the oil phase. About 20 to 30% of the newly formed triacylglycerols and about 35% of the newly formed phospholipids were secreted into the medium and were recovered as the triacylglycerol-rich particles. Labeled proteins and glycoproteins were also recovered from this fraction. Apparently, the low levels of lipid were not sufficient to drive chylomicron formation, as already noted for fasting rats studied in vivo.[26] Some studies[7] have indicated that VLDL-sized particles are formed by the intestine continuously and a transition to chylomicron production takes place only when large amounts of dietary triacylglycerol become available.

There is increasing evidence that the rat intestine also secretes characteristic forms of high-density lipoprotein into mesenteric lymph. Green et al.[73] have described a heterogeneous population of HDL particles in mesenteric lymph. It was estimated that 70% of the apoA-I in the HDL fraction of lymph originated in the intestine.[61] The HDL fraction of lymph was not derived from lymph triacylglycerol-rich lipoproteins. Recently, Forester et al.[74] have resolved mesenteric lymph HDL into discrete fractions including a discoidal fraction, a spherical HDL fraction similar in size to plasma HDL, and a small spherical particle fraction not present in plasma. Duodenal infusion of [³H]cholesterol resulted in an increase in cholesteryl ester-specific activity in lymph spherical HDL, when compared with plasma HDL. These studies provide further evidence for the concept that both surface and core components of lymph HDL may originate in the intestine. Human thoracic duct lymph contains particles similar to plasma HDL, but these particles are enriched in triacylglycerols and phospholipids and resemble HDL-2A.[36] Furthermore, chronic feeding of cholesterol to rats results in an increased proportion of lymph cholesterol carried in the LDL fraction. These particles are rich in triacylglycerol and cholesteryl ester and contain apoB and apoA-I.[41] Only small amounts of LDL are present in the mesenteric lymph under normal conditions and these are probably filtered from plasma.

Therefore, the intestine is at least a major site of synthesis for apoB of chylomicrons and of other lymph lipoproteins. There is evidence that the nature of the dietary fat affects the synthesis of these proteins and thus influences the subsequent plasma clearance and metabolism of the lipoproteins. The pertinent data are considered in subsequent sections of the discussion.

III. SATURATED FATS

The increases in plasma cholesterol and triacylglycerol levels attributed to the saturated fatty acids of dietary fats are reflected in increased VLDL and LDL and decreased HDL levels.[1,2] The effect of saturated fats on the lymph lipids is less well known and essentially all of the more detailed information has come from studies performed in experimental animals.

The saturated fats are usually of animal origin, except those derived from marine animals, fish, and shellfish. The animal fats contain high proportions of the fully saturated palmitic and stearic, and the monounsaturated oleic acids, with much lower amounts of the diunsaturated linoleic and the tetraunsaturated arachidonic acids (AA). An animal fat of special interest is butter- or milkfat, which, in addition to the saturated and monounsaturated long-chain acids, also contains high proportions of the short-chain butyric, hexanoic, and octanoic acids.

A. Lipid Composition and Particle Size

The physiological and the physicochemical properties of the synthetic saturated fats vary according to the chain length of the component fatty acids. Thus, the short-chain acids are liquid at room temperature, relatively soluble in water, and are absorbed in the free form via the portal route. The long-chain saturated fatty acids are solid at room temperature, insoluble in water, and must be reesterified before absorption via the lymphatic route. When dealing with natural fats, the distinction is less clear, but still necessary.

In both man and experimental animals butter appears to have a different biologic effect than corn oil. Beveridge et al.[75] demonstrated that experimental human diets supplemented with butter resulted in elevated cholesterol levels compared to those found when corn oil was substituted as the fat source. In rabbits tube-fed one meal of butter, the clot lysis time was significantly increased over that of animals tube-fed water; when corn oil was substituted for butter, no significant increase in clot-lysis time occurred.[76] Jones et al.[77] have shown that the different biological actions of the two fats may be accounted, in part, by the different physicochemical characteristics of the lymph lipids. Thus, in rats, the addition of large amounts of butter to a basic hypercholesterolemic diet (containing, among other ingredients, cholesterol, propyl-thiouracil, and sodium cholate) led to a different appearance of the lymphatic fat in the electron microscope when substituted for corn oil, all other conditions remaining unchanged. In butter-fed animals, the most striking morphological form was a large densely staining, fragmented mass with fairly regular borders. This form was found in the mucosal cells and within the jejunal lumen. A second form of fat found in mucosal cells was a smaller globule-shaped mass, often showing "drop-out" of lipid either centrally or peripherally. A third form of osmophilic material was composed of relatively small droplets staining fairly uniformly and densely, with irregular borders; these droplets lay in a feathery pattern. In the corn oil-fed rats, the osmophilic masses suggestive of lipid were more uniform — the droplets were generally smaller, stained less densely, did not show fragmentation or drop-out, and had smooth, regular borders. The appearance of osmophilic masses differed in the chyle of butter and corn oil-fed rats, no matter which of the various methods of preparing the chyle were used. In general, the chylomicrons from butter-fed rats on electron microscopy showed varying types of osmophilic material. The masses, which measured from 50 to 500 nm, were similar to the fat seen in the intercellular spaces of the jejunal mucosa.

A second form of osmophilic material, presumably, chylomicrons, was spherical in shape, densely staining, and measured up to 1 μm in diameter. A third form of chylomicrons seen in the visceral lymphatics of butter-fed rats was much larger in size, varying from 5 to 20 μm. The larger of these particles, called "giant chylomicrons", was very rarely seen in spun-down chyle. In the corn oil-fed rats, osmophilic material was found in two forms: as discrete round masses of fairly large size and as very small round particles lying in a powdery pattern. The discrete round masses were about 1 μm in diameter and appeared to be slightly larger than the round masses found in the chyle of butter-fed rats. No giant chylomicrons were found by electron microscopy in the chyle of corn oil-fed rats. The same difference in size of chylomicrons from butter and corn oil-fed rats was suggested in phase microscopy preparations of chyle, where the possibility of artefactual change was much less. The different physical appearances of these chylomicrons apparently reflect differences in the chemical composition of the chylomicrons and physical properties of the component triacylglycerols (see below).

Huang and Kuksis[19] compared the lipid composition and the size of the chylomicrons isolated from dog lymph following the feeding of corn oil or butter-fat meals. Although there were marked differences in the fatty acid composition of the core triacylglycerols and lesser differences in the composition of the surface phosphatidylcholines, the particle diameters calculated on the basis of the surfactant-to-core lipid ratios were about the same.[78] If the giant chylomicrons observed by Jones et al.[77] were true chylomicrons, they would have been expected to have a smaller surface area than the

Table 1
COMPOSITION OF THE MAJOR
FATTY ACIDS OF CORE
TRIACYLGLYCEROLS AND SURFACE
PHOSPHATIDYLCHOLINES OF
LYMPH CHYLOMICRONS FROM
DOGS FED EITHER BUTTERFAT OR
CORN OIL[19,78]

Fatty acids[a]	Butterfat[b] (wt %)		Corn oil[d] (wt %)	
	TG[c]	PC	TG	PC
10:0	2.4			
12:0	4.0			
14:0	13.4	0.9		
14:1ω5	4.6			
16:0	24.9	17.7	17.9	14.6
16:1ω7	4.0	2.0	1.5	
18:0	7.8	26.2	3.8	32.6
18:1ω9	32.8	22.9	35.3	11.9
18:2ω6	3.4	19.6	40.6	31.4
18:3ω3	2.7	3.4	0.9	
20:4ω6		7.3		9.5

[a]   Number of carbon atoms/number of double bonds. Omega values identify the location of last double bond with respect to methyl terminal.
[b]   70 g of butterfat fed and lymph collected over a period of 5—9 hr after feeding.
[c]   TG, triacylglycerols; PC, phosphatidylcholines.
[d]   50 g corn oil fed and lymph collected 5—9 hr after feeding.

smaller chylomicrons resulting from corn oil feeding, when compared on the basis of equal total weight. Since both types of chylomicrons possessed comparable surface-to-volume ratios, they must have been about the same size. The giant butterfat chylomicrons, therefore, must have represented aggregates of smaller particles as already suspected by these authors.[77] Why the butterfat chylomicrons should aggregate and the corn oil chylomicrons remain dispersed cannot be accounted for on the basis of the chemical analyses alone.

Table 1 gives the composition of the major fatty acids of the core triacylglycerols and the surface phosphatidylcholines of the chylomicrons of lymph as obtained from dogs fed either corn oil or butterfat.[19,78] It is obvious that the chylomicrons from the butterfat animals contain much more saturated fatty acids in the triacylglycerol cores than the chylomicrons from corn oil feeding, while the fatty acids of the phosphatidylcholines of the surface monolayer showed much less difference. Floren and Nilsson[79] have suggested that chylomicrons containing highly saturated triacylglycerols undergo partial crystallization during their isolation. The solid portion would then be expected to be enriched in the fully saturated species with a high melting point and to be less available for enzymatic action, as demonstrated following injection in hepatectomized rats. Zilversmit[80] had earlier demonstrated that the chylomicrons isolated from cream-fed dogs had a much higher melting point than those from corn oil-fed dogs. The apparent loss from cream of short-chain fatty acids, which are absorbed by the portal vein, produced chylomicrons that were semisolid at body temperature.

Parks et al.[16] have determined the physical properties of the triacylglycerol-rich lipoproteins derived from thoracic duct lymph of nonhuman primates fed a saturated fat diet. Vervet monkeys were infused via duodenal cannula with a liquid diet containing 40% of calories from butterfat. In one experiment, an animal was infused with a similar diet in which safflower oil had been substituted isocalorically for butterfat. Butterfat chylomicra and VLDL were found to have complicated phase behavior which was dependent on the thermal history of the particle. Intact particles isolated at 15°C and stored overnight at 4°C had two endothermic transitions upon heating from −10 to 60°C. It was observed that the physical state of chylomicra and VLDL triacylglycerol core can be modified by isolation temperature. The particle structure affected the crystallization, but not the melting, of chylomicra and VLDL triacylglycerols. It was also shown that chylomicron particle size does not markedly influence the physical properties of the triacylglycerol core.

Puppione et al.[81] have isolated from calf lymph, lipoprotein particles that are flat in appearance and asymmetric in shape, and possess anomalous densities, which could be explained if these lipoproteins consisted of a core of crystallized triacylglycerols encapsulated within a phospholipid monolayer. This apparently is due to the increase in density of the saturated triacylglycerols upon partial crystallization,[17] which occurs during routine lipoprotein isolation. When the lipoproteins are isolated at 37°C, the unusual structures are not observed in the intermediate density range. It is not known whether or not differences are seen in vivo in the metabolism of chylomicrons rich in saturated or unsaturated triacylglyerols.

Although coconut oil consumption also leads to elevated plasma triacylglycerol and cholesterol levels,[82] it is not known if it produces an effect upon lymph lipoproteins similar to that of milkfat consumption. Saturated triacylglycerols made up exclusively of short-chain fatty acids (MCT) are not hypercholesterolemic,[83] because these are absorbed via the portal route.

Bennet Clark et al.[84] have described in depth a study of chylomicrons and VLDL isolated from intestinal lymph of rats fed a diet designed to enrich these lipoproteins with saturated fatty acids. They have demonstrated that highly saturated triacylglycerols secreted by the intestine of the rat in native lipoproteins are present as metastable, undercooled liquids. Crystallization of triacylglycerols and concomitant structural changes in the lipoproteins are produced on cooling below the triacylglycerol nucleation temperature. After incubation at 4°C, 20 to 30% of the particles appeared flattened and polygonal. However, after incubating the chylomicrons at 58°C for 10 min the lipoprotein profile was again round and the size distribution was similar to that of the native particles. VLDL exposed to low temperature displayed structural alterations similar to those of chylomicrons. The triacylglycerols of these lipoproteins contained about 74% saturated fatty acids, primarily palmitic acid. Feldman et al.[85] have shown that the distorted particles were obtained with cooling when the saturated fatty acid concentration of triacylglycerol-rich particles exceeded 54%. Whether or not this has any physiological significance remains to be investigated. Likewise, the effects of degree of fatty acid unsaturation, the fatty acid chain length, and the triacylglycerol stereoisomerism on lipoprotein transition temperatures are not known.

Feldman et al.[85] have shown that chylomicron size increases progressively as the chylomicron lipids become less saturated. In contrast, VLDL particles were largest with the 78% saturated fatty acid diet, were intermediate with the 68% diet, and were smaller and of similar size with the two less saturated diets. Similarly, HDL particles were larger with the 78 and 68% saturated fatty acid diets than with the 38% saturated fatty acid diet. The observation of larger chylomicron particles where triacylglycerol absorption was greater is in accord with previous studies by Redgrave and Dunne[13] and Fraser.[86] The inverse relation of VLDL and HDL to chylomicron size observed by Feldman et al.[85] also occurred in an earlier study by Feldman et al.[87] on feeding diets of palmitic, stearic, oleic, and linoleic acids at greater than 78% of total fatty acids.

Lymph chylomicrons obtained from rats fed diets with 78 or 68% saturated fatty acids examined by differential scanning colorimetry were liquid when secreted and crystallized at 20 to 24°C when cooled from 37°C.

Green and Green[88] have systematically investigated the effects of dietary fat saturation on intestinal lymph lipid composition and indirectly on lymph lipoprotein size. Lipid absorption was studied in nonfasting thoracic lymph duct-cannulated rats receiving continuous intraduodenal infusions of oils varying in triacylglycerol fatty acid saturation (polyunsaturated/saturated = 4.6 or 0.2), triacylglycerol load (66 to 165 μmol/hr), and cholesterol load (0 and 3.5 μmol/hr at the low oil infusion rate, and 0 and 11 μmol/hr at the high infusion rate). Estimated average relative lymph lipoprotein size was significantly larger during infusion of the high vs. low triacylglycerol load, and of the polyunsaturated vs. the saturated oil. The proportion of the infused triacylglycerol load absorbed was significantly influenced by fat saturation, but not by total triacylglycerol load or time. The absorption of the saturated oil was 15 to 20% less efficient than that of the polyunsaturated oil. Ockner et al.[89] have suggested that absorption of saturated fatty acids occurs over a longer length of the small intestine resulting in a decreased flow of triacylglycerol through each absorptive cell in animals fed saturated fat. Thus, smaller lipoproteins would be produced when triacylglycerol load was decreased or when saturated fat was being absorbed. Lymph phospholipid flux was not significantly affected by time or fat saturation, but differed between triacylglycerol loads. Both fat saturation and triacylglycerol load significantly influenced the molar ratio of lymph phospholipids/triacylglycerols, which is an indicator of lipoprotein surface/volume ratio and thus an inverse correlate of apparent average lymph lipoprotein size.[86]

B. Apoprotein Composition

Feldman et al.[85] have also determined the apoprotein distributions in lymph VLDL and HDL samples from rats ingesting diets enriched in four different fatty acids. These data were compared in samples collected and/or stored at 4°C prior to ultracentrifugation. Within each group the distribution of apoproteins was similar in the two collections, after the gavage or while ingesting the diet. The major apoprotein for VLDL samples was apoA-I from rats fed all diets except the 78% saturated fatty acid diet. In this diet, C apoproteins predominated in VLDL. The mean apoprotein composition of VLDL samples was A-I, 40%; A-IV, 25%; C's, 18%; and E, 16%. The percent of C apoprotein in VLDL decreased progressively as percent saturated fatty acids in the diet and lymph triacylglycerols declined. VLDL apoA-IV was greatest in samples from rats fed the 38% saturated fat diet. VLDL apoE was greatest in the 68% saturated fat diet. HDL apoprotein patterns showed a progressive increase in apoA-I as the percent of saturated fatty acids declined. This was associated with a decline in apoCs of HDL. HDL apoE was relatively increased in the 68% saturated fatty acid diet. Significant amounts of apoA-IV were observed in HDL from the 48% saturated fatty acid diet. Table 2 gives the apoproteins of intestinal lymph VLDL and HDL from the different diet groups.[85] The apoproteins were separated by SDS-PAGE and scanned by densitometry of stained gels. The apoprotein composition of lymph VLDL and HDL observed by Feldman et al.[85] resembles that reported by others in the rat.[38,90] The variation in apoprotein composition of VLDL and HDL with diet suggested that the increase in apoA-I is related to increased fat absorption with the diets high in unsaturated fatty acids, which were absorbed better. In the rat, the intestine contributes more A-apoproteins to plasma than apoB or C apoproteins,[91] and increased apoA-I levels in the proximal intestine have been related to rapid absorption of fat. It should be noted that cooling of lipoprotein fractions with high concentrations of saturated fatty acids may influence their apoprotein composition (see above). The specifics of the lipid and apoprotein content of these samples are detailed elsewhere.[92]

Table 2
COMPOSITION OF APOPROTEINS OF
INTESTINAL LYMPH VLDL AND HDL
FROM GROUPS OF RATS FED DIETS
CONTAINING FATS OF DIFFERENT DEGREES
OF SATURATION[85]

Lipoprotein	Diet group % saturates	Apoprotein[a] (% optical density)			
		A-I	A-IV	C's	E
VLDL	78	24	12	52	12
	68	43	10	26	22
	48	59	4	22	15
	38	48	24	20	8
HDL	78	50	12	29	3
	68	66		20	14
	48	66	34		
	38	76	6	12	6

Apoproteins were separated by SDS-PAGE and scanned by
densitometry of stained gels. Average of determinations on two
collections each of two or three rats. Lymph samples were collected and/or stored at 4°C prior to ultracentrifugation.

Feldman et al.[93] have compared in rats the effect of different levels of dietary fatty acids on the formation of intestinal lipoproteins. Four groups of animals were fed diets with fats providing 75% of fatty acids as 16:0, 18:0, 18:1, or 18:2. The fatty acid patterns of the triacylglycerol-rich lipoproteins mirrored the diet. Exogenous cholesterol was recovered primarily in chylomicrons, except with linoleate. Cholesterol and triacylglycerol absorptions were correlated significantly and were less with the saturated fatty acid diets. Within each group the distribution of apoproteins was similar. In VLDL from rats fed the 16:0 and 18:2 diets C-apoproteins were the major components, with substantial amounts of apoA-I, apoA-IV, and apoE. In VLDL from rats fed the 18:0 diet, apoA-I was the major apoprotein with very little apoC and apoE. The apoprotein composition of lymph VLDL resembled that reported for lymph chylomicrons.[38] In HDL apoA-I was the major component with lesser amounts of apoA-IV, apoE, and apoC. There was an increase in the proportion of apoA-I to apoC in HDL with the unsaturated fat diet. The apoprotein B composition was not assessed. Clearly, a diet enriched in one specific acid has its unique effects on lymph lipoprotein formation, presumably affecting some intestinal subcellular mechanisms.

C. Cholesterolemic Effect

Feldman et al.[85] have compared in rats the effect of fat diets enriched in specific fatty acids on the lymphatic absorption of radioactive cholesterol given as a gavage. The percent of radiolabeled cholesterol absorbed into lymph was less with stearate and palmitate and greater with oleate and linoleate feeding. Radiolabeled cholesterol was recovered primarily in lymph chylomicrons with the safflower oil diet, but more in lymph VLDL with oleate and the saturated fatty acid diets. The lymph cholesterol content was greater with the unsaturated fatty acid diets compared to saturated fat diets. With the saturated fat diets cholesterol was found primarily in lymph VLDL. Plasma cholesterol levels were lower with 16:0 diets than with 18:0 or 18:2 diets, which gave comparable levels. There was also a significant increase in plasma HDL-cholesterol with the stearate diet. This increment was explained by the greater availability of cholesterol from the increased surface of the smaller chylomicron and VLDL particles. In previous studies with homogenous triacylglycerol feeding, Feldman et al.[94] have demonstrated that plasma cholesterol is highest in rats fed trilauroylglycerol. Hegsted et al.[95] had earlier shown in studies with individual fatty acids added to semipurified diets that myristic acid has a potent hypercholesterolemic effect, stearic acid has a limited effect, and palmitic acid is intermediate.

IV. UNSATURATED FATS

Several studies have demonstrated that polyunsaturated fatty acids derived from vegetable oils (principally, the ω-6 linoleic acid) will reduce plasma cholesterol levels in normal, healthy volunteers.[75,82] There is little effect of ω-6 polyunsaturated fatty acids upon plasma triacylglycerol and VLDL levels, although these data are controversial.[2] There has been an increased interest in the metabolism of ω-3 polyunsaturated fatty acids since the work of Bang and Dyerberg,[96] which implied a link between the ingestion of marine oils containing these fatty acids and the low rates of ischemic heart disease observed in Greenland Eskimos. This interest has been heightened by the finding of an inverse relation between fish consumption and 20-year mortality from coronary heart disease,[97] and the dramatic reduction of plasma lipids, lipoproteins, and apoproteins by dietary fish oils in patients with hypertriglyceridemia.[98] Furthermore, diets enriched with fish oil-derived ω-3 polyunsaturated fatty acids may exert anti-inflammatory effects by inhibiting the 5-lipoxygenase pathway in neutrophils and monocytes and inhibiting the leukotriene B-4-mediated functions of neutrophils.[99] Concern about potential adverse effects of long-term consumption of the ω-6 and ω-3 polyunsaturated fatty acids has led some investigators to examine the possibility of replacing saturated fatty acids with the ω-9 monounsaturated fatty acids in the diet.[100] The latter have been consumed in large quantities in the Mediterranean region where olive oil is used and where coronary heart disease is relatively low.[101] These acids are normally synthesized by the body and are less susceptible to oxidation. They are commonly thought of as neutral in their action on plasma cholesterol levels.

The mechanisms by which polyunsaturated ω-6 fatty acids exert their hypolipidemic effects have been extensively examined, but to date no single mechanism has been found to be operative in all instances.[102,103] A mechanism of interest to the present review is the possibility of decreased cholesterol absorption and changes in the fatty acid composition of membrane lipids, which may affect lipoprotein metabolism, including enzyme activity, and binding to cellular receptors. Although there is an enormous amount of literature on the various effects of the ω-6 fatty acids on plasma lipids and lipoproteins, only a few attempts have been made to assess the effect of dietary ω-6 fatty acids on the composition and structure of lymph lipoproteins. The lack of information on the intestinal absorption and lymphatic transport of polyunsaturated fatty acids is even greater for the ω-3 fatty acids. In the marine oils, the polyunsaturated fatty acids of the ω-3 type, such as 20:5 and 22:6, are largely found in triacylglycerols. There is evidence that ingestion of these oils does not result in a typical fat tolerance curve in man,[104] nor do the ω-3 fatty acids accumulate rapidly in tissues.[2] In vitro studies have demonstrated an ineffective lipolysis of these oils[105,106] and their resistance to chemical hydrolysis and derivatization of the ω-3 fatty acids.[106,107] Also, it has been suggested that the absorbability and route of transport (lymphatic or portal) of the unesterified polyunsaturated fatty acids may be dissimilar to that of typical fatty acids, such as the long-chain saturated and the ω-9 monounsaturated fatty acids.[1,108]

A. Lipid Composition and Particle Size

Feldman et al.[93] fed four groups of rats test diets providing 75% of fatty acids as palmitate, stearate, oleate (ω-9), and linoleate (ω-6), which were added to safflower oil as triacylglycerols. In addition, labeled triacylglycerols were given as tracers. In all instances, the fatty acid composition of triacylglycerols of whole lymph, chylomicrons, and VLDL resembled the dietary triacylglycerols fed. The closest resemblance between the dietary and absorbed fats was noted for the unsaturated diets. With the oleate diet radiolabeled triacylglycerol was absorbed predominantly into lymph chylomicrons, whereas with the saturated diets radiolabeled triacylglycerols appeared more in lymph VLDL. The lymph lipid content was greater with the oleic acid diet compared to saturated fat diets (linoleic acid effect was not determined). Chylomicron size increased as the chylomicrons became less saturated and the lipid content increased. Chylomicron diameters were largest with oleate, intermediate with linoleate, and smallest with the saturated fat diets. Larger chylomicrons were associated with smaller VLDL and HDL particles. Approximately 25% of the chylomicrons and VLDL of rats fed the saturated fat diets were polygonal rather than round. These irregular shapes were not observed in the triacylglycerol-rich lipoproteins from the oleic and linoleic acid diets. Chylomicron size was correlated significantly with radioactive triacylglycerol absorption and with triacylglycerol content in the lymph collections from 0 to 8 and 8 to 24 hr. There was a significant negative correlation between radiolabeled triacylglycerol absorption and VLDL size.

Chen et al.[109] have investigated the rate and extent of lymphatic absorption of unesterified ω-3 eicosapentaenoic acid compared to that of ω-9 octadecenoic (oleic) acid and ω-6 eicosatetraenoic (arachidonic) acid. The absorption of both the ω-6 and the ω-3 acids was as efficient as that of the ω-9 acid. The distribution of each fatty acid among the major lymph lipoproteins was similar, with 93 to 95% recovered in chylomicrons and VLDL. Distributions of absorbed cholesterol among the major lipoproteins of lymph showed minor differences. For oleic acid there was a smaller percentage of absorbed cholesterol associated with lymph chylomicrons in the rat and a greater recovery in the density (d) greater than 1.006 g/mℓ lipoproteins (LDL + HDL) compared to absorbed fatty acids. Similar results were obtained for cholesterol distributions when the infusion contained the ω-6 acid, but were less pronounced with animals receiving the ω-3 acid. The differences in cholesterol distributions among the lipoproteins for the free fatty acid groups were not of sufficient magnitude to suggest major differences in absorption and transport characteristics for cholesterol. There was a somewhat greater incorporation of the polyunsaturated fatty acids into phospholipids and phospholipid precursors in both chylomicrons and VLDL fractions, but these levels do not reflect the ultimate tissue lipid distributions of either the ω-6 or the ω-3 polyunsaturated fatty acids. It is not known whether chylomicrons or VLDL enriched with the long-chain highly unsaturated ω-3 fatty acids are larger than those carrying the more usual fatty acids.[2] When relatively small amounts of fish oils were fed to subjects on otherwise uncontrolled diets, there appeared to be a relationship between decreased plasma triacylglycerol levels and the in vitro lipolysis by lipoprotein lipase,[110] which might be taken as indirect evidence for a somewhat larger particle size of the plasma chylomicrons containing the ω-3 polyunsaturated fatty acids. Questions regarding the digestibility and absorption of oils rich in ω-3 polyunsaturated fatty acids and the metabolic fate of the resulting chylomicrons have yet to be addressed.[111]

B. Apoprotein Composition

Qualitative assessments of the apoprotein composition of lymph VLDL have been made by Riley et al.,[41] who fed 10% olive oil to rats. ApoA-I, apoA-IV, apoC, and apoE in decreasing order were found to be present as major components. Yang and

Kuksis[112] found a comparable composition of apolipoproteins for chylomicrons derived from methyl oleate feeding to rats.

Feldman et al.[93] have compared the apoprotein composition of lymph VLDL and HDL samples from rats ingesting the 16:0, 18:0, and 18:2 diets. Within each group the distribution of apoproteins was similar in the collections after the gavage or while ingesting the diet. In VLDL from rats fed the 16:0 and 18:2 diets C-apoproteins were the major components, with substantial amounts of apoA-I, apoA-IV, and apoE. In VLDL from rats fed the 18:0 diet, apoA-I was the major apoprotein with very little apoC and apoE. In HDL apoA-I was the major component with lesser amounts of apoA-IV, apoE, and apoC. There was an increase in the proportion of apoA-I to apoC in HDL with the 18:2 diet. The apoprotein composition of lymph VLDL resembled that reported for lymph chylomicrons. All diets gave discoidal HDL particles indicating intestinal synthesis. Feldman et al.[93] did not determine the apoprotein composition of the lymph lipoproteins from the oleic acid-fed rats.

There have been no studies on the effect of the ω-3 polyunsaturated fatty acids upon the apoprotein profiles of lymph or plasma lipoproteins. Reductions in the concentrations of plasma triacylglycerols and VLDL may occur from lower rates of synthesis of the apoB moieties of VLDL or may result from an increased rate of clearance of VLDL from the plasma. This could occur as a result of changes in the apoprotein content or composition of the lymph chylomicrons and VLDL.

C. Cholesterolemic Effect

Feldman et al.[93] compared the relative rates of cholesterol absorption when given as a gavage along with fat diets enriched in specific fatty acids. The percent of radiolabeled cholesterol absorbed into lymph was less with stearate and palmitate diets, and greater with oleate and linoleate feeding. Radiolabeled cholesterol was recovered primarily in lymph chylomicrons with the safflower diet and in lymph VLDL with the oleate and saturated fatty acid diets. The absorption of radiolabeled cholesterol was correlated significantly with the triacylglycerol content of lymph. Furthermore, in comparison with samples from rats that continued to be fed the stock diet (less than 5% fat) plasma cholesterol increased significantly by 50 to 75% with the 18:0, 18:1, and 18:2 diets. Plasma cholesterol levels were greatest with the 18:1 diet. Plasma triacylglycerols were significantly greater with the high palmitate and high oleate diets compared to stearate and linoleate. Plasma HDL cholesterol was significantly less with high palmitate diet compared to the other diets.

Becker et al.[113] compared the effects of saturated, monounsaturated (ω-9), and polyunsaturated (ω-6) fatty acids on levels of lipids and lipoproteins in young men. Both types of unsaturated acids, but especially polyunsaturated acids, were found to lower plasma total cholesterol and LDL cholesterol compared to saturated fatty acids. Mattson and Grundy[100] have conducted similar experiments with dietary saturated, monounsaturated (ω-9), and polyunsaturated (ω-6) fatty acids on plasma lipids and lipoproteins in man. The results show that oleic acid is as effective as linoleic acid in lowering LDL levels in normotriglyceridemic patients, but seemingly reduces HDL-cholesterol levels less frequently than does linoleic acid. Neither type of unsaturated fat had striking effects on lipoprotein levels of hypertriglyceridemic patients. In contrast to the results of Becker et al.,[113] the polyunsaturated fatty acids failed to lower plasma total and LDL cholesterol.

Feldman et al.[93] have compared the effect of feeding rats with fat diets enriched in specific fatty acids on the absorption of radioactive cholesterol administered by a gavage. The percent absorption of radiolabeled cholesterol with linoleate was similar to that obtained with oleate, and nearly double that observed with palmitate and stearate.

Plasma cholesterol levels were higher with oleate than with linoleate diet, which gave the same plasma total cholesterol level as the stearate diet. Palmitate diet resulted in the lowest plasma total cholesterol levels. In contrast, plasma triacylglycerols were significantly greater with the high palmitate and high oleate diets compared to stearate and linoleate. The lowest LDL/HDL cholesterol ratios were observed with the saturated fat diets; HDL cholesterol was relatively enriched compared to the level in rats fed the stock diet with all the high-fat diet groups except 18:2. These results are in agreement with previous work, which has shown that cholesterol absorption is increased with polyunsaturated oils high in linoleate in contrast to lesser sterol absorption with saturated fat.[114,115] Thompson et al.[116] showed that linoleic or palmitic acid perfusion increased the specific activity of lymph-free cholesterol, which was linearly correlated with a rise in lymph triacylglycerols. Thompson et al.[116] concluded that these fatty acids increased absorption of cholesterol via transport mechanisms within the intestinal wall. Uptake of cholesterol into the intestinal wall was not affected significantly.

Becker et al.[113] have compared the effects of saturated, ω-9 monounsaturated, and ω-6 polyunsaturated fatty acids on plasma lipids and lipoproteins in man. In this study 12 men consumed three different cholesterol-free formula diets in which ω-6 polyunsaturated fatty acids and saturated fats were substituted at 20% level for monoenoic fats with a constant of 40% calories from fat. The results showed that changes in dietary fat affect serum lipids, lipoproteins, and apoproteins even when consumed on a cholesterol-free diet. The ω-6 polyunsaturated fat lowered LDL cholesterol, total cholesterol, and apoB to a greater extent than monounsaturated ω-9 or saturated fat. HDL cholesterol changed very little from baseline values with any of the diets.

Baudet et al.[117] have reported the chemical composition and metabolism of human LDL and HDL following consumption of saturated (palm oil and milkfats) and ω-6 polyunsaturated fats (sunflower and peanut oil). Whereas little or no modification in the concentration and composition of lipoproteins was observed with the ω-6 polyunsaturated fats, a large increase in serum level of all lipoproteins was obtained with the milkfat diet, as reported previously by numerous investigators.[1,3,118,119] The ω-6 fatty acid diet (sunflower oil) yielded the lowest protein/lipid ratio for plasma VLDL, when compared to palm oil and peanut oil diets. With milkfat diet this ratio increased. These results are in disagreement with those of Vega et al.[118] and Kuksis et al.,[119] who found a decrease in the concentration of each component of VLDL with an ω-6 polyunsaturated fat diet as well as no change in their protein/lipid ratio. Changes observed for the LDL with the different diets in the three groups of subjects were similar to those observed for VLDL. In contrast to the results of Vega et al.[118] and Kuksis et al.,[119] a large decrease in the protein/lipid ratio was observed with the most saturated diet; a twofold decrease of this ratio was observed with the milkfat diet. In HDL a slight increase in proteins accompanied by a decrease in esterified cholesterol as compared with the ω-6 fat and palm oil diets was observed. Kuksis et al.[119] and Shepherd et al.[120] found a large decrease of the protein/lipid ratio with the ω-6 polyunsaturated fats. In regard to the apoproteins, the only important change was a large increase in apoA-I of HDL with the most saturated diets (palm oil and milkfats). Shepherd et al.[120] and Kuksis et al.[119] also observed a decrease in apoA-I with the ω-6 polyunsaturated fat diet.

Mattson and Grundy[100] have compared the effects of dietary saturated (palm oil), ω-9 monounsaturated (high oleic safflower oil), and ω-6 polyunsaturated (high linoleic safflower oil) fatty acids on plasma lipids and lipoproteins in man when supplied at 40% of total calories. After 3 to 4 weeks, ω-9 and ω-6 fatty acid diets caused statistically significant equal lowerings of plasma LDL-cholesterol, but the ω-6 polyunsaturate diet lowered HDL-cholesterol levels more frequently than did the ω-9 fatty acid diet. Neither diet changed the level of plasma triacylglycerols. The proportions of total protein and the various lipid components in isolated fractions of VLDL, IDL, LDL, and HDL were not altered by the two diets. HDL-cholesterol concentrations were low on the saturated fatty acid diet and were unaffected by either ω-9 monounsaturated or the ω-6 polyunsaturated fat diet. The decrease in plasma HDL-cholesterol noted on the ω-6 polyunsaturated fat diet is in contradiction to the report of Becker et al.,[113] who failed to demonstrate such a decrease when polyunsaturated fatty acids were substituted for saturates. Mattson and Grundy[100] have suggested that this discrepancy may reflect the way in which their data were presented rather than real differences. Apparently, the differences are best seen when the means are examined for individual changes (by the paired *t* test). Group means are less effective because of large between-individual variation.

Chen et al. compared the appearance of radioactive cholesterol in the lymph of rats given a single gastric dose of an aqueous emulsion containing radioactive oleic (ω-9), arachidonic (ω-6), and eicosapentanoic (ω-3) acids.[109] The absorption of cholesterol from each fatty acid medium was quantitatively equivalent. In earlier studies, Peifer et al.[121,122] have compared the highly unsaturated ω-3 fatty acids of tuna and menhaden oil to the ethyl esters of linolenic (ω-3), linoleic (ω-6), oleic (ω-9), and palmitic acids in hypercholesterolemic rats. Interestingly, both fish oils and the linolenic acid produced equivalent cholesterol lowering. Linoleic acid was the next most effective, followed by oleate and palmitate. In studies with hypercholesterolemic chickens the fish oils were more hypocholesterolemic than either linolenic or linoleic acids.[123,124] Thus, in some species, different chain length acids of the ω-3 family may be equally hypercholesterolemic, while in other species the fatty acids with the longest chain length are more hypocholesterolemic.[125]

V. DIETARY CHOLESTEROL

Stange and Dietschy[126] have investigated the origin of cholesterol in the mesenteric lymph of the rat by quantitating the amounts of total and newly synthesized cholesterol secreted into the mesenteric lymph of the rat. It was concluded that in the absence of fat absorption, sterol synthesized in the intestinal mucosa was incorporated predominantly into cell membranes and did not enter intestinal lymph to any significant degree. However, during fat absorption, a fraction of this newly formed sterol pool was incorporated into lipoproteins and was delivered through intestinal lymph to the body pools of cholesterol. Such newly synthesized cholesterol was found predominantly in the free form and accounted for 27% of the total sterol found in lymph at the end of the experiment. Stange et al.[127] have shown that feeding corn oil to rats results in a pronounced stimulation of sterol synthesis in the mid and lower villus of the jejunum. The infusion of corn oil[126] used in the present study had an even greater effect upon local cholesterol synthesis.

A. Lipid Composition and Particle Size

Cholesterol feeding has been shown to produce abnormal plasma lipoproteins in a variety of experimental animals[128] and man.[129] Riley et al.[41] have examined the role of the intestine in the pathogenesis of plasma lipoprotein changes induced by chronic cholesterol feeding in the rat by analyzing the composition of the mesenteric lymph lipoproteins of animals fed a diet containing 1% cholesterol and 10% olive oil. The studies showed that chronic cholesterol feeding in the rat results in significant changes in apoprotein and lipoprotein secretion by the intestine. Specifically, this was associated with the appearance in mesenteric lymph of lipoproteins enriched in cholesterol.

In addition, intermediate density particles (d = 1.006 to 1.030 g/ml) were prominent and carried 19% of total lymph cholesterol. This lipoprotein subfraction was rich in apoA-I compared to plasma IDL, where apoE was the major apoprotein. Control animals chronically fed 10% olive oil had similar apoA-I, cholesterol, and triacylglycerol outputs in mesenteric lymph, and similar apoA-I and cholesterol distribution in lipoprotein subfractions compared to the chow-fed controls. While all groups elevated the lymph triacylglycerols to similar levels with lipid feeding, there was a marked reduction in hourly lymph cholesterol and apoA-I secretion in chronically cholesterol-fed animals compared to both sets of controls. There was a marked decrease in apoA-I secretion in the lymph d less than 1.006 g/ml lipoproteins, as well as a marked decrease in total lymph apoA-I secretion in the chronically cholesterol-fed rat, and it has been suggested[41] that this may be in part responsible for the decreased HDL levels observed in the chronically cholesterol-fed rat.

B. Apoprotein Composition

Swift et al.[130] have characterized the nascent lipoproteins retrieved from Golgi apparatus-rich fractions of intestinal epithelial cells from chow-fed control and diet-induced hypercholesterolemic rats. It was shown that the intestine of the fasted control rat synthesizes a triacylglycerol-rich VLDL containing only apoproteins B-240, A-IV, and A-I. In addition, it was demonstrated that the intestine of the hypercholesterolemic rat synthesizes a cholesteryl ester-enriched VLDL that contains relatively more apoA-IV and less apoB-240 compared with Golgi VLDL as determined by radioisotope incorporation studies. These particles contained less triacylglycerol and more cholesterol compared to control Golgi VLDL. These findings of differences in the nascent Golgi VLDL from hypercholesterolemic and control rats differ from the results reported by Riley et al.,[41] who studied mesenteric lymph lipoproteins in rats made hypercholesterolemic by feeding diets containing 1% cholesterol and 10% olive oil. They found no differences in lipid composition of lymph VLDL from cholesterol-fed and control animals. Furthermore, after infusion of olive oil and cholesterol, the relative composition of d less than 1.006 g/ml lymph lipoproteins (VLDL and chylomicrons) from cholesterol-fed and control rats was comparable. Swift et al.[130] have suggested that the differences may be the result of the diets used. The diet used by Swift et al.[130] contained taurocholate to facilitate intestinal absorption of fat, and propylthiouracil, which produces higher serum cholesterol levels than the diets used by Riley et al.[41] The differences in the two studies may also be related to the dilution of the lymph lipoproteins by plasma lipoproteins in the study of Riley et al.[41] Approximately 60% of the total apoprotein radioactivity was found in apoA-I in both preparations.[130] However, the qualitative and quantitative significance of these alterations with regard to the marked hypercholesterolemia observed in these animals remains to be determined.

It should be noted that the apoprotein composition of Golgi VLDL from control rats as determined by the radioisotope incorporation studies by Swift et al.[130] is quite different from the apoprotein composition of lymph VLDL reported previously.[38,42,131]

Sloop et al.[132] have determined the distribution, chemical, and apoprotein composition of plasma and peripheral lymph lipoproteins in control and cholesterol-fed dogs. In both groups of animals, the agarose electrophoretic patterns of plasma and lymph lipoproteins were similar. In hypercholesterolemic dogs, β-VLDL, β-migrating intermediate density lipoprotein, and HDLc were major components both in plasma and lymph. On sodium dodecyl sulfate-polyacrylamide gel electrophoresis (SDS-PAGE), the apolipoprotein composition of lymph lipoproteins was similar to their plasma counterparts with the exception of HDL, especially in the cholesterol-fed animals. In plasma HDL, only apoA-I could be detected, whereas in peripheral lymph HDL, a significant enrichment in apoE and apoA-IV was observed. A significant proportion of lymph HDL obtained from cholesterol-fed animals was in the form of disc-shaped particles stacked in roleau structures. It was concluded that these particles are assembled *de novo* in the periphery as a crucial stage of reverse cholesterol transport to the liver.

C. Cholesterolemic Effect

The effect of two different levels of dietary cholesterol (0.16 mg/kcal and 0.79 mf/cal) on the composition of thoracic duct lymph lipoproteins isolated from nonhuman primates has been studied by Klein and Rudel,[133] with the diets being infused intraduodenally. No dietary cholesterol-induced differences in the apoprotein patterns of lymph chylomicrons and VLDL were apparent. The apoprotein distributions of chylomicrons and VLDL were qualitatively similar during infusions of both diets. The higher level of dietary cholesterol resulted in an increase in the amount of cholesteryl ester in lymph chylomicrons and VLDL.

Numerous investigations have established that diets high in saturated or animal fats increase levels of circulating cholesterol, while the polyunsaturated vegetable or fish oils have an opposite effect.[1-4] There is general agreement that increased dietary intakes of polyunsaturated vegetable oils lower the concentrations of plasma LDL cholesterol.[1,2,4] Recent studies lend support to the claim that no change in plasma HDL levels occurs during polyunsaturated fat feeding.[1,113] Dietary fats vary in their digestion and absorption. Variations in lipid solubilization and absorption influence cholesterol homeostasis. Other possible mechanisms for these phenomena include changes in formation and secretion of lipoproteins with specific lipid and apoprotein composition.

The relative importance of the saturated fatty acids in cholesterol homeostasis varies with the level of fat consumption and the dietary intake of cholesterol.[134,135] A relationship exists between test period endogenous cholesterol flux and exogenous cholesterol load. Thus, Sylven and Borgstrom[134] found that the contribution of endogenous cholesterol to lymph total cholesterol varied substantially with dietary load in lymph duct-cannulated rats given pulse doses of trioleoylglycerol containing increasing amounts of dietary cholesterol. Klein and Rudel[135] have also found that absorption of dietary cholesterol leads to a decrease in endogenous cholesterol flux in thoracic duct lymph of monkeys. As pointed out in Section III, the studies of Green and Green[88] have shown that the lymph cholesterol flux during infusion of oils of various degrees of unsaturation only averaged 3.2 μmol/hr and was not significantly influenced by dietary fat saturation or triacylglycerol load. The initial rapid rise in lymph cholesterol after addition of cholesterol (mass + tracer) to the oils was due primarily to endogenous cholesterol. Thus, absorption of dietary cholesterol resulted in an initial displacement of endogenous cholesterol into the lymph. However, in samples collected 18 to 24 hr after addition of cholesterol to the oil, 87 to 100% of the increased lymph cholesterol flux was exogenous (labeled). This increased flux was not significantly influenced by fat saturation of the infused oil and averaged 5.6 μmol/hr for rats in the low oil/lower cholesterol infusion groups, and 9.5 μmol/hr for those in the higher infusion groups. The percent esterified cholesterol in lymph of rats in the low oil infusion groups was significantly higher in the S vs. P animals during both infusion of oil and of oil + cholesterol.

An effect of dietary cholesterol on endogenous cholesterol flux is related to the disagreement between chemical and isotopic estimates of cholesterol absorption reported by Green and Green[88] and by Green.[136] Valid use of isotopic exogenous cholesterol as a monitor of the effect of dietary cholesterol on total intestinal cholesterol input requires that endogenous cholesterol input not be influenced by dietary load. The data of Green and Green[88] and of Green[136] demonstrate that the use of isotopic cholesterol to measure cholesterol absorption is limited even in carefully balanced studies and show that disagreement between isotope and mass data may depend on experimental conditions.

VI. METABOLIC PATHWAYS OF TRIACYLGLYCEROL BIOSYNTHESIS

The stereochemical course of triacylglycerol biosynthesis and the intramolecular structure of dietary fat may have implications for chylomicron formation, secretion, and subsequent metabolism. The positional distribution and molecular association of the fatty acids in the triacylglycerols affect the rate of hydrolysis by pancreatic lipase[137] and determine the nature of the fatty acid composition of the monoacylglycerols and free fatty acids that become available for absorption and for resynthesis into chylomicron triacylglycerols. The relative proportions of the short- and long-chain fatty acids released from the dietary triacylglycerols determine the composition of the fat being absorbed via the portal and the lymphatic routes.[46] Furthermore, the amounts of the monoacylglycerols and the free fatty acids entering the villus cells determine the relative proportions of the triacylglycerols that are resynthesized via the monoacylglycerol and the phosphatidic acid pathways, as well as the overall rate of fat absorption.[138-140] There appears to be no inhibition of the phosphatidic acid pathway by 2-monoacylglycerols in vivo,[138-141] contrary to an earlier claim based on in vitro work.[142] Using the incorporation of glucose into lipid glycerol as an estimate of phosphatidic acid pathway, it was found that this pathway is stimulated by both fat and carbohydrate feeding.[141] Likewise, the positional distribution and molecular association of the fatty acids in the dietary glycerophospholipids are known to vary with the dietary fat. Since much of the luminal lysophospholipid becomes absorbed, reacylated, and incorporated into the chylomicron surface monolayer,[18,143] it would be expected to exert different metabolic effects depending on the diet. There have been very few experiments to examine these theoretical possibilities experimentally.

A. Triacylglycerol Structure and Particle Size

Under normal conditions the bulk of the dietary fat is resynthesized via the 2-monoacylglycerol pathway and leads to an essentially random reacylation of the *sn*-1- and *sn*-3-positions with the common 18:1 and 18:2 fatty acids.[144] There is evidence that this reacylation proceeds preferentially through the *sn*-1,2-diacylglycerol intermediates,[145,146] but the *sn*-2,3-diacylglycerols have also been isolated[147,148] and shown[149] to serve as acceptors of acyl groups in vitro. As a result, the chylomicron triacylglycerol structure would be expected to differ significantly from that of the dietary fat. However, certain dietary fats of very pronounced enantiomeric nature have been shown to retain much of their original structure during the digestion and absorption process. Thus, mustard seed oil, which contains much of the monounsaturated 22:1 (erucic) acid in the *sn*-3-position, retains it largely in the *sn*-3-position following digestion and absorption[150] and after incorporation into heart triacylglycerols.[151] Table 3 shows the positional distribution of the fatty acids among the three positions of the glycerol molecule in the dietary and chylomicron triacylglycerols following feeding mustard seed oil to rats.[150] Although the biochemical basis of this specificity is not known, it may be rationalized that the *sn*-3-position of the triacylglycerol molecule becomes acylated last and receives the fatty acids which are least readily activated and utilized. Alternatively, it is possible that the *sn*-3-monoacylglycerols have been absorbed and have served as effective substrates for triacylglycerol biosynthesis, although in vitro experiments have

Table 3

POSITIONAL DISTRIBUTION OF FATTY ACIDS IN THE
TRIACYLGLYCEROLS OF DIETARY MUSTARD SEED OIL
AND IN LYMPH CHYLOMICRONS[150]

Fatty acids[a]	Mustard seed oil (mol %)			Chylomicrons (mol %)		
	Pos. 1[b]	Pos. 2	Pos. 3	Pos. 1	Pos. 2	Pos. 3
14:0				0.23	0.50	0.13
15:0				0.11	0.12	0.06
16:0	4.33	1.99	2.41	8.91	8.35	4.64
16:1ω7	0.12	0.32	—	0.95	0.94	0.90
17:0				0.20	0.10	0.20
18:0	1.87	1.28	0.29	3.90	0.91	1.99
18:1ω9	25.88	38.42	5.3	24.98	32.96	22.20
18:2ω6	10.66	32.26	2.78	13.31	31.04	13.71
18:3ω3	4.42	27.74	6.96	8.88	17.54	11.08
20:0	0.61	0.24	1.57	0.91	0.16	0.82
20:1ω9	14.33	0.20	18.31	14.07	2.86	11.94
20:4ω6				0.45	0.95	1.45
22:0				0.34	0.27	0.40
22:1ω9	36.11	0.50	59.03	21.37	3.00	27.38
22:6ω3				0.33	0.22	0.94
24:1ω9	1.86	—	3.34	1.06	0.08	2.16

[a]   Number of carbon atoms/number of double bonds. Omega values identify the
      location of the last double bond from the methyl terminal.
[b]   Stereospecific numbering of glycerol carbons.

indicated[152] that sn-3-monoacylglycerols are not utilized. In contrast, bovine milk fat triacylglycerols, which contain butyric and other short-chain fatty acids in the sn-3-position, are resynthesized largely into symmetrical triacylglycerols[153] because the short-chain fatty acids are not activated and become transported to the liver in the free form via the portal route. However, short-chain fatty acids located at the 2-position of a dietary triacylglycerol may survive digestion and may become absorbed and reacylated to triacylglycerols appearing in the lymph as a combination of one short-chain and two long-chain fatty acids. Figure 1 shows the carbon number distribution of the triacylglycerols recovered from the ascitic fluid of a baby receiving medium-chain length triacylglycerols.[154] Four distinct populations of triacylglycerols can be recognized from the maxima in the elution pattern: three short-chain acids, two short-chain and one long-chain acid, one short-chain and two long-chain acids, and three long-chain acids. Since the short-chain acids are not activated in the intestinal mucosa, the triacylglycerols containing short-chain fatty acids must have been absorbed intact, as have some diacylglycerols and monoacylglycerols, which then became acylated by long-chain fatty acids. If esterified at the 2-position of the glycerol molecule, stearic acid is absorbed better than free stearate.[155] Filer et al.[156] have shown that natural lard, which contains much of its palmitate in the 2-position, is more readily absorbed than randomized lard, in which the fatty acids are distributed equally among the three glycerol positions. They have attributed this effect to the greater extent of micellarization of the fat due to the yield of greater amounts of 2-palmitoylglycerol from lipolysis of the native lard. In contrast, behenic acid was not well absorbed either as the 2-monoacylglycerol or the free acid.[157] From micellar solutions free stearic acid is absorbed at a rate comparable to that of palmitic, oleic, and linoleic acids.[158]

When free fatty acids or their methyl or ethyl esters are fed in the absence of 2-monoacylglycerols, the chylomicron triacylglycerols are formed via the phosphatidic

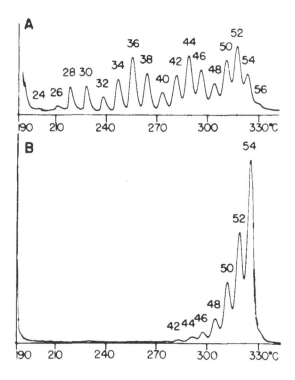

FIGURE 1. GLC resolution of triacylglycerols from chylous ascites fluid of an infant. (A) After feeding medium-chain triacylglycerols; (B) after feeding trioleoylglycerol. Peaks identified by total
number of acyl carbon atoms. Instrument, Beckman GC-4 gas
chromatograph with special on-column inlet heater; column,
stainless steel tube, 50 cm × 2 mm i. d., packed with 3% w/w JXR
on 100 to 120 mesh Gas Chrom Q. Nitrogen flow, 120 mℓ/min.
On column injector, 280°C; detector, 340°C; column temperature
programmed as shown. Sample, 1 μℓ of 1% solution of triacylglycerols in chloroform.[154] Attenuation 5 × 10².

acid pathway, which is characterized by the preferential incorporation of the saturated
fatty acids in the *sn*-1-position and of the unsaturated fatty acids in the *sn*-2-posi-
tion.[153] The *sn*-3-position of the triacylglycerol molecule is apparently filled by polyunsaturated and long-chain fatty acids which are less readily utilized in phosphatidic acid
biosynthesis. Aside from yielding enantiomeric triacylglycerols, this pathway is much
less efficient and results in considerably slower uptake of dietary fatty acids and in
formation of chylomicrons of a significantly smaller diameter than those resulting
from the operation of the monoacylglycerol pathway utilizing the same fatty acids.
Yang and Kuksis[14,112] have demonstrated that feeding corn oil fatty acids as the methyl
esters yields chylomicrons about one half the size of those resulting from corn oil
feeding. The chylomicrons resulting from the feeding of the methyl esters of teaseed
oil, which are made up largely of oleate, were also smaller than those resulting from
corn oil feeding, and were similar to those resulting from feeding the corn oil fatty acid
methyl esters.[14,112]

B. Apoprotein Composition

Yang and Kuksis[14,112] have compared the apoprotein composition of the chylomicrons containing triacylglycerols formed via the phosphatidic acid (fatty acid methyl
esters) and the monoacylglycerol (corn oil) pathway in the rat. No difference was ob-

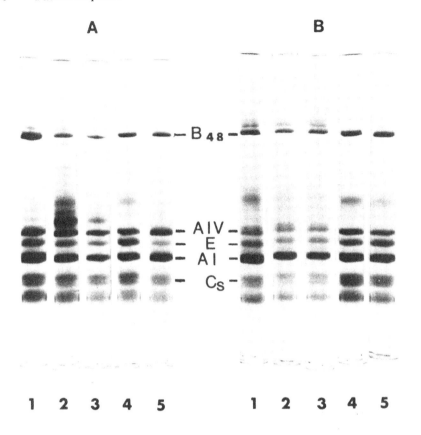

FIGURE 2.   SDS-PAGE resolution of apoproteins of lymph chylomicrons as obtained from rats fed (A) corn oil or (B) corn oil fatty acid methyl esters. Numbers 1 to 5 represent the initial 24-hr samples from each rat. 4% polyacrylamide SDS gels; Coomassie brilliant blue staining. The apoprotein mixtures were run in parallel with appropriate molecular weight standards.[14,122]

served in the proportions of the different apoprotein classes (Figure 2). It was calculated that in both instances only one molecule of apoprotein B occurred per particle. However, because of the smaller size, the chylomicrons arising from the phosphatidic acid pathway carried into the bloodstream twice as much apoprotein B and other apolipoproteins than did those arising via the monoacylglycerol pathway. Interestingly, freshly collected and worked-up chylomicrons appeared to contain a proportion of the total apoprotein B in the high molecular weight form (B-100), as previously reported by Lee et al.[30] There was evidence that the larger chylomicrons had a slightly higher proportion of apoprotein A-IV than the smaller chylomicrons.[112] In other studies it was observed that the differences resulting from the feeding of free fatty acids or corn oil to rats produced apoprotein profiles very similar to those arising from mustard seed oil.[14]

Cholesterol feeding results in the appearance of HDLc in plasma[159] and of an HDL fraction rich in cholesterol in lymph.[160] The latter, however, is believed[132,161] to originate largely from tissues other than the intestinal mucosa and may represent the reverse transport of cholesterol from tissues to the liver. Both of these HDL types have been shown to be rich in apoE.[132,159] It is not known if HDL rich in cholesterol is also formed by the villus cells.

C. Cholesterolemic Effect

There have been no systematic studies on the effect of dietary triacylglycerol structure on the absorption and lymphatic transport of cholesterol, although it has been demonstrated that unsaturated fats generally carry more cholesterol than saturated fats in the form of chylomicrons.[94] Furthermore, certain saturated fats contain cholesterol and are more atherogenic than unsaturated fats because of the sterol content. Thus, natural butterfat is more hypercholesterolemic than butterfat distillates, which are low in cholesterol.[162] The hypercholesterolemic effect of natural lard could be attributed to its high content of palmitic acid in the 2-position of the triacylglycerol molecules. Upon lipolysis, the 2-palmitoylglycerol would become available for enhanced absorption of both fatty acids and cholesterol.[156] Randomized lard is less readily absorbed, less hypercholesterolemic, and therefore presumably less atherogenic. Comparative studies of different unsaturated fats have also shown differences in their relative degree of atherogenicity. Yamamoto et al.[163] studied the relative hypocholesterolemic effects of animal and plant fats including lard, soybean oil, and hydrogenated coconut oil. They suggested that the differences observed by them could be due to the positional distribution of essential fatty acids. Essential fatty acids at the 2-position conferred a greater hypocholesterolemic potential than essential fatty acids at the 1- and 3-positions. Kritchevsky et al.[164] have shown that native peanut oil is more atherogenic than interesterified peanut oil. A stereospecific analysis of native and rearranged peanut oils showed significant differences in the positional distribution and molecular association of the long-chain fatty acids (20:0, 22:0, and 24:0) in the two oils.[165,166] As a result of the rearrangement more of the longer-chain acids had become associated with the *sn*-1- and *sn*-2-positions than in the native oil, which contained these acids largely in the *sn*-3-position. The most atherogenic of the native peanut oils contained the highest proportions of linoleic acid in the *sn*-2-position, which in the rearranged oils became distributed randomly among all three positions. The presence of the essential fatty acid (linoleic) in the 2-position appeared to confer a greater hypercholesterolemic or at least atherogenic effect, which was opposite to the effect observed by Yamamoto et al.[163] The significance of the positional distribution of the long-chain saturated and the unsaturated fatty acids in the peanut oils in regard to the mechanism of atherogenesis is not known, but the relative accessibility of the fatty acids from the different positions in the triacylglycerol molecule may have some effect. Thus, lipoprotein lipase has been shown to preferentially attack the *sn*-1-position of triacylglycerols,[167,168] while the hepatic lipase has shown preference for the *sn*-3-position.[168] Consideration should also be given to the possibility of altered distribution of the ω-3 fatty acids (18:3ω3), which might become more readily available as a result of the rearrangement of the fatty acids. Members of the ω-3 family of fatty acids are believed to be potent inhibitors of lipoxygenases.[99] Since the natural and rearranged oils possess identical fatty acid composition, the atherogenicity cannot be attributed to differences in fatty acid composition and must be related to the positional distribution and/or molecular association of the fatty acids. It is therefore possible that the atherogenic effect is less marked with natural fats and oils which possess more random distribution for their fatty acids. In other studies,[169] it has been shown that the molecular species of cocoa butter, which is less atherogenic than some other oils of comparable degree of fatty acid saturation, possess an essentially 1-random-2-random-3-random structure. Independent studies by Hung et al.[170] have shown that the feeding of randomized rapeseed oils containing either high or very low levels of erucic acid results in low growth rates of the animals, which could account for the lower hypercholesterolemia and lesser atherosclerosis.

In all instances the feeding of the atherogenic oils resulted in increased plasma cholesterol and triacylglycerol levels and in increased levels of plasma LDL. Since the structure of the triacylglycerols was not determined, it is not possible to conclude whether or not it possessed any abnormalities. However, structural analyses have been made of plasma triacylglycerols from different states of hyperlipemia[171-173] and marked differences have been noted in the stereochemical distribution of the fatty acids when compared to a normolipemic state. A detailed examination of the structure of the molecular species of the triacylglycerols of plasma chylomicrons and VLDL, however, has revealed that the apparent discrepancies in the fatty acid composition of the three positions are due to differences in the fatty acid composition between the subject groups.[174,175] The positional distribution and molecular association of the fatty acids in these triacylglycerols in all instances could be shown to approximate closely the 1-random 2-random 3-random noncorrelative distribution. It was therefore not possible to attribute either the hyperlipemic or the atherogenic effect to any abnormalities in the molecular structure of the plasma triacylglycerols.

The possibility must be considered that the cholesterolemic and atherogenic effects of the dietary fats are due to abnormalities in the composition, positional distribution, and molecular association of the fatty acids in the glycerophospholipids of the surface monolayer of the chylomicrons and of other lymph or plasma lipoproteins. There is evidence that much of the surfactant glycerophospholipid is derived from dietary and biliary sources by partial hydrolysis in the lumen and a reacylation in the villus cells.[18,143] In this process the fatty acids in the *sn*-2-position of the luminal glycero-phospholipids are removed by pancreatic phospholipase A-2, while the fatty acids in the *sn*-1-position remain unchanged. Following absorption, these lysophospholipids are reacylated largely by dietary fatty acids and are secreted into the lymph chylomicrons and VLDL. The lymph chylomicrons contain significant amounts of both phosphatidylcholine and phosphatidylethanolamine, while the plasma lipoproteins are usually very low in phosphatidylethanolamine. The surfactant composition of the lymph chylomicrons would be expected to vary with the composition and content of the dietary phospholipids. There have been no systematic studies on the effect of the phospholipid class composition of the surface monolayer of the chylomicrons on the apolipoprotein composition of the chylomicrons, although it is known that both particle size and surface composition affect the binding of the apolipoproteins to the lipid particles. Likewise, there have been no systematic studies on the effect of the fatty acid composition of the surface phospholipids on the apoprotein binding to lipid particles, or on the clearance and metabolism of lymph chylomicrons from plasma, although the role of diet fat in subcellular structure and function is no longer disputed.[176] Since the glycerophospholipids of plant origin contain unsaturated and those of animal origin saturated fatty acids in the *sn*-1-position, significant differences would be expected to be manifested in their physicochemical and physiological behavior on the chylomicron surfaces. There have been no systematic studies of these potential differences.

VII. SUMMARY AND CONCLUSIONS

The intestinal villus cell is a major synthetic site of lymph and plasma VLDL in the fasting state and of the chylomicrons and VLDL in the fed state. In addition, a discoid nascent HDL particle is believed to originate in the intestine. Each lipoprotein isolated from intestinal lymph contains a characteristic complement of apoproteins, some of which are synthesized in the gut while others are adsorbed from the lymph serum following secretion of the lipid particles. There is agreement that both apoB and apoA-I are synthesized in the intestine. There is also agreement that apoB is formed in rapid response to dietary fat, but the intestinal synthesis of apoA-I in response to fat feeding is controversial.

Lymph chylomicron and lipid composition are clearly affected by the nature of the dietary fat. Lymph chylomicron size and lipid content are greater with unsaturated than with saturated fat diets, and there is some evidence that a diet rich in a particular fatty acid can bring about a specific effect upon the lymph lipoproteins. There is no evidence, however, that the diet-induced changes in lymph lipids would be directly related to the changes in plasma lipids and lipoproteins known to be associated with the feeding of different dietary fats. Thus, ω-6 fatty acid-rich diets lead to lowered plasma, but increased lymph lipoprotein cholesterol levels, while diets rich in ω-3 fatty acids lead to decreased plasma, but increased lymph triacylglycerol levels.

Similarly, the intramolecular structure of the dietary triacylglycerols and the stereochemical course of triacylglycerol resynthesis in the villus cells appear to have no effect on the apoprotein composition of the chylomicrons, although changes in the size and other physical properties of the resulting chylomicrons have been demonstrated. Furthermore, the lymph apoprotein formation, partition of cholesterol, and triacylglycerols into lymph lipoproteins are affected by the absorbability of dietary fats and cholesterol via the intestine. The degree to which these changes affect the subsequent clearance and metabolism of lymph lipoproteins from plasma and atherosclerosis remains to be established.

ACKNOWLEDGMENTS

The studies by the author and collaborators referred to in this review were supported by funds from the Medical Research Council of Canada, Ottawa, Ontario and from the Ontario Heart Foundation, Toronto, Ontario.

REFERENCES

1. Morrisett, J. D., Pownall, H. J., Jackson, R. L., Segura, R., Gotto, A. M., and Taunton, O. D., Effects of polyunsaturated and saturated fat diets on the chemical composition and therotropic properties of human plasma lipoproteins, in *Polyunsaturated Fatty Acids,* Kunau, W. H. and Holman, R. T., Eds., American Oil Chemists' Society, Champaign, Ill., 1977, 139.
2. Goodnight, S. H., Jr., Harris, W. S., Connor, W. E., and Illingworth, D. R., Polyunsaturated fatty acids, hyperlipidemia and thrombosis, *Arteriosclerosis,* 2, 87, 1982.
3. Spritz, N. and Mishkel, M. A., Effects of dietary fats on plasma lipids and lipoproteins: an hypothesis for the lipid lowering effect of unsaturated fatty acids, *J. Clin. Invest.,* 48, 78, 1969.
4. Horrobin, D. F. and Manku, M. S., How do polyunsaturated fatty acids lower plasma cholesterol levels?, *Lipids,* 18, 558, 1983.
5. Cortese, B., Levy, Y., Janus, E. D., Turner, P. R., Rao, S. N., Miller, N. E., and Lewis, B., Modes of action of lipid-lowering diets in man: studies of apolipoprotein B kinetics in relation to fat consumption and dietary fatty acid composition, *Eur. J. Clin. Invest.,* 13, 79, 1983.
6. Goldberg, A. C. and Schonfeld, G., Effect of diet on lipoprotein metabolism, *Ann. Rev. Nutr.,* 5, 195, 1985.
7. Tytgat, G. N., Rubin, C. E., and Saunders, D. R., Synthesis and transport of lipoprotein particles by intestinal absorptive cells in man, *J. Clin. Invest.,* 50, 2065, 1971.
8. Redgrave, T. G., Formation and metabolism of chylomicrons, in *Gastrointestinal Physiology IV, International Review of Physiology,* Vol. 28, Young, J. A., Ed., University Park Press, Baltimore, 1983, 103.
9. Bisgaier, C. L. and Glickman, R. M., Intestinal synthesis, secretion, and transport of lipoproteins, *Ann. Rev. Physiol.,* 45, 625, 1983.
10. Shaikh, N. A. and Kuksis, A., Expansion of phospholipid pool size of rat intestinal villus cells during fat absorption, *Can. J. Biochem.,* 60, 444, 1982.
11. Shaikh, N. A. and Kuksis, A., Further evidence for enhanced phospholipid synthesis by rat jejunal villus cells during fat absorption, *Can. J. Biochem. Cell Biol.,* 61, 370, 1983.
12. Zilversmit, D. B., Sisco, P. H., and Yokoyama, A., Size distribution of thoracic duct lymph chylomicrons from rats fed cream and corn oil, *Biochim. Biophys. Acta,* 125, 129, 1966.

13. Redgrave, T. G. and Dunne, K. B., Chylomicron formation and composition in unanaesthetized rabbits, *Atherosclerosis*, 22, 389, 1975.
14. Yang, L.-Y. and Kuksis, A., Apoprotein profiles of lymph chylomicrons of rats fed corn oil or fatty acid methyl esters, in *Abstracts, Int. Symp. on Lipoprotein Deficiency Syndromes*, Vancouver, Canada, May 1985.
15. Glickman, R. M. and Kirsch, K., Lymph chylomicron formation during the inhibition of protein synthesis, *J. Clin. Invest.*, 52, 2910, 1973.
16. Parks, J. S., Atkinson, D., Small, D. M., and Rudel, L. L., Physical characterization of lymph chylomicron and very low density lipoproteins from nonhuman primates fed saturated dietary fat, *J. Biol. Chem.*, 256, 12992, 1981.
17. Landman, W., Feuge, R. O., and Lovegren, N. V., Melting and dilatometric behaviour of 2-oleopalmitostearin and 2-oleodistearin, *J. Am. Oil Chem. Soc.*, 37, 638, 1960.
18. Arvidson, G. A. E. and Nilsson, A., Formation of lymph chylomicron phosphatidylcholines in the rat during the absorption of safflower oil or triolein, *Lipids*, 7, 344, 1971.
19. Huang, T. C. and Kuksis, A., A comparative study of the lipids of chylomicron membrane and fat core, and of the lymph serum of dogs, *Lipids*, 2, 443, 1967.
20. Hamilton, J. A. and Small, D. M., Solubilization and localization of triolein in phosphatidylcholine bilayers: a ^{13}C NMR study, *Proc. Natl. Acad. Sci. U.S.A.*, 78, 6878, 1981.
21. Janiak, M. T., Loomis, C. R., Shipley, G. G., and Small, D. M., The ternary phase diagram of lecithin, cholesteryl linoleate and water: phase behavior and structure, *J. Mol. Biol.*, 86, 325, 1974.
22. Janiak, M. J., Small, D. M., and Shipley, G. G., Interactions of cholesteryl esters with phospholipids: cholesteryl myristate and dimyrisoyl lecithin, *J. Lipid Res.*, 20, 183, 1979.
23. Kuksis, A., Breckenridge, W. C., Myher, J. J., and Kakis, G., Replacement of endogenous phospholipids in rat plasma lipoproteins during intravenous infusion of an artificial lipid emulsion, *Can. J. Biochem.*, 56, 630, 1978.
24. Redgrave, T. G. and Carlsson, L. A., Changes in plasma very low density and low density lipoprotein content, composition, and size after a fatty meal in normo- and hypertriglyceridemic man, *J. Lipid Res.*, 20, 217, 1979.
25. Kuksis, A., Myher, J. J., Geher, K., Breckenridge, W. C., Jones, G. J. L., and Little, J. A., Lipid class and molecular species interrelationships among plasma lipoproteins of normolipemic subjects, *J. Chromatogr. Biomed. Appl.*, 224, 1, 1981.
26. Ockner, R. K., Hughes, F. B., and Isselbacher, K. J., Very low density lipoproteins in intestinal lymph: origin, composition, and role in lipid transport in the fasting state, *J. Clin. Invest.*, 48, 2079, 1969.
27. Mahley, R. W., Bennett, B. D., Morre, D. J., Gray, M. E., Thistlethwaite, W., and LeQuire, V. S., Lipoproteins associated with the Golgi apparatus isolated from epithelial of rat small intestine, *Lab. Invest.*, 25, 435, 1971.
28. Krishnaiah, K. V., Walker, L. F., Boresztajn, J., Schonfeld, G., and Getz, G. S., Apoprotein B variant from rat intestine, *Proc. Natl. Acad. Sci. U.S.A.*, 77, 3806, 1980.
29. Kane, J. P., Hardman, D. A., and Paulus, H. A., Heterogeneity of apolipoprotein B: isolation of a new species from human chylomicrons, *Proc. Natl. Acad. Sci. U.S.A.*, 77, 2465, 1980.
30. Lee, D. M., Koren, E., Singh, S., and Mok, T., Presence of B-100 in rat mesenteric chyle, *Biochem. Biophys. Res. Commun.*, 123, 1149, 1984.
31. Lusis, A. J., et al., Cloning and expression of apolipoprotein B, the major protein of low and very low density lipoproteins, *Proc. Natl. Acad. Sci. U.S.A.*, 82, 4597, 1985.
32. Bhattacharya, S. and Redgrave, T. G., The content of apolipoprotein B in chylomicron particles, *J. Lipid Res.*, 22, 820, 1981.
33. Glickman, R. M., Kilgore, A., and Khorana, J., Chylomicron apoprotein localization within intestinal epithelium: studies of normal and impaired lipid absorption, *J. Lipid Res.*, 19, 260, 1978.
34. Windmueller, H. G. and Wu, A.-L., Biosynthesis of plasma apolipoproteins by rat small intestine without dietary or biliary fat, *J. Biol. Chem.*, 256, 3012, 1981.
35. Imaizumi, K., Havel, R. J., Fainaru, M., and Vigne, J.-L., Origin and transport of the A-I and arginine-rich apolipoproteins in mesenteric lymph of rats, *J. Lipid Res.*, 19, 1038, 1978.
36. Anderson, D. W., Schaefer, E. J., Bronzert, T. J., Lindgren, F. T., Forte, T., Starzl, T. E., Niblack, G. D., Zech, L., and Brewer, H. B., Jr., Transport of apolipoprotein A-I and A-II by human thoracic duct lymph, *J. Clin. Invest.*, 67, 857, 1981.
37. Zannis, V., Breslow, J. L., and Katz, A. J., Isoprotein of human apolipoprotein A-I demonstrated in plasma and intestinal organ culture, *J. Biol. Chem.*, 255, 8612, 1980.
38. Imaizumi, K., Fainaru, M., and Havel, R. J., Composition of proteins of mesenteric lymph chylomicrons in the rat and alterations produced by exposure of chylomicrons to blood serum and serum lipoproteins, *J. Lipid Res.*, 19, 712, 1978.
39. Wu, A.-L. and Windmueller, H. G., Relative contributions by liver and intestine to individual plasma apolipoproteins in the rat, *J. Biol. Chem.*, 254, 7316, 1979.

40. Green, P. H. R., Lefkowitch, J. H., Glickman, R. M., Riley, J. W., Quinet, E., and Blum, C. B., Apolipoprotein localization and quantitation in the human intestine, *Gastroenterology*, 83, 1223, 1982.
41. Riley, J. W., Glickman, R. M., Green, P. H. R., and Tall, A. R., The effect of chronic cholesterol feeding on intestinal lipoproteins in the rat, *J. Lipid Res.*, 21, 942, 1980.
42. Wu, A.-L. and Windmueller, H. G., Identification of circulating apolipoproteins synthesized by rat small intestine *in vivo*, *J. Biol. Chem.*, 253, 2525, 1978.
43. Krause, B. R., Sloop, C. H., Castle, C. K., and Roheim, P. S., Mesenteric lymph apolipoproteins in control and ethinyl estradiol-treated rats: a model for studying apolipoproteins of intestinal origin, *J. Lipid Res.*, 22, 610, 1981.
44. Palay, S. C. and Karlin, L. J., An electron microscopic study of the intestinal villus. II. The pathway of fat absorption, *J. Biophys. Biochem. Cytol.*, 5, 373, 1959.
45. Reaven, E. P. and Reaven, G. M., Distribution and content of microtubules in relation to the transport of lipid. An ultrastructural quantitative study of the absorptive cell of the small intestine, *J. Cell. Biol.*, 75, 559, 1977.
46. Bezard, J. and Bugaut, M., Absorption of glycerides containing short, medium and long chain fatty acids, in *Fat Absorption*, Vol. 1, Kuksis, A., Ed., CRC Press, Boca Raton, Fla., 1986.
47. Stremmel, W., Lotz, G., Strohmeyer, G., and Berk, P. D., Identification, isolation, and partial characterization of a fatty acid binding protein from rat jejunal membranes, *J. Clin. Invest.*, 75, 1068, 1985.
48. Ockner, R. K. and Manning, J. A., Fatty acid binding protein in small intestine. Identification, isolation, and evidence for its role in cellular fatty acid transport, *J. Clin. Invest.*, 54, 326, 1974.
49. O'Doherty, P. J. A. and Kuksis, A., Stimulation of triacylglycerol synthesis by Z protein in rat liver and intestinal mucosa, *FEBS Lett.*, 60, 256, 1975.
50. Sabesin, S. M. and Isselbacher, K. J., Protein synthesis inhibition: mechanism for the production of impaired fat absorption, *Science*, 147, 1149, 1965.
51. Glickman, R. M., Kirsch, K., and Isselbacher, K. J., Fat absorption during inhibition of protein synthesis: studies of lymph chylomicrons, *J. Clin. Invest.*, 51, 356, 1972.
52. Redgrave, T. G. and Zilversmit, D. B., Does puromycin block release of chylomicrons from the intestine?, *Am. J. Physiol.*, 217, 336, 1969.
53. O'Doherty, P. J. A. and Kuksis, A., Effect of puromycin *in vitro* on protein and glycoprotein biosynthesis in isolated epithelial cells of rat intestine, *Int. J. Biochem.*, 6, 434, 1975.
54. Yousef, I. M., O'Doherty, P. J. A., Whitter, E. F., and Kuksis, A., Ribosome structure and chylomicron formation in intestinal mucosa, *Lab. Invest.*, 34, 256, 1976.
55. Friedman, H. I. and Nylund, B., Intestinal fat digestion, absorption and transport, *Am. J. Clin. Nutr.*, 33, 1108, 1980.
56. Rachmilewitz, D., Albers, J. J., and Saunders, D. R., Apoprotein B in fasting and postprandial human jejunal mucosa, *J. Clin. Invest.*, 57, 530, 1976.
57. Renner, F., Samuelsson, A., Rogers, M., and Glickman, R. M., Effect of saturated and unsaturated lipid on the composition of mesenteric triglyceride-rich lipoproteins in the rat, *J. Lipid Res.*, 27, 72, 1986.
58. Alpers, D. H., Lock, D. R., Lancaster, N., Poksay, K., and Schonfeld, G., Distribution of apolipoproteins A-I and B among intestinal lipoproteins, *J. Lipid Res.*, 26, 1, 1985.
59. Ghiselli, G., Schaefer, E. J., Light, J. A., and Brewer, H. B., Jr., Apolipoprotein A-I isoforms in human lymph: effect of fat absorption, *J. Lipid Res.*, 24, 731, 1983.
60. Davidson, N. O. and Glickman, R. M., Apolipoprotein A-I synthesis in rat small intestine: regulation by dietary triglyceride and biliary lipid, *J. Lipid Res.*, 26, 368, 1985.
61. Bearnut, H. R., Glickman, R. M., Weinberg, L., and Green, P. H. R., Effect of biliary diversion on rat mesenteric lymph apolipoprotein-I and high density lipoprotein, *J. Clin. Invest.*, 69, 210, 1982.
62. Gordon, J. I., Strauss, A. W., Smith, D. P., Ockner, R., and Alpers, D. H., Identification of the abundant mRNA species which accumulate in the enterocyte and cloning of their cDNA's, *Gastroenterology*, 82(Abstr.), 1071, 1982.
63. Strauss, E. W. and Jacob, J. S., Some factors affecting the lipid secretory phase of fat absorption by intestine *in vitro* from golden hamster, *J. Lipid Res.*, 22, 147, 1981.
64. Redgrave, T. G., The role in chylomicron formation of phospholipase activity of intestinal Golgi membranes, *Aust. J. Exp. Biol. Med.*, 51, 427, 1973.
65. Malaisse, W. J., Malaisse-Lagae, F., Van-Obbengeren, E., Somers, G., Davis, G., Ravazzola, M., and Orci, L., Role of microtubules in the phasic pattern of insulin release, *Ann. N.Y. Acad. Sci.*, 253, 630, 1975.
66. Arreaza-Plaza, C. A., Bosch, U., and Otayek, M. A., Lipid transport across the intestinal epithelial cell. Effect of colchicine, *Biochim. Biophys. Acta*, 431, 297, 1976.

67. Glickman, R. M., Perrotto, J. L., and Kirsch, K., Intestinal lipoprotein formation: effect of colchicine, *Gastroenterology*, 70, 347, 1976.

68. Hoffman, A. G. D., Child, P., and Kuksis, A., Synthesis and release of lipids and lipoproteins by isolated rat jejunal enterocytes in the presence of sodium taurocholate, *Biochim. Biophys. Acta*, 665, 283, 1981.

69. Redman, C. M., Banerjee, K., Howell, K., and Palade, G. E., The step at which colchicine blocks the secretion of plasma protein by rat liver, *Ann. N.Y. Acad. Sci.*, 253, 780, 1975.

70. Sabesin, S. M. and Frase, S., Electron microscopic studies of the assembly, intracellular transport and secretion of chylomicrons by rat intestine, *J. Lipid Res.*, 18, 496, 1977.

71. Yousef, I. M. and Kuksis, A., Release of chylomicrons by isolated cells of rat intestinal mucosa, *Lipids*, 7, 380, 1972.

72. O'Doherty, P. J. A., Yousef, I. M., and Kuksis, A., Effect of puromycin on protein and glycerolipid biosynthesis in isolated mucosal cells, *Arch. Biochem. Biophys.*, 156, 586, 1973.

73. Green, P. H. R., Tall, A. R., and Glickman, R. M., Rat intestine secretes discoid high density lipoprotein, *J. Clin. Invest.*, 61, 528, 1978.

74. Forester, G. P., Tall, A. R., Bisgaier, C. L., and Glickman, R. M., Rat intestine secretes spherical high density lipoproteins, *J. Biol. Chem.*, 258, 5938, 1983.

75. Beveridge, J. M. R., Connell, W. F., and Mayer, G. A., Dietary factors affecting the level of plasma cholesterol in humans: the role of fat, *Can. J. Biochem. Physiol.*, 34, 441, 1956.

76. Scott, R. F. and Thomas, W. A., Methods of comparing effects of various fats on fibrinolysis, *Proc. Soc. Exp. Biol. Med.*, 96, 24, 1957.

77. Jones, R., Thomas, W. R., and Scott, R. F., Electron microscopy study of chyle from rats fed butter or corn oil, *Exp. Mol. Pathol.*, 1, 65, 1962.

78. Huang, T. C., A Comparative Study of the Structure of Milk Fat Globules and Chylomicrons, Ph.D. thesis, Queen's University, Kingston, Ontario, Canada, 1965.

79. Floren, C.-H. and Nilsson, A., Effects of fatty acid unsaturation on chylomicron metabolism in normal and hepatectomized rats, *Eur. J. Biochem.*, 77, 23, 1977.

80. Zilversmit, D. B., The composition and structure of lymph chylomicrons in dog, rat and man, *J. Clin. Invest.*, 44, 1610, 1965.

81. Puppione, D. L., Kunitake, S. T., Hamilton, R. L., Philips, M. L., Schumaker, V. N., and Davis, L. D., Characterization of unusual intermediate density lipoproteins, *J. Lipid Res.*, 23, 283, 1982.

82. Ahrens, E. H., Insull, W., Blomstrand, R., Hirsch, J., Tsaltas, T. T., and Peterson, M. L., The influence of dietary fats on serum-lipid levels in man, *Lancet*, 1, 943, 1957.

83. Frederickson, D. S., Levy, R. I., Jones, E., Bonnell, M., and Ernst, N., The Dietary Management of Hyperlipiproteinemias: A Handbook for Physicians, revised NIH Publ. No. 73-110, Public Health Service, U.S. Department of Health, Education, and Welfare, Washington, D.C., 1973.

84. Bennet Clark, S., Atkinson, D., Hamilton, J. A., Forte, T., Russel, B., Feldman, E. B., and Small, D. M., Physical studies of d<1.006 g/ml lymph lipoproteins from rats fed palmitate-rich diets, *J. Lipid Res.*, 23, 28, 1982.

85. Feldman, E. B., Russell, B. S., Chen, R., Johnson, J., Forte, T., and Bennet Clark, S., Dietary saturated fatty acid content affects lymph lipoproteins: studies in the rat, *J. Lipid Res.*, 24, 967, 1983.

86. Fraser, R., Size and lipid composition of chylomicrons of different Svedberg units of flotation, *J. Lipid Res.*, 11, 60, 1970.

87. Feldman, E. B., Russell, B. S., Hawkins, G. B., and Forte, T., Fatty acid composition of the diet and intestinal lipoproteins, *Arteriosclerosis*, 1, 84, 1981.

88. Green, M. H. and Green, J. B., Effects of dietary fat saturation and triglyceride and cholesterol load on lipid absorption in the rat, *Atherosclerosis*, 46, 181, 1983.

89. Ockner, R. K., Pittman, J. P., and Yager, J. L., Differences in the intestinal absorption of saturated and unsaturated long chain fatty acids, *Gastroenterology*, 62, 981, 1972.

90. Green, P. H. R. and Glickman, R. M., Intestinal lipoprotein metabolism, *J. Lipid Res.*, 22, 1153, 1981.

91. Holt, P. R., Wu, A. L., and Bennet Clark, S., Apoprotein composition and turnover in rat intestinal lymph during steady-state triglyceride absorption, *J. Lipid Res.*, 20, 494, 1979.

92. Feldman, E. B., Chen, R., Russell, B. S., Forte, T., and Bennet Clark, S., Cooling affects intestinal lymph lipoproteins of rats fed saturated fats, *Clin. Res.*, 30, 185A, 1982.

93. Feldman, E. B., Russell, B. S., Hawkins, C. B., and Forte, T., Intestinal lymph lipoproteins in rats fed diets enriched in specific fatty acids, *J. Nutr.*, 113, 2323, 1983.

94. Feldman, E. B., Russell, B. S., Schnare, F. H., Morretti-Rojas, I., Miles, B. C., and Doyle, E. A., Effects of diets of homogeneous saturated triglycerides on cholesterol balance in rats, *J. Nutr.*, 109, 2237, 1979.

95. Hegsted, D. M., McGandy, R. B., Myers, M. L., and Stare, F. J., Quantitative effects of dietary fat on serum cholesterol in man, *Am. J. Clin. Nutr.*, 17, 281, 1965.

96. Bang, H. O. and Dyerberg, J., Lipid metabolism and ischemic heart disease in Greenland Eskimos, *Adv. Nutr. Res.*, 3, 1, 1980.
97. Kromhout, D., Bosschieter, E. B., and Coulander, C.-L., The inverse relation between fish consumption and 20-year mortality from coronary heart disease, *N. Engl. J. Med.*, 312, 1205, 1985.
98. Phillipson, B. E., Rothrock, D. W., Connor, W. E., Harris, W. S., and Illingworth, D. R., Reduction of plasma lipids, lipoproteins, and apoproteins by dietary fish oils in patients with hypertriglyceridemia, *N. Engl. J. Med.*, 312, 1210, 1985.
99. Lee, T. H., Hoover, R. L., Williams, J. D., Sperling, R. I., Ravalese, J., III, Spur, B. W., Robinson, D. R., Corey, E. J., Lewis, R. A., and Austen, K. F., Effect of dietary enrichment with eicosapentaenoic and docosahexaenoic acids on in vitro neutrophil and monocyte leukotriene generation and neutrophil function, *N. Engl. J. Med.*, 312, 1217, 1985.
100. Mattson, F. H. and Grundy, S. M., Comparison of effects of dietary saturated, monounsaturated, and polyunsaturated fatty acids on plasma lipids and lipoproteins in man, *J. Lipid Res.*, 26, 194, 1985.
101. Keys, A., Ed., Coronary heart disease in seven countries, *Circulation*, 44(Suppl. 1), 1970.
102. Illingworth, D. R., Present status of polyunsaturated fats in the prevention of cardiovascular disease, in *Nutrition and Food Science*, Vol. 3, Santos, W., Lopus, N., Barbosa, J. J., Chavez, D., and Valente, J. C., Eds., Plenum Press, New York, 1980, 365.
103. Ramesha, P. R. and Ganguly, J., On the mechanism of hypocholesterolemic effects of polyunsaturated lipids, *Adv. Lipid Res.*, 17, 155, 1979.
104. Harris, W. S., Connor, W. E., and Goodnight, S. H., Jr., Dietary fish oils, plasma lipids and platelets in man, *Prog. Lipid Res.*, 20, 75, 1981.
105. Brockerhoff, H., Hoyle, R. J., and Huang, P. C., Positional distribution of fatty acids in the fats of a polar bear and a seal, *Can. J. Biochem.*, 44, 1519, 1966.
106. Bottino, N. R., Vandenberg, G. A., and Reiser, R., Resistance of certain long-chain polyunsaturated fatty acids of marine oils to pancreatic lipase hydrolysis, *Lipids*, 2, 489, 1967.
107. Yurkowski, M. and Brockerhoff, H., Fatty acid distribution of triglycerides determined by deacylation with methyl magnesium bromide, *Biochim. Biophys. Acta*, 125, 55, 1966.
108. Willis, A. L., Nutritional and pharmacological factors in eicosanoid biology, *Nutr. Rev.*, 39, 289, 1981.
109. Chen, I. S., Subramaniam, S., Cassidy, M. M., Sheppard, J. J., and Vahouny, G. V., Intestinal absorption and lipoprotein transport of omega-3 eicosapentaenoic acid, *J. Nutr.*, 115, 219, 1985.
110. Engleberg, H., Mechanisms involved in the reduction of serum triglycerides in man upon adding unsaturated fats to the normal diet, *Metabolism*, 15, 796, 1966.
111. Illingworth, D. R., Harris, W. S., and Connor, W. E., Inhibition of low density lipoprotein synthesis by dietary w-3-fatty acids in humans, *Arteriosclerosis*, 4, 270, 1984.
112. Yang, Y.-L. and Kuksis, A., Differences in apoprotein composition of chylomicrons containing triacylglycerols derived from different biosynthetic pathways, unpublished results, 1985.
113. Becker, N., Illingworth, D. R., Alaupovic, P., Connor, W. E., and Sundberg, E. E., Effects of saturated, monounsaturated, and omega-6 polyunsaturated fatty acids on plasma lipids, lipoproteins, and apoproteins in humans, *Am. J. Clin. Nutr.*, 37, 355, 1982.
114. Vahouny, G. V. and Treadwell, C. R., Comparative effects of dietary fatty acids and glycerides on lymph lipids in the rat, *Am. J. Physiol.*, 196, 881, 1959.
115. Bloomfield, D. K., Cholesterol metabolism. III. Enhancement of cholesterol absorption and accumulation in safflower oil-fed rats, *J. Lab. Clin. Med.*, 64, 613, 1964.
116. Thompson, G. R., Ockner, R. K., and Isselbacher, K. J., Effect of mixed micellar lipid on the absorption of cholesterol and vitamin D$_3$ into lymph, *J. Clin. Invest.*, 48, 87, 1969.
117. Baudet, M. F., Dachet, C., Lasserre, M., Esteva, O., and Jacotot, B., Modification in the composition and metabolic properties of human low density and high density lipoproteins by different dietary fats, *J. Lipid Res.*, 25, 456, 1984.
118. Vega, G. L., Groszek, E., Wolf, R., and Grundy, S. M., Influence of polyunsaturated fats on composition of plasma lipoproteins and apolipoproteins, *J. Lipid Res.*, 23, 811, 1982.
119. Kuksis, A., Myher, J. J., Geher, K., Jones, G. J. L., Shepherd, J., Packard, C. J., Morrisett, J. D., Taunton, O. D., and Gotto, A. M., Effect of saturated and unsaturated fat diets on lipid profiles of plasma lipoproteins, *Atherosclerosis*, 41, 221, 1982.
120. Shepherd, J., Packard, C. J., Patsch, J. R., Gotto, A. M., Jr., and Taunton, O. D., Effects of dietary polyunsaturated and saturated fats on the properties of high density lipoproteins and the metabolism of apolipoprotein A-I, *J. Clin. Invest.*, 61, 1582, 1978.
121. Peifer, J. J., Jansen, F., Ahn, P., Cox, W., and Lundberg, W. O., Studies on the distribution of lipids in hypercholesterolemic rats. I. The effect of feeding palmitate, oleate, linoleate, linolenate, menhaden and tuna oils, *Arch. Biochem. Biophys.*, 80, 302, 1960.

122. Peifer, J. J., Lundberg, W. O., Ishio, S., and Warmanen, E., Studies of the distribution of lipids in hypercholesterolemic rats. III. Changes in hypercholesterolemia and tissue fatty acids induced by dietary fats and marine oil fractions, *Arch. Biochem. Biophys.*, 110, 270, 1965.

123. Kahn, S. G., Vandeputte, J., Wind, S., and Yacowitz, H., A study of the hypocholesterolemic activity of the ethyl esters of the polyunsaturated fatty acids of cod liver oil in the chicken. I. Effect on total serum cholesterol, *J. Nutr.*, 80, 403, 1963.

124. Dam, H., Kristensen, G., Nielsen, G. K., and Sondergaard, E., Influence of dietary cholesterol, cod liver oil and linseed oil on cholesterol and polyenoic fatty acids in tissues from fasted and non-fasted chicks, *Acta Physiol. Scand.*, 45, 31, 1959.

125. Ruiter, A., Jongbloed, A. N., VanGent, C. M., Danse, L. H. J. C., and Metz, S. H. M., The influence of dietary mackerel oil on the condition of organs and on blood lipid composition in the young growing pig, *Am. J. Clin. Nutr.*, 31, 2159, 1978.

126. Stange, E. F. and Dietschy, J. M., The origin of cholesterol in the mesenteric lymph of the rat, *J. Lipid Res.*, 26, 175, 1985.

127. Stange, E. F., Suckling, K. E., and Dietschy, J. M., Synthesis and coenzyme A dependent esterification of cholesterol in rat intestinal epithelium. Differences in cellular localization and mechanisms of regulation, *J. Biol. Chem.*, 258, 12868, 1983.

128. Lasser, N. L., Roheim, P. S., Edelstein, D., and Eder, H. A., Serum lipoproteins of normal and cholesterol fed rats, *J. Lipid Res.*, 14, 1, 1973.

129. Mahley, R. W., Alterations in plasma lipoproteins induced by cholesterol feeding in animals including man, in *Disturbances in Lipid and Lipoprotein Metabolism,* Dietschy, J. M., Gotto, A. M., Jr., and Ontko, J. A., Eds., American Physiological Society, Bethesda, Md., 1978, 181.

130. Swift, L. L., Soule, P. D., Gray, M. E., and LeQuire, V. S., Intestinal lipoprotein synthesis. Comparison of nascent Golgi lipoproteins from chow-fed and hypercholesterolemic rats, *J. Lipid Res.*, 25, 1, 1984.

131. Fidge, N. H. and McCullagh, P. J., Studies on the apoproteins of rat lymph chylomicrons: characterization and metabolism of a new chylomicron-associated apoprotein, *J. Lipid Res.*, 22, 138, 1981.

132. Sloop, C. H., Dory, L., Hamilton, R., Krause, B. R., and Roheim, P. S., Characterization of dog peripheral lymph lipoproteins: the presence of a disc-shaped "nascent" high density lipoprotein, *J. Lipid Res.*, 24, 1429, 1983.

133. Klein, R. L. and Rudel, L. L., Effect of dietary cholesterol level on the composition of thoracic duct lymph lipoproteins isolated from nonhuman primates, *J. Lipid Res.*, 24, 357, 1983.

134. Sylven, C. and Borgstrom, B., Absorption and lymphatic transport of cholesterol in the rat, *J. Lipid Res.*, 9, 596, 1968.

135. Klein, R. L. and Rudel, L. L., Cholesterol absorption and transport in thoracic duct lymph lipoproteins of nonhuman primates. Effect of dietary cholesterol level, *J. Lipid Res.*, 24, 343, 1983.

136. Green, M. H., Chemical and isotopic measurement of cholesterol absorption in the rat, *Atherosclerosis*, 37, 343, 1980.

137. Brockerhoff, H. and Jensen, R. G., *Lipolytic Enzymes,* Academic Press, New York, 1978.

138. O'Doherty, P. J. A. and Kuksis, A., Glycerolipid biosynthesis in isolated rat intestinal epithelial cells, *Can. J. Biochem.*, 53, 1010, 1975.

139. Breckenridge, W. C. and Kuksis, A., Triacylglycerol biosynthesis in everted sacs of rat intestinal mucosa, *Can. J. Biochem.*, 53, 1184, 1975.

140. Johnston, J. M., Rao, G. A., and Lowe, P. A., The separation of the alpha-glycerophosphate and monoglyceride pathways in the intestinal biosynthesis of triglycerides, *Biochim. Biophys. Acta*, 137, 578, 1967.

141. Wilson, A. C., Goldstein, R. C., Conn, A. R., and Kuo, P. T., Effect of high fat and high sucrose feeding on lipogenesis by isolated rat intestinal cells, *Can. J. Biochem. Cell Biol.*, 61, 340, 1983.

142. Kuksis, A. and Myher, J. J., Analytical methodology in the fat absorption area, in *Fat Absorption,* Vol. 1, Kuksis, A., Ed., CRC Press, Boca Raton, Fla., 1986.

143. Scow, R. O., Stein, Y., and Stein, O., Incorporation of dietary lecithin and lysolecithin into lymph chylomicrons in the rat, *J. Biol. Chem.*, 212, 4919, 1967.

144. Mills, S. C., Cook, L. J., Scott, T. W., and Nestel, P. J., Effect of dietary fat supplementation on the composition and positional distribution of fatty acids in ruminant and porcine glycerides, *Lipids*, 11, 49, 1976.

145. Johnston, J. M., Paltauf, F., Schiller, C. M., and Schultz, L. D., The utilization of the alpha-glycerophosphate and monoglyceride pathways for phosphatidylcholine biosynthesis in the intestine, *Biochim. Biophys. Acta*, 218, 124, 1970.

146. O'Doherty, P. J. A. and Kuksis, A., Microsomal synthesis of di- and triacylglycerols in rat liver and Ehrlich ascites cells, *Can. J. Biochem.*, 52, 514, 1974.

147. Breckenridge, W. C. and Kuksis, A., Stereochemical course of diacylglycerol formation in rat intestine, *Lipids*, 7, 256, 1972.

148. Bugaut, M., Myher, J. J., and Kuksis, A., An examination of the stereochemical course of acylation of 2-monoacylglycerols by rat intestinal villus cells using [^{2}H$_3$]palmitic acid, *Biochim. Biophys. Acta*, 792, 254, 1984.

149. O'Doherty, P. J. A., Kuksis, A., and Buchnea, D., Enantiomeric diglycerides as stereospecific probes in triglyceride synthesis *in vitro*, *Can. J. Biochem.*, 50, 881, 1972.

150. Myher, J. J., Kuksis, A., and Yang, Y.-L., Stereochemical course of pancreatic lipolysis and mucosal resynthesis of mustard seed oil triacylglycerols in the rat, *Can. J. Biochem. Cell Biol.*, submitted.

151. Myher, J. J., Kuksis, A., Vasdev, S. C., and Kako, K. J., Acylglycerol structure of mustard seed oil and of cardiac lipids of rats during dietary lipidosis, *Can. J. Biochem.*, 57, 1315, 1979.

152. Paltauf, F., Metabolism of the enantiomeric 1-O-alkylglycerol ethers in the rat intestinal mucosa *in vivo*; incorporation into 1-O-alkyl and 1-O-alk-1'-enyl glycerol lipids, *Biochim. Biophys. Acta*, 239, 38, 1971.

153. Breckenridge, W. C., Yeung, S. K. F., and Kuksis, A., Biosynthesis of triacylglycerols by rat intestinal mucosa *in vivo*, *Can. J. Biochem.*, 54, 170, 1976.

154. Hamilton, J. R., Kuksis, A., and Jeejeebhoy, K. N., Neonatal chylous ascites. Response to dietary triglyceride, *Clin. Res.*, 18(Abstr.), 725, 1970.

155. Mattson, F. H., Nolen, G. A., and Webb, M. R., The absorbability by rats of various triglycerides of stearic and oleic acid and the effect of dietary calcium and magnesium, *J. Nutr.*, 109, 1682, 1979.

156. Filer, L. J., Jr., Mattson, F. H., and Fomon, S. J., Triglyceride configuration and fat absorption by the human intact, *J. Nutr.*, 99, 293, 1970.

157. Mattson, F. H. and Streck, J. A., Effect of consumption of glycerides containing behenic acid on the lipid content of the heart of weanling rats, *J. Nutr.*, 104, 483, 1974.

158. Parsons, H. G. and Emken, E. A., Simultaneous measurement of the assimilation of four long-chain fatty acids in human chylomicron triglycerides and cholesteryl esters, *Fed. Proc.*, 43(Abstr. 1965), 621, 1984.

159. Mahley, R. W. and Holcombe, K. S., Alterations of the plasma lipoproteins and apoproteins following cholesterol feeding in the rat, *J. Lipid Res.*, 18, 314, 1977.

160. Reichl, D., Myant, N. B., Rudra, D. N., and Pflug, J. J., Evidence for the presence of tissue-free cholesterol in low density and high density lipoproteins of human peripheral lymph, *Atherosclerosis*, 37, 489, 1980.

161. Dory, L., Sloop, C. H., Boquet, L. M., Hamilton, R. L., and Roheim, P. S., LCAT-mediated modification of discoidal peripheral lymph HDL: a possible mechanism of HDLc formation in cholesterol-fed dogs, *Proc. Natl. Acad. Sci. U.S.A.*, 80, 3489, 1983.

162. Beveridge, J. M. R., A second report to the students of Queen's University on a co-operative study of dietary factors affecting plasma lipid levels, *Queen's Univ. Med. Rev.*, 8, 2, 1960.

163. Yamamoto, I., Sugano, M., and Wada, M., Hypocholesterolemic effect of animal and plant fats in rats, *Atherosclerosis*, 13, 171, 1971.

164. Kritchevsky, D., Tepper, S. A., Vesselinovitch, D., and Wissler, R. W., Cholesterol vehicle in experimental atherosclerosis. XIII. Randomized peanut oil, *Atherosclerosis*, 17, 225, 1973.

165. Myher, J. J., Marai, L., Kuksis, A., and Kritchevsky, D., Acylglycerol structure of peanut oils of different atherogenic potential, *Lipids*, 12, 775, 1977.

166. Manganaro, F., Myher, J. J., Kuksis, A., and Kritchevsky, D., Triacylglycerol structure of genetic varieties of peanut oils of varying atherogenic potential, *Lipids*, 16, 508, 1981.

167. Morley, N. H. and Kuksis, A., Positional specificity of lipoprotein lipase, *J. Biol. Chem.*, 247, 6389, 1972.

168. Paltauf, F., Esfandi, F., and Holasek, A., Stereo-specificity of lipases. Enzymatic hydrolysis of enantiomeric alkyldiglycerides by lipoprotein lipase, lingual lipase and pancreatic lipase, *FEBS Lett.*, 40, 119, 1974.

169. Pind, S., Kuksis, A., Myher, J. J., Marai, L., and Kritchevsky, D., Stereospecific analysis of molecular species of triacylglycerols from cocoa butter and coconut oil, unpublished results.

170. Hung, S., Umemura, T., Yamashiro, S., Slinger, S. J., and Holub, B. J., The effects of original and randomized rapeseed oils containing high and very low levels of erucic acid on cardiac lipids and myocardial lesions in rats, *Lipids*, 12, 215, 1977.

171. Parijs, J., De Weerdt, G. A., Beke, R., and Barbier, F., Stereospecific analysis of human plasma triglycerides, *Lipids*, 9, 937, 1974.

172. Gordon, D. T., Pitas, R. E., and Jensen, R. G., Effects of diet and type IIa hyperlipoproteinemia upon structure of triacylglycerols and phosphatidylcholines from human plasma lipoproteins, *Lipids*, 10, 270, 1975.

173. Parijs, J., De Weerdt, G. A., Beke, R., and Barbier, F., Stereospecific distribution of fatty acids in human plasma triglycerides, *Clin. Chim. Acta*, 66, 43, 1976.

174. Myher, J. J., Kuksis, A., Breckenridge, W. C., and Little, J. A., Studies of triacylglycerol structure of very low density lipoproteins of normolipemic subjects and patients with type III and type IV hyperlipoproteinemia, *Lipids*, 19, 683, 1984.

175. Myher, J. J., Kuksis, A., Breckenridge, W. C., McGuire, V., and Little, J. A., Comparative studies on triacylglycerol structure of very low density lipoproteins and chylomicrons of normolipemic subjects and patients with type II hyperlipoproteinemia, *Lipids,* 20, 90, 1985.

176. Clandinin, M. T., Field, C. J., Hargreaves, L., Morson, L., and Zsigmond, E., Role of diet fat in subcellular structure and function, *Can. J. Physiol. Pharmacol.,* 63, 546, 1985.

Chapter 7

THE EFFECT OF FAT-SOLUBLE XENOBIOTICS ON INTESTINAL LIPID, APOPROTEIN, AND LIPOPROTEIN SYNTHESIS AND SECRETION*

Frank P. Bell

TABLE OF CONTENTS

---

* This chapter is dedicated to Dr. F. W. Quackenbush, whose inspiration, support, and guidance in the early years will always be remembered.

I. INTRODUCTION

Intestinal tissue in man and experimental animals has a wide variety of metabolic capabilities which are associated with maintaining the integrity of the tissue and with performing the functions of nutrient absorption and transport. The intestinal wall also possesses a high capacity for the biotransformation of xenobiotics.[1,2] Such biotransformations, which may be synthetic (e.g., acetylation, glucuronic acid conjugation) or nonsynthetic (e.g., oxidation, reduction) are commonly called detoxification reactions; the reactions do not necessarily yield products with reduced toxicity potential, however.[3] A major function of the small intestine (duodenum and proximal jejunum) is the absorption of digested lipids from the intestinal lumen and their transport into the circulation via the intestinal lymphatics or the portal venous system. In the case of lipids dependent upon the lymphatic route of transport, a complex series of biochemical events takes place in the intestinal wall; these include resynthesis of triglycerides, phospholipids, and cholesteryl esters that were hydrolyzed during digestion by lipases as well as the synthesis of apoproteins necessary for the formation of chylomicron particles.[4,5] The assembly of lipids (which include the fat-soluble vitamins A and E) and apoproteins to form the chylomicra and their subsequent secretion into the lymphatics ultimately completes the transferral of digested lipid from the intestinal lumen to the systemic circulation.

In addition to resynthesizing triglycerides, phospholipids, and cholesteryl esters from the absorbed products of luminal fat digestion (i.e., cholesterol, fatty acids, monoglycerides), the intestine also synthesizes phospholipids[6,7] and cholesterol[8-10] *de novo* and is, in fact, regarded as an important production site for plasma cholesterol.[8-10] More recently, the intestine has also become recognized as an important production site of plasma very low-density lipoproteins (VLDL)[4,11,12] and high-density lipoproteins (HDL)[5,13-15] as well.

The purpose of the present chapter is to review our knowledge concerning the effect of fat-soluble xenobiotics* on the ability of the intestine to synthesize and secrete lipids and lipoproteins. This review is a particularly important undertaking, not for the reason that it will summarize a great body of literature (for it does not exist), but for the reason that it will, hopefully, identify an important area of research that has only been marginally investigated.

As one examines the biological literature on xenobiotics, it becomes clear that the intestine has been extensively examined in its role as a route of entry of the xenobiotics into the blood circulation, with considerably less being published on the effects of xenobiotics on intestinal metabolism, per se. Our need to understand how xenobiotics can affect major aspects of lipid and lipoprotein metabolism in the intestine becomes important in view of the fact that man's exposure to medicinal xenobiotics (drugs) and to environmental xenobiotics (chemicals) is likely to increase rather than to decrease in the future.[16,17]

II. NORMAL LIPID ABSORPTION, RESYNTHESIS, AND SECRETION BY THE INTESTINE

A. Overview

The processes of intralumenal digestion of ingested fats (lipids) have been the subject of several excellent reviews[18,19] and will not be discussed in detail here. Figure 1, however, summarizes the metabolic fate of the absorbed products which arise from digestion (hydrolysis) of triglycerides, phospholipids, and cholesteryl esters (the major classes of dietary lipids) by pancreatic lipase, pancreatic phospholipase A_2, and pancreatic cholesterol ester hydrolase. These products are primarily fatty acids, 2-monoglycerides, lysophosphatides, and cholesterol; fat-soluble vitamins A, D, E, and K are likely to be present as well. Esters of vitamins A, D, and E are hydrolyzed by a nonspecific pancreatic carboxylic ester hydrolase which may be identical to cholesterol ester hydrolase.[19,20] Once within the mucosa, most products of fat digestion, including vitamin A, become substrates for reesterification reactions in the endoplasmic reticulum (ER) which prepare them for incorporation into chylomicron particles[4,5] for secretion into the lymphatics. Vitamin K (natural), as well as vitamins D and E, are also transported predominantly via the lymph.[21] Although vitamin D,[21] but apparently not

---

* Other xenobiotics are occasionally discussed to illustrate specific points.

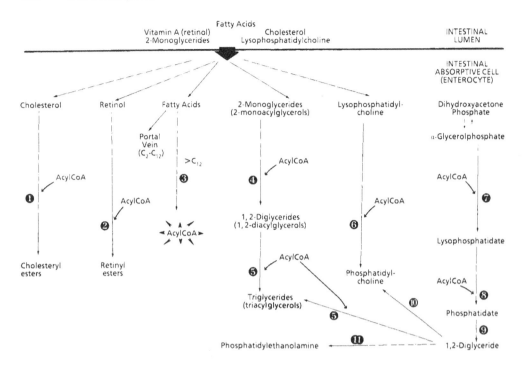

FIGURE 1.   Lipid metabolism in the small intestine. Numbers shown in the metabolic scheme correspond with the various enzymes listed as follows: (1) AcylCoA: cholesterol acyltransferase (E.C. 2.3.1.26) and/or cholesterol esterase (E.C. 3.1.1.13); (2) acylCoA: retinol acyltransferase; (3) acylCoA synthetase (E.C. 6.2.1.3); (4) monoacylglycerol acyltransferase (E.C. 2.3.1.22); (5) diacylglycerol acyltransferase (E.C. 2.3.1.20); (6) lysophosphatidylcholine acyltransferase (E.C. 2.3.1.23); (7) glycerol phosphate acyltransferase (E.C. 2.3.1.15); (8) lysophosphatidate acyltransferase; (9) phosphatidate phosphohydrolase (E.C. 3.1.3.4); (10) choline phosphotransferase (E.C. 2.7.8.2); (11) ethanolamine phosphotransferase (E.C. 2.7.8.1).

vitamin E,[22,23] can be esterified in the mucosa, esterification is not a prerequisite for its secretion into lymph.[21] Lipid digestion products which are not dependent on the lymphatic route of transport are the short-chain (C_2 to C_6) and medium-chain (C_8 to C_{12}) fatty acids[24,25] and some molecular species of fat-soluble vitamins (e.g., vitamin K_3, synthetic)[26,27] which are transferred unaltered into the portal venous blood for direct delivery to the liver.

B. Fatty Acids

The long-chain fatty acids are activated to their acylCoA esters by acylCoA synthetase (E.C. 6.2.1.3) in the ER and are thought to be carried to that site by a cytosolic protein known as fatty acid binding protein (FABP),[28,29] formerly known as the Z-protein.

C. Triglycerides

Triglycerides are synthesized either by the acylation of 2-monoglyceride with two molecules of acylCoA (predominant pathway, 80%) or by acylating α-glycerol phosphate (via phosphatidic acid) with three molecules of acylCoA.[5,19] This latter route to triglycerides depends upon the generation of trioses (dihydroxyacetone phosphate and α-glycerol phosphate) from carbohydrate metabolism in the mucosa.[5]

D. Phospholipids

Phospholipids such as lecithin (phosphatidylcholine) are synthesized by direct acylation of absorbed lysolecithin. *De novo* synthesis of lecithin from α-glycerol phosphate also occurs, but appears to be more important in providing phospholipid for membrane synthesis rather than for transport purposes.[6,7]

E. Cholesterol and Cholesteryl Esters

Approximately 70 to 90% of the cholesterol absorbed by the intestinal mucosa is reesterified prior to secretion into lymph in chylomicrons;[30,31] the remainder is secreted into lymph as the unesterified sterol. Three enzymes capable of esterifying cholesterol in the mucosa have been described.[32] These are acylCoA: cholesterol acyltransferase (ACAT, E.C. 2.3.1.26) of the ER,[33-35] pancreatic cholesterol esterase (E.C. 3.1.1.13, a soluble enzyme) that is taken up from the intestinal lumen,[36-38] and cholesterol ester synthetase.[39] The role of each of these in the formation of chylomicron cholesteryl esters is controversial at present.[32,35,40,41]

F. Vitamin A (Retinol)

Esterification of absorbed vitamin A with fatty acid in the mucosa is thought to be performed by acylCoA: retinol acyltransferase in the ER.[42] The enzyme has a number of properties in common with ACAT,[42] but appears to be a separate enzyme. About 90% of lymph vitamin A is present in the esterified form.[43] β-Carotene (provitamin A) is absorbed from the lumen intact and split in the mucosa to yield 2 molecules of vitamin A,[26] which then become esterified with fatty acid.

G. Chylomicron Apoproteins and Lipids and Chylomicron Formation

Chylomicrons are synthesized during fat absorption and are the principal transport vehicle for the transference of exogenous lipid from the intestinal lumen to the blood circulation via the lymphatic system. The protein portion of chylomicrons isolated from intestinal lymph of man and the rat represents from 1 to 2% of the total mass of the chylomicra[5,44] and is comprised of a number of apoproteins synthesized by the intestine. These apoproteins are apoB, apoA-I, apoA-IV, and a small amount of apoC-II; in addition, man synthesizes apoA-II[5,25] which is a minor apoprotein in the rat.[5] ApoE and some apoC species are also acquired by the chylomicrons upon their exposure to plasma,[5,45] although recent evidence of a functional apoE m-RNA in rat intestine does allow for the possibility that some chylomicron apoE is derived from the intestine.[46]

The primary lipid constituents of chylomicrons are triglyceride (80 to 90%), phospholipid (8 to 13%), and cholesterol (2 to 3%)[25,44] which reflect the nature of the major dietary fats. The bulk of the triglyceride, cholesteryl esters, and about one third of the unesterified cholesterol are located in a central core of the chylomicron particle. The outer surface of the chylomicron is comprised of phospholipid, apoproteins, unesterified cholesterol, and some saturated triglyceride.[47] Failure of the intestine to synthesize chylomicron apoproteins results in triglyceride droplet accumulation within the intestinal cells.[48,49] The significance of apoB to chylomicron formation and triglyceride transport is demonstrated in patients with the genetic disorder of abetalipoproteinemia. The inability of the intestine to produce apoB in such patients is coupled to an inability to synthesize triglyceride-rich lipoproteins and results in intestinal accumulation of triglyceride.[5,50,51]

An understanding of the mechanisms by which apoproteins and lipids are assembled into chylomicrons and the details of the secretion process for chylomicrons are still incomplete. However, through combined biochemical and ultrastructural studies of the small intestine of man and experimental animals, it has been possible to gain substantial information about these processes.[25,48,52-60]

Briefly, the following sequence is suggested.[52,59] Absorbed lipid first appears as triglyceride droplets in the smooth endoplasmic reticulum (SER) in the apical region of enterocytes (absorptive cells). The droplets are then transported through reticular channels to the Golgi apparatus. As the droplets accumulate in the Golgi, the Golgi becomes more prominent. Golgi-derived secretory vesicles carrying nascent chylomicrons migrate to, and fuse with, the lateral plasmalemma of the enterocytes; the nascent chylomicrons are thereby discharged into the intercellular space by reverse pinocytosis (exocytosis). The chylomicrons then pass into the lamina propria and finally into the lacteal lumina which constitute the intestinal lymphatics.

Since the SER, RER (rough endoplasmic reticulum), and the Golgi represent a contiguous tubular network, it is thought that products of protein synthesis in the RER can merge with products of lipid synthesis in the SER,[56,59] and that protein glycosylation (apoB is a glycoprotein)[61] and final assembly of nascent chylomicrons take place in the Golgi.[56,62]

III. XENOBIOTIC EFFECTS ON INTESTINAL LIPID AND LIPOPROTEIN METABOLISM

A. Contributing Factors to Xenobiotic Effects

The ability of a xenobiotic to affect any aspect of intestinal lipid or lipoprotein metabolism will depend, in part, upon factors which determine its rate and extent of absorption, its rate of biotransformation by intestinal enzymes, and/or its rate of clearance from the tissue. The amount absorbed is influenced by such events as intestinal motility,[17,63] xenobiotic metabolism by intestinal microflora,[64-66] the physical and chemical properties of the xenobiotic,[67,68] the nature of the diet and coingestion of competing substances,[69-74] species differences,[17,65] age,[2,75,76] and intestinal blood flow,[77] to name a few. Intestinal biotransformation of xenobiotics[3,17,78,79] may be enhanced through induction of intestinal enzymes.[80-82]

B. Modification of Triglyceride Synthesis and Secretion

1. Ethanol

Triglyceride synthesis in rat intestinal tissue is increased in animals pretreated orally with ethanol and in normal intestine which is preincubated with ethanol;[83] increases under both sets of circumstances are 30% or greater. The increased synthesis of triglyceride correlates with increases in lymph triglyceride secretion observed after intraduodenal administration of ethanol in the rat.[84,85] The mechanism of the stimulation of triglyceride synthesis is attributed to increased activity in the microsomal enzymes acylCoA synthetase (E.C. 6.2.1.3) and monoacylglycerol acyltransferase (E.C. 2.3.1.22).[86]

2. Colchicine

Colchicine, which is known to inhibit microtubule polymerization, results in the accumulation of exogenous[53,54,87] and endogenous[54,58] lipid in the intestinal wall, an observation which has suggested a functional significance of microtubules in lipoprotein secretion.[53,54,87] In addition to blocking lipoprotein secretion, however, there is now evidence that colchicine preferentially directs fatty acid esterification into triglycerides, thereby reducing the percentage of absorbed fatty acid that becomes incorporated into phospholipids.[58]

3. 2-Ethyl-n-Caproic Acid

In contrast to ethanol, 2-ethyl-*n*-caproic acid inhibits triglyceride synthesis by everted sacs of rat small intestine.[88] The inhibition appears to arise from an inhibition of long-chain fatty acid activation by acylCoA synthetase (E.C. 6.2.1.3) in the intestinal mucosa.[88]

4. Fenfluramine (N-ethyl-α-m-[trifluoromethyl]phenethylamine)

Inhibition of triglyceride synthesis in rat intestinal mucosa has also been demonstrated both in vivo and in vitro with the anorexic compound, fenfluramine; in this case inhibition appears to result from inhibition of fatty acylCoA transfer to monoglyceride by monoacylglycerol acyltransferase (E.C. 2.3.1.22).[89,90]

5. 2,6-Di-tert-butylamino-3-acetyl-4-methylpyridine

Recent studies with an experimental glucose transport inhibitor (2,6-di-*tert*-butylamino-3-acetyl-4-methylpyridine) in the rat have revealed an unusual type of chemically induced lipidosis in the rat small intestine. The agent results in a coalescence of chylomicra in the intercellular spaces between epithelial cells in the duodenum, and they subsequently become engulfed by macrophages as they reach the lamina propria.[91] The factors responsible for chylomicron coalescence remain speculative, but may be related to structural or compositional changes in the particles as a result of changes in triglyceride (or protein) synthesis.

6. TCDD (2,3,7,8-Tetrachloro-dibenzo-p-dioxin)

TCDD is an extremely toxic contaminant of herbicides such as 2,4,5-trichlorophenoxyacetic acid (2,4,5,-T) and agent orange (a 1:1 mixture of 2,4,5-T and 2,4-D which is 2,4-dichlorophenoxyacetic acid). An intestinal receptor for TCDD has been detected in rat intestinal mucosa[92] and is presumably present in man as well. In rats previously exposed to orally administered TCDD (80 μg/kg), oral dosing with 1 m*l* olive oil led to a progressive accumulation of fat in the intestinal epithelial cells. The fat (triglyceride) accumulated in abnormally large droplets which were delayed in passage into the lymph.[93] The accumulation seems likely to reflect an impairment in chylomicron formation secondary to inhibition of protein synthesis.[94] The relationship, if any, between these intestinal effects and the subsequent hypertriglyceridemia which follows oral exposure of rats[95] and rabbits[96] to TCDD is unknown.

7. α-Hexachlorocyclohexane

In addition to the studies reported above, there is a number of studies in which changes in triglyceride following xenobiotic exposure have been attributed to liver metabolism which could, in part, reflect a combined effect of the xenobiotics on the intestine as well. For example, there is a progressive rise in serum VLDL and triglycerides (five- and threefold, respectively) within a 48-hr period following a single oral dose (200 mg/kg) of α-hexachlorocyclohexane (α-isomer of the insecticide lindane) in the rat.[97] Despite the fact that there was no direct study made of intestinal triglyceride or lipoprotein secretion, it was concluded that the VLDL and triglyceride elevations were likely of hepatic origin.[97]

8. Polychlorinated Biphenyls (PCBs)

PCBs (refers to over 200 possible chlorobiphenyl isomers)[98] are quantitatively absorbed by the intestine[99] and transported in the circulation in association with the plasma lipoproteins (VLDL, LDL, HDL).[100] PCBs are also readily taken up by lipoproteins in vitro.[101,102] Various reports in the literature indicate disturbances of triglyceride metabolism that are associated with PCBs. None of the studies, however, examined intestinal tissue. The incorporation of [14C]acetate into triglycerides by cultured human skin fibroblasts was increased over twofold when the PCB 2,4,5,2′,4′,5′-hexachlorobiphenyl (HCB) was present at a level of 5.5 μM in the culture medium.[103] In addition, cellular triglyceride content was elevated 29% as a result of 24-hr exposure to HCB (13.8 μM).[103] In liver, PCB exposure results in proliferation of SER, disorientation of RER, and triglyceride accumulation.[104-107] The triglyceride accumulation appears to reflect a decreased clearance (rather than increased synthesis) that derives from an impaired delivery of VLDL particles from the ER to the Golgi and an impaired release of VLDL from the Golgi.[108] Whether or not similar events occur in the intestine (i.e., triglyceride accumulation, impaired release of VLDL, or chylomicrons) with PCB exposure appears to be an open question.

9. Aflatoxins

Xenobiotics with effects on intestinal lipid and lipoprotein metabolism may not necessarily be restricted to the man-made variety, since aflatoxin B_1 (one of a group of toxic, carcinogenic furanocoumarins) produced by *Aspergillus parasiticus* has been shown to inhibit lipid synthesis from [14C]acetate in mink intestine in vitro.[109] Although the triglyceride fraction was not specifically isolated from the tissue lipid extract, parallel studies done with liver specifically showed a reduced incorporation of [14C]acetate into the triglyceride fraction.[109]

C. Modification of Phospholipid Synthesis
1. DDT(1,1,1-Trichloro-2,2-bis[p-chlorophenyl]ethane)

The insecticide DDT, which is readily absorbed by the intestine and incorporated into chylomicrons,[110,111] has recently been shown to increase (approximately twofold) the phospholipid content of intestinal mucosa and brush border membranes in the Rhesus monkey.[112,113] In vitro studies with intestinal mucosa from treated monkeys (150 mg/kg body weight for 2 days) also showed a twofold increase in the incorporation of [14C]acetate into phospholipids.[113] To date, no data are available as to which mucosal enzymes of phospholipid biosynthesis are affected by DDT. The stimulatory effect of DDT on phospholipid synthesis is somewhat selective in that triglyceride and cholesteryl ester synthesis are unaffected.[113]

2. Halothane(2-Bromo-2-chloro-1,1,1-trifluoroethane)

The inhalation anesthetic halothane has been reported to inhibit the base-exchange reaction by which phosphatidylserine is formed from phosphatidylethanolamine in guinea pig ileum.[114] Whether or not this inhibition represents a selective effect is not known, since the study dealt specifically with phosphatidylserine.[114]

3. PCBs

The PCBs are known to affect phospholipid synthesis, but it appears that studies thus far have been conducted in tissues other than the intestine. The effect of PCBs on phospholipid synthesis in tissues and cells is somewhat enigmatic. Phospholipid synthesis in liver microsomes from rats exposed to PCB may be increased[115] or decreased,[116,117] depending upon experimental conditions, and may be accompanied by increases in liver phospholipid levels.[116,118] In cultured human skin fibroblasts, low levels (5.5 μM) of HCB stimulate [14C]acetate incorporation into phospholipids, whereas high levels (>27 μM) are inhibitory.[103] When [14C]glycerol-3-phosphate is the precursor, inhibition of phospholipid synthesis is >50% at 5.5 μM, with no stimulation being observed at higher HCB concentrations.[103] Since the pathways of phospholipid synthesis in intestine and liver are similar, there is the possibility that PCBs can also affect intestinal phospholipid synthesis. In view of the widespread distribution of PCBs in the environment and in the food supply,[98,119,120] studies of the effects of PCBs on intestinal lipid metabolism are needed.

D. Modification of Cholesterol and Cholesteryl Ester Synthesis
1. Clofibrate(Ethyl-p-chlorophenoxyisobutyrate)

Xenobiotics which affect intestinal cholesterol synthesis include the well-known lipid-lowering drug, clofibrate.[121] Inhibition of sterologenesis has been observed in hamster jejunum[122] and in rat jejunal villous cells (but not crypt cells)[123] following clofibrate treatment. Studies in man also indicate inhibition of sterologenesis in intestinal mucosa.[121] This inhibition, however, has not been observed in intestinal tissue of the dog[124] or pig,[125] nor has it been consistently observed in the rat[126] where differences in experimental design may be of major importance.[126]

2. Phenobarbital, Ethanol, and AOMA

Intestinal sterol synthesis is also increased with some xenobiotics. Among these are phenobarbital,[127] ethanol,[128] and the experimental compound AOMA (a copolymer of maleic acid and an α-olefin chain of 18 carbons).[126] Phenobarbital treatment of rats (80 mg/kg/day i.p.) for 5 days resulted in a twofold increase in the incorporation of [14C]acetate into cholesterol by the small intestine in vitro.[127] In the case of ethanol, a single oral dose (7.5 g/kg) 18 hr prior to experimentation was sufficient to increase cholesterol synthesis from [14C]acetate by a factor of 2 to 4 in the small intestine, in vitro.[128] The mechanisms of the phenobarbital and ethanol effects have not been elucidated. In the case of AOMA, however, the stimulation in intestinal sterologenesis (fourfold) is thought to result from lowering feedback inhibition on the biosynthetic pathway as a result of decreased cholesterol absorption by the intestinal mucosa.[126]

3. Compounds 57-118 and 58-035

The recent report of several inhibitors of intestinal ACAT (acylCoA: cholesterol acyltransferase, E.C. 2.3.1.26) has permitted evaluation of the role of this microsomal enzyme in the esterification of absorbed cholesterol in the intestine.[35,41] Compound 57-118 (N-[1-oxo-9-octadecenyl]-DL-tryptophan) (z) ethyl ester inhibits cholesterol absorption in the cholesterol-fed rabbit in a dose-dependent fashion and has been demonstrated to inhibit mucosal microsomal ACAT both in vitro and ex vivo.[35] Similarly, compound 58-035 (3-[decyldimethylsilyl]-N-[2-(4-methylphenyl)-1-phenyl-ethyl]propanamide), which inhibits rat intestinal ACAT in vitro[41] and decreases lymphatic secretion of cholesteryl esters in the lymph-fistula rat,[41] also inhibits cholesterol absorption in the intact rat.[129] The studies with both of these compounds suggests that mucosal ACAT plays an important role in the intestinal transport of cholesterol.

E. Modification of Fatty Acid Synthesis

1. Cyclopropenoid Fatty Acids

Naturally occurring cyclopropenoid fatty acids found in plant lipids such as cottonseed oil and sterculia foetida oil are inhibitors of long-chain fatty acid desaturation in liver of various species.[130-132] Whether or not these fatty acids affect any aspect of lipid metabolism in the intestine is unknown.

F. Modification of Apoprotein and Lipoprotein Synthesis and Secretion

The preceding sections have made mention of the ability of the intestine to synthesize and secrete chylomicrons, VLDL, and HDL into the lymphatic system.[4,5,13,15] Of these, it is only the chylomicron that is a unique product of intestinal tissue; VLDL and HDL are produced by the liver as well.[44] In man and the rat, intestinal VLDL (particle size 200 to 800 Å) resembles chylomicrons (particle size 750 to 6000 Å) in apoprotein composition and lipid composition, although the apoproteins and lipids are in somewhat different proportions.[4,5,44,61] The role of the chylomicrons is to transport lipid of exogenous origin, whereas the VLDL transport lipid of endogenous origin.[5,44] The role of intestinal HDL is unclear, but may offer an alternate transport vehicle for fat-soluble vitamins and other apolar lipids from the intestine.[133] Relationships between intestinal exposure to xenobiotics and changes in lipoprotein (or apoprotein) synthesis and secretion have been mentioned earlier in connection with ethanol,[84,85] colchicine,[53,54,87] 2,6-di-*tert*-butylamino-3-acetyl-4-methylpyridine,[91] and TCDD.[93] Theoretically, there are a variety of mechanisms through which xenobiotics can (or do) operate to modify intestinal apoprotein and lipoprotein synthesis and secretion; these can likely be characterized by specific events which would fall within the broad categories of (1) changes in apoprotein synthesis, (2) alterations in lipoprotein assembly, and (3) alterations in lipoprotein secretion (release).

1. Changes in Apoprotein Synthesis

The significance of intestinal apoprotein synthesis to lipid transport was first evident in patients with abetalipoproteinemia[4,134,135] and has since been demonstrated in animal models with protein synthesis inhibitors such as puromycin.[48,49,136-138]

a. DDT

A number of fat-soluble xenobiotics are known to affect amino acid uptake by the intestine and have, therefore, the potential to alter intestinal apoprotein synthesis. For example, a single oral dose of DDT (150 mg/kg) to rhesus monkeys has been shown to inhibit the uptake of L-alanine and L-phenylalanine by the small intestine under in vitro conditions.[139] A single oral dose of DDT has also been shown to reduce the uptake of [^{14}C]-labeled L-glycine and L-leucine in the pigeon jejunum by 19 and 7%, respectively.[140] Although a direct addition of DDT (final concentration 1.0 m*M*) to pigeon jejunum incubations did not alter amino acid uptake according to the same investigators,[140] L-tyrosine uptake by everted sacs of rat small intestine has been shown to be inhibited by DDT and its metabolite DDE, at concentrations of 10^{-4} to 10^{-5} *M*.[141]

b. TCDD

TCDD has also been shown to inhibit L-[^{14}C]leucine uptake in everted sacs of small intestine from rats previously pretreated with a single oral dose of the compound (100 µg/kg);[142] the effects were not observed, however, until 2 to 3 weeks postadministration. This lengthy time lag in development of the TCDD effect may explain why L-leucine uptake by intestinal rings of rats which were examined 7 days postadministration of TCDD (80 µg/kg orally) did not differ from controls.[95]

c. Dieldrin and Malathion

Inhibition of L-leucine uptake (approximately 50%) by small intestine of the rhesus monkey has also been reported 24 hr postdosing with the pesticide dieldrin (20 mg/kg orally).[143] Additionally, the pesticide malathion has been found to decrease (approximately 35%) L-glycine uptake by rat small intestine following a single oral dose (1 g/kg) of the pesticide.[144]

The doses of xenobiotics used in the studies cited above are greatly in excess of what one might expect from environmental exposure and as such, may reflect exaggerated responses of host tissue. Aside from L-leucine and L-phenylalanine, studies with the other essential amino acids appear to be lacking.

d. Miscellaneous

Apoprotein synthesis could also be affected for reasons other than problems with amino acid transport. ApoB, a major component of intestinal VLDL and chylomicrons, is a glycoprotein with about 5% carbohydrate; apoA-I also appears to have a small component of carbohydrate.[61] A study by Kessler et al.[62] in apical cell suspensions from rat small intestine indicates that the protein and sugar moieties of apoproteins are synthesized in the rough endoplasmic reticulum and the Golgi apparatus, respectively, and then assembled in the Golgi along with lipid to make lipoprotein. At the time of this writing, there do not appear to be any reports of intestinal apoprotein or lipoprotein synthesis being modified by xenobiotics that alter the sugar (carbohydrate) moiety. The possibility that xenobiotics could modify a glycoprotein like apoB does exist, since inhibition of synthesis of other intestinal glycoproteins has been reported with a number of xenobiotics such as salicylate,[145] phenylbutazone,[146] indomethacin,[147] and the β-adrenergic blocker, practolol.[148]

Recent studies by Gordon et al.[149,150] have demonstrated that apoA-IV mRNA levels increase in rat intestine after fat feeding and that the major translation product of apoA-I mRNA in rat intestine is a high molecular weight preprotein. The possibility that fat-soluble xenobiotics may have effects on intestinal protein synthesis at the transcriptional and translational levels should be considered.

2. Alterations in Lipoprotein Assembly
a. 2,6-Di-tert-butylamino-3-acetyl-4-methylpyridine

Several studies have suggested that some problems of intestinal lipid transport are related to alterations in lipoprotein assembly. For example, the intestinal lipidosis observed in rats treated with 2,6-di-*tert*-butylamino-3-acetyl-4-methylpyridine appeared to result from the coalescence of normally appearing chylomicrons to form abnormally large lipoprotein particles.[91] While the mechanism responsible for particle coalescence is unclear, modifications in assembly of lipid components, apoproteins, or carbohydrate moieties may have altered surface characteristics of the chylomicra in a way which favored coalescence.

b. PCBs

The presence of abnormally large lipoprotein particles has also been reported for VLDL isolated from livers of rats injected with PCBs;[108] the larger particles were considered to be the possible result of interference in the assembly of lipid components with the apoproteins. The possibility that PCBs affect the intestinal VLDL in a similar fashion has not been investigated.

c. Pluronic L-81, a Hydrophobic Detergent

The most unusual effects on intestinal lipoprotein assembly have emerged from a study in which rats received an intraduodenal infusion of the hydrophobic detergent pluronic L-81[151] along with a lipid emulsion. Subsequent examination of jejunum by transmission electron microscopy and of lymph lipoproteins by negative staining indicated that VLDL formation and VLDL release into lymph was operative, but that chylomicron particles were absent from the tissue and the lymph. Since VLDL and chylomicrons contain common lipid and apoproteins,[4,5,44,61] it would appear that the chylomicron assembly processes had been altered by the detergent as opposed to a failure to synthesize apoprotein and lipid components.

3. Lipoprotein and Apoprotein Secretion
a. Colchicine, Vinblastine, and N-Methylated Neomycin

Studies employing antimicrotubule agents such as colchicine and vinblastine provide evidence that microtubules are intimately involved in the process of lipoprotein secretion by enterocytes[53,54,87] as well as by liver.[152-154] Biochemical, radiochemical, and ultrastructural studies in the intestine and lymph from fed and fasted rats treated with colchicine[53,54,58,87] and in organ cultures of pig duodenojejunal mucosa exposed to colchicine[155] have demonstrated a diminished release of chylomicrons (exogenous lipid) and VLDL (endogenous lipid) and an accumulation of tissue lipid. The secretion of apoA-I by intestinal mucosa is not entirely via lipoprotein, but appears to also be in a free form not dependent upon microtubule integrity.[155] Whether or not the inhibition of intestinal lipoprotein secretion by colchicine can be explained solely on the basis of the antimicrotubular activity of this xenobiotic is still an open question.[58] At the present time, the literature is devoid of studies in which an antimicrotubular effect of any fat-soluble xenobiotic has been reported.

In contrast to colchicine, calcium ion[156] and N-methylated neomycin[157] have been reported to increase the secretion of chylomicrons and LDL plus HDL, respectively, in in vitro preparations of hamster and rat intestine. Whether or not the action mechanism of Ca^{2+} and N-methylated neomycin could be mimicked or duplicated by a fat-soluble xenobiotic is unknown.

IV. PHTHALATES

Phthalates (phthalic acid esters) are ubiquitous xenobiotics which have become widely distributed throughout the environment as a result of their widespread industrial use as solvents and as constituents of products such as coatings, films, adhesives, and polyvinyl chloride plastic goods.[158,159] Although this class of compounds has been found to affect lipid accumulation[160,161] and lipid synthesis[160] in animal tissues and to alter plasma lipoprotein compositions,[162] an examination of the possible effect of phthalates on intestinal lipid and lipoprotein metabolism has not been examined to date. There are, however, several studies whose results should direct attention to the intestine. For example, the binding capacity of FABP (fatty acid binding protein) is increased six- to sevenfold in the liver of rats receiving 2 to 4% di-(2-ethyl-hexyl)phthalate (DEHP) in the diet.[163] Whether or not similar changes occur in intestinal FABP is unknown. It has also been shown that DEHP can inhibit hepatic cholesterolgenesis[160,164] and cholesterol esterification by ACAT.[165] Since both of these enzyme systems are present in the intestine as well, the possibility of their inhibition by DEHP seems probable and may be linked to the plasma cholesterol lowering effect of DEHP in several species.[160,166]

Phthalate esters are also known to affect Zn metabolism.[167-170] Studies in rodents have shown that some phthalate esters (e.g., DEHP, mono(2-ethylhexyl)phthalate, and di-n-pentyl phthalate) can cause a decrease in Zn levels in testes,[167-170] despite the presence of adequate levels of the mineral in plasma and liver[171] and unimpaired intestinal absorption of Zn.[171] Whether or not phthalates also affect Zn levels necessary for intestinal cell function is unknown. The importance of Zn to the secretion of absorbed lipid has been documented in Zn-deficient rats.[172] Such a deficiency results in an accumulation of lipid droplets (consisting primarily of triglyceride) in the intestinal mucosa and is accompanied by ultrastructural changes in the RER and Golgi apparatus;[172,173] the lipid accumulation is probably related to a failure of the mucosal cells to synthesize protein for chylomicron formation.[172]

V. METABOLISM OF XENOBIOTICS BY LIPID METABOLIZING ENZYMES AND ITS IMPLICATIONS IN INTESTINAL LIPID AND LIPOPROTEIN SYNTHESIS AND SECRETION

A number of xenobiotics have been shown to undergo metabolism through the action of enzymes which are usually associated with lipid biochemistry.[174]

A. Carboxylic Acids and Their Esters
1. 4-Benzyloxybenzoic Acid and Zardex (Hexadecylcyclopropane Carboxylate)
Studies with 4-benzyloxybenzoic acid in rat jejunum in vitro showed that the com-

pound mimics aliphatic fatty acids in that it is incorporated into triglyceride in the tissue.[175] Information on an interrelationship between lipid metabolism and the metabolism of other xenobiotics comes mainly from studies not specifically focusing on the intestine. For example, the miticide cycloprate (Zardex, hexadecylcyclopropane carboxylate), when administered orally as [carboxyl-^{14}C]cycloprate to the rat,[176] dog,[177] and bovine,[178] is hydrolyzed to the carboxylic acid. Chain elongation of the carboxylic acid to form "hybrid fatty acids" occurs subsequently,[174] and has been demonstrated by identification of a homologous series of saturated fatty acids having a terminal cyclopropyl group;[176] chain elongation probably proceeds by stepwise 2-carbon additions, which characterizes fatty acid elongation in tissues and which involves fatty acid activation with CoA.[176] The major tissue fatty acid generated from cycloprate is 13-cyclopropyltridecanoic acid.[176,177] These unusual fatty acids reside in host tissue, primarily in the esterified form in triglycerides, phospholipids, and carnitine esters, and to some extent are esterified with cholesterol;[176,177] they are also secreted into milk as triglycerides and carnitine esters.[178]

2. 2-Tetradecylglycidic Acid and 5-(Tetradecyloxy)-2-furoic Acid

The hypoglycemic agent 2-tetradecylglycidic acid[179] and the hypolipidemic agent 5-(tetradecyloxy)-2-furoic acid[180] are known to form acylCoA thioesters in hepatic cells.

B. Alcohols
1. Cannabinoid Metabolites, Dipyridamole, and Etofenamate

The xenobiotics bearing alcoholic hydroxyl groups have also been reported to undergo esterification with fatty acids in vivo; examples include metabolites of cannabinoids,[181,182] the vasodilator dipyridamole [2,2',2'',2'''-(4,8-dipiperidinopyrimido[5,4-d]pyrimidine-2,6-diyldinitrilo) tetraethanol][174] and the antiinflammatory agent etofenamate [N-(α,α,α-trifluoro-m-tolyl) anthranilic acid 2-(2-hydroxy-ethoxy)ethylester].[183]

C. Physiologic Implications

The physiologic consequences of a dual utilization of lipid metabolizing enzymes (and their co-factors) for lipid synthesis (or catabolism) as well as for xenobiotic metabolism have scarcely been addressed. A hint of possible problems is suggested, however, in the current literature. For example, the formation of tofylCoA, the CoA ester of 5-(tetradecyloxy)-2-furoic acid, in hepatocytes is paralleled by reductions in cellular levels of CoA.[180] Xenobiotics which might affect the intestine in a similar fashion could complicate the activation of fatty acids for triglyceride synthesis or steryl ester formation. Additionally, xenobiotic metabolism could reduce substrate availability for normal lipid metabolism (e.g., the esterification of hydroxyl-bearing xenobiotics with fatty acids) or yield hybrid lipid molecules (e.g., triglycerides and phospholipids) which have different requirements for transport than native molecular species or which could alter chylomicron size or stability,[91] thus, indirectly modifying lipid transport. Another possibility is the formation of xenobiotic lipid hybrids that resist subsequent metabolism by the tissue. The ability of the intestine to produce xenobiotic lipid hybrids that resist normal metabolism has already been demonstrated, i.e., the 4-benzyloxybenzoic acid-containing triglycerides formed in rat jejunum in vitro are resistant to the action of lipoprotein lipase.[175] The resistance of hybrid lipids to further metabolism could lead to problems of accumulation of the lipid in tissues or plasma.[184]

Another potential problem for the intestine (and other tissues) is that hybrid lipid, once synthesized, could replace or substitute for specific lipids in intracellular membranes of the intestine (e.g., hybrid phospholipids). The insertion of such lipids (or the xenobiotic itself) into membrane could indirectly affect the activity of membrane-bound enzymes. Under physiologic conditions, the activity of a variety of membrane-bound enzymes is influenced by the fluidity of the membrane lipids which in turn is determined largely by the length and degree of unsaturation of fatty acyl chains and the presence of sterol.[185,186] Alterations in membrane fluidity resulting from compositional change or changes in bilayer thickness can affect the conformation of particulate enzymes and thereby modify their rates of catalysis.[187,188] Xenobiotics demonstrating membrane effects include insecticides (e.g., parathion and azinphos) which have been shown to increase membrane permeability to nonelectrolytes[189] and to induce molecular disorder in model lipid bilayers.[190] Changes in cellular membrane composition and stability have also been reported in rats exposed to cyclopropenoid fatty acids[191] and brominated oils[192,193] in the diet. PCB exposure in rats also alters cellular membrane structure (e.g., endoplasmic reticulum).[118,194]

VI. CONCLUDING REMARKS

The remarkable processes involved in the absorption of digested lipids by the enterocyte, lipid resynthesis, apoprotein synthesis, and the assembly of lipids and apoproteins into lipoproteins for secretion by exocytosis are only now beginning to be understood at the molecular and ultrastructural levels. The integrity of such complex processes in modern man is under continuous assault by the wide variety of man-made xenobiotics which enter the body directly through food and water supplies or indirectly through inhaled air or by absorption through skin. Unfortunately, we know all too little about the biochemical effects of most xenobiotics on the intestine; hopefully the opportunity for research on this important organ will be more fully recognized.

ACKNOWLEDGMENT

The author is grateful to Dr. R. Virkhaus for his valuable contribution to information gathering for this chapter and to Ms. J. Obreiter for excellent secretarial assistance in preparing the manuscript. The generosity of Dr. J. E. MacNintch and Mrs. James Burwell in providing reference material is also very much appreciated.

REFERENCES

1. Hartiala, K., Metabolism of foreign substances in the gastrointestinal tract, in *Handbook of Physiology Section 9*, Lee, D. H. K., Falk, H. L., Murphy, S. D., and Geiger, S. R., Eds., American Physiological Society, Bethesda, Md., 1976, 375.
2. Klinger, W., Pharmacology of the developing digestive system, *Pharm. Ther.*, 22, 41, 1983.
3. Hartiala, K., Metabolism of hormones, drugs and other substances by the gut, *Physiol. Rev.*, 53, 496, 1973.
4. Green, P. H. R. and Glickman, R. M., Intestinal lipoprotein metabolism, *J. Lipid Res.*, 22, 1153, 1981.
5. Bisgaier, C. L. and Glickman, R. M., Intestinal synthesis, secretion, and transport of lipoproteins, *Ann. Rev. Physiol.*, 45, 625, 1983.
6. Mansbach, C. M., II, Complex lipid synthesis in hamster intestine, *Biochim. Biophys. Acta*, 296, 386, 1973.
7. Mansbach, C. M., II, The origin of chylomicron phosphatidylcholine in the rat, *J. Clin. Invest.*, 60, 411, 1977.
8. Dietschy, J. M. and Siperstein, M. D., Cholesterol synthesis by the gastrointestinal tract: localization and mechanisms of control, *J. Clin. Invest.*, 44, 1311, 1965.
9. Lindsey, C. A., Jr. and Wilson, J. D., Evidence for a contribution by the intestinal wall to the serum cholesterol of the rat, *J. Lipid Res.*, 6, 173, 1965.

10. Dietschy, J. M. and Wilson, J. D., Regulation of cholesterol metabolism, *N. Engl. J. Med.*, 282, 1128, 1970.

11. Ockner, R. K., Hughes, F. B., and Isselbacher, K. J., Very low density lipoproteins in intestinal lymph: origin, composition, and role in lipid transport in the fasting state, *J. Clin. Invest.*, 48, 2079, 1969.

12. Keim, N. L. and Marlett, J. A., Intestinal secretion of lipids and lipoproteins during carbohydrate absorption in the rat, *J. Nutr.*, 110, 1354, 1980.

13. Green, P. H. R., Tall, A. R., and Glickman, R. M., Rat intestine secretes discoid high density lipoprotein, *J. Clin. Invest.*, 61, 528, 1978.

14. Bearnot, H. R., Glickman, R. M., Weinberg, L., Green, P. H. R., and Tall, A. R., Effect of biliary diversion on rat mesenteric lymph apolipoprotein-I and high density lipoprotein, *J. Clin. Invest.*, 69, 210, 1982.

15. Forester, G. P., Tall, A. R., Bisgaier, C. L., and Glickman, R. M., Rat intestine secretes spherical high density lipoproteins, *J. Biol. Chem.*, 258, 5938, 1983.

16. Office of Technological Assessment, Environmental contamination of food, in Environmental Contaminants in Food, OTA Report No. OTA-F-103 (U.S. Printing Office Stock No. 052-003-00724-0), Washington, D.C., 1979, 15.

17. Chhabra, R. S. and Eastin, W. C., Jr., Intestinal absorption and metabolism of xenobiotics in laboratory animals, in *Intestinal Toxicology*, Schiller, C. M., Ed., Raven Press, New York, 1984, 145.

18. Carey, C. M., Small, D. M., and Bliss, C. M., Lipid digestion and absorption, *Ann. Rev. Physiol.*, 45, 651, 1983.

19. Kuksis, A., Intestinal digestion and absorption of fat-soluble environmental agents, in *Intestinal Toxicology*, Schiller, C. M., Ed., Raven Press, New York, 1984, 69.

20. Hyun, J., Kothari, H., Herm, E., Motenson, J., Treadwell, C. R., and Vahouny, G. V., Purification and properties of pancreatic juice cholesterol ester hydrolase, *J. Biol. Chem.*, 244, 1937, 1969.

21. Clark, M. L. and Harries, J. T., Absorption of lipids, in *Intestinal Absorption in Man*, McColl, I. M. and Sladen, G. E., Eds., Academic Press, New York, 1975, 187.

22. Gallo-Torres, H. E., Obligatory role of bile for the intestinal absorption of vitamin E, *Lipids*, 5, 379, 1970.

23. Pudelkiewicz, W. J. and Nakiya, M., Some relationships between plasma, liver and excreta tocopheral in chicks fed graded levels of alpha-tocopheryl acetate, *J. Nutr.*, 97, 303, 1969.

24. Jackson, M. J., Transport of short chain fatty acids, in *Biomembranes*, Vol. 4B, Smyth, D. H., Ed., Plenum Press, New York, 1974, 673.

25. Riley, J. W. and Glickman, R. M., Fat malabsorption: advances in our understanding, *Am. J. Med.*, 67, 980, 1979.

26. Forth, W. and Rummel, W., Absorption of fat-soluble vitamins, in *Pharmacology of Intestinal Absorption: Gastrointestinal Absorption of Drugs*, Forth W. and Rummel, W., Eds., Pergamon Press, Oxford, 1975, 447.

27. Hollander, D., Intestinal absorption of vitamins A, E, D, and K, *J. Lab. Clin. Med.*, 97, 449, 1981.

28. Ockner, R. K. and Manning, J. M., Fatty acid binding protein in small intestine: identification, isolation, and evidence for its role in cellular fatty acid transport, *J. Clin. Invest.*, 54, 326, 1974.

29. Ockner, R. K. and Manning, J. M., Fatty acid binding protein. Role in esterification of absorbed long chain fatty acid in rat intestine, *J. Clin. Invest.*, 58, 632, 1975.

30. Treadwell, C. R. and Vahouny, G. V., Cholesterol absorption, in *Handbook of Physiology Section 6*, Vol. 3, Code, C. F. and Heidel, W., Eds., Williams & Wilkins, Baltimore, 1968, 1407.

31. Goodman, D. S., Cholesterol ester metabolism, *Physiol. Rev.*, 45, 747, 1965.

32. Gallo, L. L., Myers, S., and Vahouny, G. V., Rat intestinal acylcoenzyme A: cholesterol acyltransferase. Properties and localization, *Proc. Soc. Exp. Biol. Med.*, 177, 188, 1984.

33. Drevon, C. A., Lilljeqvist, A. C., Shreiner, B., and Norum, K., Influence of cholesterol/fat feeding on cholesterol esterification and morphological structures in intestinal mucosa from guinea pigs, *Atherosclerosis*, 34, 207, 1979.

34. Norum, K. R., Helgerud, P., and Lilljeqvist, A. C., Enzymatic esterification of cholesterol in rat intestinal mucosa catalyzed by acylCoA:cholesterol acyltransferase, *Scand. J. Gastroenterol.*, 16, 401, 1981.

35. Heider, J. G., Pickens, C. E., and Kelly, L. A., Role of acylCoA:cholesterol acyltransferase in cholesterol absorption and its inhibition by 57-118 in the rabbit, *J. Lipid Res.*, 24, 1127, 1983.

36. Swell, L., Byron, J. E., and Treadwell, C. R., Cholesterol esterase. IV. Cholesterol esterase of rat intestinal mucosa, *J. Biol. Chem.*, 186, 543, 1950.

37. Gallo, L., Newbill, T., Hyun, J., Vahouny, G. V., and Treadwell, C. R., Role of pancreatic cholesterol esterase in the uptake and esterification of cholesterol by isolated intestinal cells, *Proc. Soc. Exp. Biol. Med.*, 156, 277, 1977.

38. Gallo, L., Chiang, Y., Vahouny, G. V., and Treadwell, C. R., Localization and origin of rat intestinal cholesterol esterase determined by immunocytochemistry, *J. Lipid Res.*, 21, 537, 1980.

39. Watt, S. M. and Simmonds, W. J., The effect of pancreatic diversion on lymphatic absorption and esterification of cholesterol in the rat, *J. Lipid Res.*, 22, 157, 1981.

40. Field, F. J., Cooper, A. D., and Erickson, S. K., Regulation of rabbit intestinal acylcoenzyme A — cholesterol acyltransferase in vivo and in vitro, *Gastroenterology*, 83, 873, 1982.

41. Clark, S. B. and Tercyak, A. M., Reduced cholesterol transmucosal transport in rats with inhibited mucosal acylCoA:cholesterol acyltransferase and normal pancreatic function, *J. Lipid Res.*, 25, 148, 1984.

42. Norum, K. R., Helgerud, P., Petersen, L. B., Groot, P. H. E., and DeJonge, H. R., Influence of diets on acylCoA:cholesterol acyltransferase and on acylCoA:retinol acyltransferase in villous and crypt cells from rat small intestinal mucosa and in the liver, *Biochim. Biophys. Acta*, 751, 153, 1983.

43. Huang, H. S. and Goodman, D. S., Vitamin A and carotenoids. I. Intestinal absorption and metabolism of ^{14}C-labeled vitamin A alcohol and β-carotene in the rat, *J. Biol. Chem.*, 240, 2839, 1965.

44. Eisenberg, S., Lipoprotein metabolism and hyperlipemia, *Atherosclerosis Rev.*, 1, 23, 1976.

45. Imaizumi, K., Fainaru, M., and Havel, R. J., Composition of proteins of mesenteric lymph chylomicrons in the rat and alterations produced upon exposure of chylomicrons to blood serum and serum proteins, *J. Lipid Res.*, 19, 712, 1978.

46. Tanaka, Y., Lin-Lee, Y.-C., Lin-Su, M.-H., and Chan, L., Intestinal biosynthesis of apolipoproteins in the rat: Apo E and Apo A-I mRNA translation and regulation, *Metabolism*, 31, 861, 1982.

47. Zilversmit, D. B., The composition and structure of lymph chylomicrons in dog, rat, and man, *J. Clin. Invest.*, 44, 1610, 1965.

48. Sabesin, S. M. and Isselbacher, K. J., Protein synthesis inhibition: mechanism for the production of impaired fat absorption, *Science*, 147, 1149, 1965.

49. Glickman, R. M., Kirsch, K., and Isselbacher, K. J., Fat absorption during inhibition of protein synthesis: studies of lymph chylomicrons, *J. Clin. Invest.*, 51, 356, 1972.

50. Glickman, R. M., Green, P. H. R., Lees, R. S., Lux, S. E., and Kilgore, A., Immunofluorescence studies of apolipoprotein B in intestinal mucosa. Absence in abetalipoproteinemia, *Gastroenterology*, 76, 288, 1979.

51. Dobbins, W. O., An ultrastructural study of the intestinal mucosa in congenital β-lipoprotein deficiency with particular emphasis upon the intestinal absorptive cell, *Gastroenterology*, 50, 195, 1966.

52. Tytgat, G. N., Rubin, C. E., and Saunders, D. B., Synthesis and transport of lipoprotein particles by intestinal absorptive cells in man, *J. Clin. Invest.*, 50, 2065, 1971.

53. Glickman, R. M., Perrotto, J. L., and Kirsch, K., Intestinal lipoprotein formation: effect of cholchicine, *Gastroenterology*, 70, 347, 1976.

54. Reaven, E. P. and Reaven, G. M., Distribution and content of microtubules in relation to the transport of lipid. An ultrastructural quantitative study of the absorptive cell of the small intestine, *J. Cell Biol.*, 75, 559, 1977.

55. Glickman, R. M., Green, P. H. R., Lees, R. S., and Tall, A. R., Inheritance of apolipoprotein C-II deficiency with hypertriglyceridemia and pancreatitis, *N. Engl. J. Med.*, 299, 1421, 1978.

56. Rubin, C. E., Perkins, W. D., Surawicz, C. M., McDonald, G. B., and Albers, J. J., Ultrastructural apoprotein B localization within human jejunal absorptive cells during fat absorption, *Gastroenterology*, 78, 1248, 1980.

57. Thomson, A. B. R. and Dietschy, J. M., Intestinal lipid absorption: major extracellular and intracellular events, in *Physiology of the Gastrointestinal Tract*, Vol. 2, Johnson, L. R., Ed., Raven Press, New York, 1981, 1147.

58. Pavelka, M. and Gangl, A., Effects of colchicine on the intestinal transport of endogenous lipid. Ultrastructural, biochemical, and radiochemical studies in fasting rats, *Gastroenterology*, 84, 544, 1983.

59. Sabesin, S. M. and Frase, S., Electronmicroscopic studies of the assembly, intracellular transport, and secretion of chylomicrons by rat intestine, *J. Lipid Res.*, 18, 496, 1977.

60. Strauss, E. S., Electron microscopic study of intestinal fat absorption, *J. Lipid Res.*, 7, 307, 1966.

61. Schaefer, E. J., Eisenberg, S., and Levy, R. I., Lipoprotein apoprotein metabolism, *J. Lipid Res.*, 19, 667, 1978.

62. Kessler, J. I., Narcessian, P., and Mauldin, D. P., Biosynthesis of lipoproteins by intestinal epithelium. Site of synthesis and sequence of association of lipid, sugar, and protein moieties, *Gastroenterology*, 68, 105, 1975.

63. Serafin, J. A., Avian species differences in the intestinal absorption of xenobiotics (PCB, dieldrin, Hg^{2+}), *Comp. Pharmacol. Toxicol.*, 78C, 491, 1984.

64. Nelson, D. V. M. and Mata, L. J., Bacterial flora associated with the human gastrointestinal mucosa, *Gastroenterology*, 58, 56, 1970.

65. Miranda, C. L. and Chhabra, R. S., Species differences in stimulation of intestinal and hepatic mixed-function oxidase enzymes, *Biochem. Pharmacol.*, 29, 1161, 1980.

66. Goldman, P., Transformation of xenobiotics by the intestinal microflora, in *Microbial Transformations of Bioactive Compounds*, Vol. 2, Rosazza, J. P., Ed., CRC Press, Boca Raton, Fla., 1982, 43.

67. Houston, J. B., Upshall, D. G., and Bridges, J. W., Studies using carbamate esters as model compounds to investigate the role of lipophilicity in the gastrointestinal absorption of foreign compounds, *J. Pharmacol. Exp. Ther.*, 195, 67, 1975.
68. Ahdaya, S. M., Monroe, R. J., and Guthrie, F. E., Absorption and distribution of intubated insecticides in fasted mice, *Pest. Biochem. Physiol.*, 16, 38, 1981.
69. Morgan, D. P., Dotson, T. B., and Lin, L. I., Effectiveness of activated charcoal, mineral oil, and castor oil in limiting gastrointestinal absorption of chlorinated hydrocarbon pesticide, *Clin. Toxicol.*, 11, 61, 1977.
70. Chhabra, R. S., Intestinal absorption and metabolism of xenobiotics, *Environ. Health Perspect.*, 33, 61, 1979.
71. Keller, W. C. and Yeary, R. A., A comparison of the effects of mineral oil, vegetable oil, and sodium sulfate on the intestinal absorption of DDT in rodents, *Clin. Toxicol.*, 16, 223, 1980.
72. Volpenhein, R. A., Webb, D. R., and Jandacek, R. J., Effect of a nonabsorbable lipid, sucrose polyester, on the absorption of DDT by the rat, *J. Toxicol. Environ. Health*, 6, 679, 1980.
73. Withey, J. R., Collins, B. T., and Collins, P. G., Effect of vehicle on the pharmacokinetics and uptake of four halogenated hydrocarbons by the gastrointestinal tract of the rat, *J. Appl. Toxicol.*, 3, 249, 1983.
74. Manara, L., Coccia, P., and Croci, T., Prevention of TCDD toxicity in laboratory rodents by addition of charcoal or cholic acids to chow, *Food Chem. Toxicol.*, 22, 815, 1984.
75. Shah, P. V. and Guthrie, F. E., Penetration of insecticides through isolated sections of the mouse digestive system. Effects of age and region of intestine, *Toxicol. Appl. Pharmacol.*, 25, 621, 1973.
76. Stohs, S. J., Al-Turk, W. A., and Hassing, J. M., Altered drug metabolism in hepatic and extrahepatic tissues in mice as a function of age, *Age*, 3, 88, 1980.
77. Winne, D., Influence of blood flow on intestinal absorption of xenobiotics, *Pharmacology*, 21, 1, 1980.
78. Hietanen, E. and Lang, M., Control of glucuronide biosynthesis in the gastrointestinal mucosa, in *Conjugation Reactions in Drug Biotransformation*, Aitio, A., Ed., Elsevier/North-Holland, Amsterdam, 1978, 399.
79. Hoensch, H. P., Hutt, R., and Hartmann, F., Biotransformation of xenobiotics in human intestinal mucosa, *Environ. Health Perspect.*, 33, 71, 1979.
80. Lake, B. G., Collins, M. A., and Gangolli, S. D., The effect of treatment with polychlorinated-biphenyl mixture on hepatic and extrahepatic xenobiotic metabolism in the rat and ferret, *Biochem. Soc. Trans.*, 6, 1251, 1978.
81. Stohs, S. J., Graftstrom, R. C., Burke, M. D., and Orrenius, S., Xenobiotic metabolism and enzyme induction in isolated rat intestinal microsomes, *Drug Metab. Dispos.*, 4, 517, 1976.
82. Chhabra, R. S., Effect of dietary factors and environmental chemicals on intestinal drug metabolizing enzymes, *Toxicol. Environ. Chem.*, 3, 173, 1981.
83. Carter, E. A., Drummey, G. D., and Isselbacher, K. J., Ethanol stimulates triglyceride synthesis by the intestine, *Science*, 174, 1245, 1971.
84. Mistilis, S. P. and Ockner, R. K., Alcohol-induced fatty liver: importance of endogenous intestinal lipoproteins, *J. Clin. Invest.*, 49, 66a, 1970.
85. Mistilis, S. P. and Ockner, R. K., Effects of ethanol (ETOH) on intestinal metabolism of endogenous lipids, *Clin. Res.*, 19, 398, 1971.
86. Rodgers, J. B. and O'Brien, R. J., The effect of acute ethanol treatment on lipid-reesterifying enzymes of the rat small bowel, *Am. J. Dig. Dis.*, 20, 354, 1975.
87. Arreaza-Plaza, C. A., Bosch, U., and Otayek, M. A., Lipid transport across the intestinal epithelial cell. Effect of colchicine, *Biochim. Biophys. Acta*, 431, 297, 1976.
88. Vahouny, G. V., Nelson, J., and Treadwell, C. R., Inhibition of triglyceride synthesis in everted intestinal sacs, *Proc. Soc. Exp. Biol. Med.*, 128, 495, 1968.
89. Dannenburg, W. N., Fenfluramine and triglyceride synthesis by microsomes of intestinal mucosa in the rat, *Vie Med. Can. Fr.*, 2, 41, 1973.
90. Dannenburg, W. N., Kardian, B. C., and Norrell, L. Y., Fenfluramine and triglyceride synthesis by microsomes of intestinal mucosa in the rat, *Arch. Int. Pharmacodyn. Ther.*, 201, 115, 1973.
91. Visscher, G. E., Robison, R. L., and Hartman, H. A., Chemically induced lipidosis of small intestinal villi in the rat, *Toxicol. Appl. Pharmacol.*, 55, 535, 1980.
92. Johansson, G., Gillner, M., Högberg, B., and Gustafsson, J. Å., Detection of the TCDD-receptor in rat intestinal mucosa using isoelectric focusing in polyacrylamide gel, in *Developments in Biochemistry*, Vol. 13, Gustafsson, J.-Å., Carlstedt-Duke, J., Mode, A., and Rafter, J., Eds., Elsevier/North-Holland, Amsterdam, 1980, 243.
93. McConnell, E. E. and Shoaf, C. R., Studies on the mechanism of 2,3,7,8-tetrachlorodibenzo-*p*-dioxin (TCDD)toxicity-lipid assimilation. I. Morphology, *Pharmacologist*, 23, 176(342a), 1981.
94. Shoaf, C. R. and Schiller, C. M., Studies on the mechanism of 2,3,7,8-tetrachlorodibenzo-*p*-dioxin (TCDD) toxicity-lipid assimilation. II. Biochemistry, *Pharmacologiest*, 23, 176(341a), 1981.

95. Schiller, C. M., Walden, R., and Shoaf, C. R., Studies on the mechanism of 2,3,7,8-tetrachlorodi-benzo-*p*-dioxin toxicity: nutrient assimilation, *Fed. Proc.*, 41, 1426, 1982.
96. Lovati, M. R., Galbussera, M., Franceschini, G., Weber, G., Resi, L., Tanganelli, P., and Sirtori, C. R., Increased plasma and aortic triglycerides in rabbits after acute administration of 2,3,7,8-tetrachlorodibenzo-p-dioxin, *Toxicol. Appl. Pharmacol.*, 75, 91, 1984.
97. Grajewski, O. and Oberdisse, E., Increase in serum very low density lipoproteins in rats after admin-istration of *a*-hexachlorocyclohexane, *Naunyn-Schmiedeberg's Arch. Pharmacol.*, 298, 129, 1977.
98. Rosenman, K. D., Chemical contamination episodes, in *Environmental and Occupational Medicine*, Rom, W. N., Ed., Little, Brown, Boston, 1983, 595.
99. Morales, N. M., Tuey, D. B., Colburn, W. A., and Matthews, H. B., Pharmacokinetics of multiple oral doses of selected polychlorinated biphenyls in mice, *Toxicol. Appl. Pharmacol.*, 48, 397, 1979.
100. Spindler-Vomachka, M., Vodicnik, M. J., and Lech, J. J., Transport of 2,4,5,2′,4′,5′-hexachlorobi-phenyl by lipoproteins in vivo, *Toxicol. Appl. Pharmacol.*, 74, 70, 1984.
101. Maliwal, B. P., and Guthrie, F. E., In vitro uptake and transfer of chlorinated hydrocarbons among human lipoproteins, *J. Lipid Res.*, 23, 474, 1982.
102. Vomachka, M. S., Vodicnik, M. J., and Lech, J. J., Characteristics of 2,4,5,2′,4′,5′-hexachlorobi-phenyl distribution among lipoproteins in vitro, *Toxicol. Appl. Pharmacol.*, 70, 350, 1983.
103. Beranek, S. R., Becker, M. M., Kling, D., and Gamble, W., Phospholipid and glyceride biosynthesis in 2,4,5,2′,4′,5′-hexachlorobiphenyl-treated human skin fibroblasts, *Environ. Res.*, 34, 103, 1984.
104. Kimbrough, R. D., Linder, R. E., and Gaines, T. B., Morphological changes in livers of rats fed polychlorinated biphenyls. Light microscopy and ultrastructure, *Arch. Environ. Health*, 25, 354, 1972.
105. Koller, L. D. and Zinkl, J. G., Pathology of polychlorinated biphenyls in rabbits, *Am. J. Pathol.*, 70, 363, 1973.
106. Kasza, L., Weinberger, M. A., Carter, C., Hinton, D. E., Trump, B. F., and Brouwer, E. A., Acute, subacute and residual effects on polychlorinated biphenyl (PCB) in rats. II. Pathology and electron microscopy in liver and serum enzyme study, *J. Toxicol. Environ. Health*, 1, 689, 1976.
107. Lipsky, M. M., Klaunig, J. E., and Hinton, D. E., Comparison of acute response to polychlorinated biphenyl in liver of rat and channel catfish: a biochemical and morphological study, *J. Toxicol. Environ. Health*, 4, 107, 1978.
108. Sandberg, P.-O. and Glaumann, H., Studies on the cellular toxicity of polychlorinated biphenyls (PCBs) partial block and alteration of intracellular migration of lipoprotein particles in rat liver, *Exp. Mol. Pathol.*, 32, 1, 1980.
109. Chou, C. C. and Marth, E. H., Incorporation of [2-¹⁴C]acetate into lipids of mink *(Mustela vison)* liver and intestine during in vitro and in vivo treatment with aflatoxin B1, *Appl. Microbiol.*, 30, 946, 1975.
110. Popcock, D. M.-E. and Vost, A., DDT absorption and chylomicron transport in the rat, *Lipids*, 9, 374, 1974.
111. Sieber, S. M., The entry of foreign compounds in the thoracic duct lymph of the rat, *Xenobiotica*, 4, 265, 1974.
112. Mahmood, A., Sanyal, S., Agarwal, N., and Subrahmanyam, D., Lipid composition of monkey intestinal brush border membrane: effect of DDT administration, *Ind. J. Biochem. Biophys.*, 16, 175, 1979.
113. Mahmood, A., Agarwal, N., Sanyal, S., Dudeja, P. K., and Subrahmanyam, D., Effect of DDT (chlorophenotane) administration on lipid metabolism in intestinal epithelium of Rhesus monkeys, *Ind. J. Exp. Biol.*, 18, 660, 1980.
114. Paton, W. D. M. and Wing, D. R., Effects of halothane on the incorporation of [¹⁴C]serine into phospholipid in the guinea-pig ileum, *Br. J. Pharmacol.*, 72, 393, 1981.
115. Dzogbefia, V. P., Kling, D., and Gamble, W., Polychlorinated biphenyl in vivo and in vitro modifi-cations of phospholipid and glyceride biosynthesis, *J. Environ. Pathol. Toxicol.*, 1, 841, 1978.
116. Ishidate, K. and Nakazawa, Y., Effect of polychlorinated biphenyls (PCBs) administration on phos-pholipid synthesis in rat liver, *Biochem. Pharmacol.*, 25, 1255, 1976.
117. Hinton, D. E., Glaumann, H., and Trump, F., Studies on the cellular toxicity of polychlorinated biphenyls (PCBs), *Virchows Arch. B*, 27, 279, 1978.
118. Kohli, K. K., Gupta, B. N., Albro, P. W., Mukhtar, H., and McKinney, T. D., Biochemical effects of pure isomers of hexachlorobiphenyl: fatty livers and cell structure, *Chem. Biol. Interact.*, 25, 139, 1979.
119. Lo, M.-T. and Sandi, E., Polycyclic aromatic hydrocarbons (polynuclears) in foods, *Residue Rev.*, 69, 35, 1978.
120. Wasserman, M., Wasserman, D., Cucos, S., and Miller, H. J., World PCBs map; storage and effects in man and his biological environment in the 1970s, *Ann. N.Y. Acad. Sci.*, 320, 69, 1979.

121. Grundy, S. M., Ahrens, E. H., Jr., Salen, G., Schreibman, P. H., and Nestel, P. J., Mechanisms of action of clofibrate on cholesterol metabolism in patients with hyperlipidemia, *J. Lipid Res.*, 13, 531, 1972.

122. Cheng, C. Y. and Feldman, E. B., Clofibrate, inhibitor of intestinal cholesterogenesis, *Biochem. Pharmacol.*, 20, 3509, 1971.

123. Strandberg, T. E., Kuusi, T., Tilvis, R. S., and Miettinen, T. A., Clofibrate decreases jejunal cholesterol synthesis and activity of postheparin plasma lipoprotein lipase in the rat, *Pharmacology*, 26, 290, 1983.

124. Gans, J. H. and Cater, M. R., Metabolic effects of clofibrate and of cholestyramine administration to dogs, *Biochem. Pharmacol.*, 20, 3321, 1971.

125. Rogers, D. H., Kim, D. N., Lee, K. T., Reiner, J. M., and Thomas, W. A., Circadian variation of 3-hydroxy-3-methyl-glutaryl coenzyme A reductase activity in swine liver and ileum, *J. Lipid Res.*, 22, 811, 1981.

126. Turley, S. D. and Dietschy, J. M., Effects of clofibrate, cholestyramine, Zanchol, Probucol and AOMA feeding on hepatic and intestinal cholesterol metabolism and on biliary lipid secretion in the rat, *J. Cardiovasc. Pharmacol.*, 2, 281, 1980.

127. Middleton, W. R. J. and Isselbacher, K. J., The stimulation of intestinal cholesterogenesis in the rat by phenobarbital, *Proc. Soc. Exp. Biol. Med.*, 131, 1435, 1969.

128. Middleton, W. R. J., Carter, E. A., Drummey, G. D., and Isselbacher, K. J., Effect of oral ethanol administration on intestinal cholesterogenesis in the rat, *Gastroenterology*, 60, 880, 1971.

129. Clark, S. B. and Maranhao, R., Cholesterol absorption and inhibition of intestinal acylCoA:cholesterol acyltransferase, *Fed. Proc.*, 42, 327, 1983.

130. Johnson, A. R., Pearson, J. A., Shenstone, F. S., and Fogerty, A. C., Inhibition of the desaturation of stearic to oleic acid by cyclopropene fatty acids, *Nature (London)*, 214, 1244, 1967.

131. Raju, P. K. and Reiser, R., Inhibition of fatty acyl desaturase by cyclopropene fatty acids, *J. Biol. Chem.*, 242, 379, 1967.

132. Pande, S. V. and Mead, J. F., Inhibition of the stearylCoA desaturase system by sterculate, *J. Biol. Chem.*, 245, 1856, 1970.

133. Deckelbaum, R. J., The intestine and new high density lipoprotein formation, *Gastroenterology*, 86, 1619, 1984.

134. Isselbacher, K. J., Scheig, R. L., Plotkin, G. R., and Caulfield, J. B., Congenital β-lipoprotein deficiency: an hereditary disorder involving a defect in the absorption and transport of lipids, *Medicine (Baltimore)*, 43, 347, 1964.

135. Kostner, G. M., Apo β-deficiency (abeta lipoproteinemia): a model for studying the lipoprotein metabolism in *Lipid Absorption: Biochemical and Clinical Aspects*, Rommel, K., Goebell, H., and Böhmer, R., Eds., University Park Press, Baltimore, 1976, 203.

136. Kayden, H. J. and Medick, M., The absorption and metabolism of short and long chain fatty acids in puromycin-treated rats, *Biochim. Biophys. Acta*, 176, 37, 1969.

137. Glickman, R. M. and Kirsch, K., Lymph chylomicron formation during the inhibition of protein synthesis. Studies of chylomicron apoproteins, *J. Clin. Invest.*, 52, 2910, 1973.

138. Glickman, R. M., Kilgore, A., and Khorana, J., Chylomicron apoprotein localization within rat intestinal epithelium: studies of normal and impaired lipid absorption, *J. Lipid Res.*, 19, 260, 1978.

139. Mahmood, A., Agarwal, N., Dudeja, P. K., Sanyal, S., Mahmood, R., and Subrahmanyam, D., Effect of a single oral dose of DDT on intestinal uptake of nutrients and brush border enzymes in protein-caloric-malnourished monkeys, *Pest. Biochem. Physiol.*, 15, 143, 1981.

140. Patil, S. D., Prakash, K., and Hegde, S. N., Effects of a single oral dose of DDT on D-glucose and amino acid uptake and on brush border enzymes in pigeon intestine, *Ind. J. Exp. Biol.*, 20, 904, 1982.

141. Iturri, S. J. and Wolff, D., Inhibition of the active transport of D-glucose and L-tyrosine by DDT and DDE in the rat small intestine, *Comp. Biochem. Physiol.*, 71C, 131, 1982.

142. Ball, L. M. and Chhabra, R. S., Intestinal absorption of nutrients in rats treated with 2,3,7,8-tetrachlorodibenzo-p-dioxin (TCDD), *J. Toxicol. Environ. Health*, 8, 629, 1981.

143. Mahmood, A., Agarwal, N., Sanyal, S., Dudeja, P. K., and Subrahmanyam, D., Acute dieldrin toxicity: effect on the uptake of glucose and leucine and on brush border enzymes in monkey intestine, *Chem. Biol. Interact*, 37, 165, 1981.

144. Chowdhury, J. S., Dudeja, P. K., Mehta, S. K., and Makomood, A., Effect of a single oral dose of malathion on D-glucose and glycine uptake and on brush border enzymes in rat intestine, *Toxicol. Lett.*, 6, 411, 1980.

145. Lukie, B. E. and Forstner, G. G., Synthesis of intestinal glycoproteins. Inhibition of [1-¹⁴C]glucosamine incorporation by sodium salicylate in vitro, *Biochim. Biophys. Acta*, 273, 380, 1972.

146. Lukie, B. E. and Forstner, G. G., Synthesis of intestinal glycoproteins. Inhibition of [1-¹⁴C]glucosamine incorporation by phenylbutazone in vitro, *Biochim. Biophys. Acta*, 338, 345, 1974.

147. Rainsford, K. D., The effects of aspirin and other non-steroid antiinflammatory/analgesic drugs on gastro-intestinal mucus glycoprotein biosynthesis in vivo: relationship to ulcerogenic actions, *Biochem. Pharmacol.*, 27, 877, 1978.

148. Okine, L. K. N. A., Ioannides, C., and Parke, D. V., Inhibition of gastrointestinal mucosal glycoprotein synthesis by the β-adrenergic blocking drug, practolol, *Biochem. Pharmacol.*, 31, 2263, 1982.

149. Gordon, J. I., Strauss, A. W., Smith, D. P., Ockner, R., and Alpers, D. H., Identification of the abundant mRNA species which accumulate in the enterocyte and cloning of their cDNAs, *Gastroenterology*, 82, 1071, 1982.

150. Gordon, J. I., Smith, D. P., Andy, R., Alpers, D. H., Schonfeld, G., and Strauss, A. W., The primary translation product of rat intestinal apolipoprotein A-I mRNA is an unusual preprotein, *J. Biol. Chem.*, 257, 971, 1982.

151. Tso, P., Balint, J. A., Edmonds, R. H., and Rodgers, J. B., The effect of hydrophobic detergent in lipoprotein formation and transport in the intestine, *Gastroenterology*, 78, 1280, 1980.

152. Stein, O. and Stein, Y., Colchicine-induced inhibition of very-low density lipoprotein release by rat liver in vivo, *Biochim. Biophys. Acta*, 306, 142, 1973.

153. LeMarchand, Y., Singh, A., Assimacopoulos-Jeannet, F., Orci, L., Rouiller, C., and Jeanrenaud, B., A role for the microtubular system in the release of very-low density lipoproteins by perfused mouse livers, *J. Biol. Chem.*, 248, 6862, 1973.

154. Reaven, E. P. and Reaven, G. M., Evidence that microtubules play a permissive role in hepatocyte very low density lipoprotein secretion, *J. Cell Biol.*, 84, 28, 1980.

155. Mougin-Schutz, A., Vacher, D., and Girard-Globa, A., Synthesis and secretion of apolipoproteins by pig intestinal mucosa in organ culture. Lack of inhibition of apolipoprotein A-I secretion by colchicine, *Biochim. Biophys. Acta*, 754, 208, 1983.

156. Strauss, E. W. and Jacob, J. S., Some factors affecting the lipid secretory phase of fat absorption by intestine in vitro from the golden hamster, *J. Lipid Res.*, 22, 147, 1981.

157. Kessler, J. I., Sehgal, A. K., and Turcotte, R., Effect of neomycin on amino acid uptake and on synthesis and release of lipoproteins by rat intestine, *Can. J. Physiol. Pharmacol.*, 56, 420, 1978.

158. Peakall, D. B., Phthalate esters: occurrence and biological effects, *Residue Rev.*, 54, 1, 1975.

159. Thomas, J. A. and Thomas, M. J., Biological effects of di-(2-ethylhexyl) phthalate and other phthalic acid esters, *CRC Crit. Rev. Toxicol.*, 13, 283, 1984.

160. Bell, F. P., Effects of phthalate esters on lipid metabolism in various tissues, cells and organelles in mammals, *Environ. Health Perspect.*, 45, 41, 1982.

161. Stein, M. S., Caasi, P. I., and Nair, P. P., Influence of dietary fat and di-2-ethylhexyl phthalate on tissue lipids in rats, *J. Nutr.*, 104, 187, 1974.

162. Bell, F. P., Patt, C. S., Brundage, B., Gillies, P. J., and Phillips, W. A., Studies on lipid biosynthesis and cholesterol content of liver and serum lipoproteins in rats fed various phthalate esters, *Lipids*, 13, 66, 1978.

163. Kawashima, Y., Nakagawa, S., and Kozuka, H., Effects of some hypolipidemic drugs and phthalic acid esters on fatty acid binding protein in rat liver, *J. Pharm. Dyn.*, 5, 771, 1982.

164. Bell, F. P., Inhibition of hepatic sterol and squalene biosynthesis in rats fed di-2-ethylhexyl phthalate, *Lipids*, 11, 769, 1976.

165. Bell, F. P. and Buthala, D. A., Biochemical change in liver of rats fed the plasticizer di (2-ethylhexyl) phthalate, *Bull. Environ. Contam. Toxicol.*, 31, 177, 1983.

166. Reddy, J. K., Moody, D. E., Azarnoff, D. L., and Rao, M. S., Di (2-ethylhexyl) phthalate: an industrial plasticizer induces hypolipidemia and enhances hepatic catalase and carnitine acetyltransferase activities in rats and mice, *Life Sci.*, 18, 941, 1976.

167. Oishi, S. and Hiraga, K., Testicular atrophy induced by phthalic acid esters: effect on testosterone and zinc concentrations, *Toxicol. Appl. Pharmacol.*, 53, 35, 1980.

168. Foster, P. M. D., Foster, J. R., Cook, M. W., Thomas, L. V., and Gangolli, S. D., Changes in ultrastructure and cytochemical localization of zinc in rat testis following the administration of di-*n*-pentyl phthalate, *Toxicol. Appl. Pharmacol.*, 63, 120, 1982.

169. Foster, P. M. D., Thomas, L. V., Cook, M. W., and Gangolli, S. D., Study of the effects and changes in zinc excretion produced by some *n*-alkyl phthalates in the rat, *Toxicol. Appl. Pharmacol.*, 54, 392, 1980.

170. Oishi, S. and Hiraga, K., Testicular atrophy induced by phthalic acid monoesters: effects of zinc and testosterone concentrations, *Toxicology*, 15, 197, 1980.

171. Oishi, S. and Hiraga, K., Testicular atrophy induced by di-2-ethylhexyl phthalate: effect of zinc supplement, *Toxicol. Appl. Pharmacol.*, 70, 43, 1983.

172. Koo, S. I. and Turk, D. E., Effect of zinc deficiency on intestinal transport of triglyceride in the rat, *J. Nutr.*, 107, 909, 1977.

173. Koo, S. I. and Turk, D. E., Effect of zinc deficiency on the ultrastructure of the pancreatic acinar cell and intestinal epithelium in the rat, *J. Nutr.*, 107, 896, 1977.

174. Caldwell, J. and Marsh, M. V., Interrelationships between xenobiotic metabolism and lipid biosynthesis, *Biochem. Pharmacol.*, 32, 1667, 1983.

175. Fears, R., Baggaley, K. H., Alexander, R., Morgan, B., and Hindley, R. M., The participation of ethyl 4-benzyloxybenzoate (BRL 10894) and other aryl-substituted acids in glycerolipid metabolism, *J. Lipid Res.*, 19, 3, 1978.

176. Quistad, G. B., Staiger, L. E., and Schooley, D. A., Environmental degradation of the miticide cycloprate (hexadecylcyclopropanecarboxylate). I. Rat metabolism, *J. Agric. Food Chem.*, 26, 60, 1978.

177. Quistad, G. B., Staiger, L. E., and Schooley, D. A., Environmental degradation of the miticide cycloprate (hexadecylcyclopropanecarboxylate). IV. Beagle dog metabolism, *J. Agric. Food Chem.*, 26, 76, 1978.

178. Quistad, G. B., Staiger, L. E., and Schooley, D. A., Environmental degradation of the miticide cycloprate (hexadecylcyclopropanecarboxylate). III. Bovine metabolism, *J. Agric. Food Chem.*, 26, 71, 1978.

179. Tutwiler, G. F. and Dellevigne, P., Action of the oral hypoglycemic agent 2-tetra decylglycidic acid on hepatic fatty acid oxidation and gluconeogenesis, *J. Biol. Chem.*, 254, 2935, 1979.

180. McCune, S. A. and Harris, R. A., Mechanism responsible for 5-(tetradecyloxy)-2-furoic acid inhibition of hepatic lipogenesis, *J. Biol. Chem.*, 254, 10095, 1979.

181. Leighty, E. G., Metabolism and distribution of cannabinoids in rats after different methods of administration, *Biochem. Pharmacol.*, 22, 1613, 1973.

182. Leighty, E. G., Fentiman, A. F., Jr., and Flotz, R. L., Long-retained metabolites of Δ⁹- and Δ⁸-tetrahydrocannabinols identified as novel fatty acid conjugates, *Res. Commun. Chem. Pathol. Pharmacol.*, 14, 13, 1976.

183. Dell, H.-D., Fiedler, J., Kamp, R., Gau, W., Kurz, J., Weber, B., and Wuensche, C., Etofenamate fatty acid esters. An example of a new route of drug metabolism, *Drug Metab. Dispos.*, 10, 55, 1982.

184. Anon., Information section, Brominated oil residues and toxicity, *Food Chem. Toxicol.*, 22, 319, 1984.

185. Chapman, D., Lipid dynamics in cell membranes, in *Cell Membranes: Biochemistry, Cell Biology and Pathology*, Weissmann, G. and Clairborne, R., Eds., HP Publishing, New York, 1975, 13.

186. Brasitus, T. A. and Schachter, D., Cholesterol biosynthesis and the modulation of lipid fluidity in rat intestinal microvillus membranes, *Gastroenterology*, 78, 1144, 1980.

187. Bell, F. P., The dynamic state of membrane lipids: the significance of lipid exchange and transfer reactions to biomembrane composition structure, function and cellular lipid metabolism, in *Biomembranes*, Vol. 12, Kates, M. and Manson, L. A., Eds., Plenum Press, New York, 1984, 543.

188. Johannsson, A., Smith, G. A., and Metcalfe, J. C., The effect of bilayer thickness on the activity of (Na⁺ + K⁻)-ATPase, *Biochim. Biophys. Acta*, 641, 416, 1981.

189. Antunes-Madeira, M. C. and Madeira, V. M. C., Interaction of insecticides with lipid membranes, *Biochim. Biophys. Acta*, 550, 384, 1979.

190. Antunes-Madeira, M. C., Carvalho, A. P., and Madeira, V. M. C., Effects of insecticides on thermotropic lipid phase transitions, *Pestic. Biochem. Physiol.*, 14, 161, 1980.

191. Nixon, J. E., Eisele, T. A., Wales, J. H., and Sinnhuber, R. O., Effect of subacute toxic levels of dietary cyclopropenoid fatty acids upon membrane function and fatty acid composition in the rat, *Lipids*, 9, 314, 1974.

192. Jones, B. A., Tinsley, I. J., and Lowry, R. R., Bromine levels in tissue lipids of rats fed brominated fatty acids, *Lipids*, 18, 319, 1983.

193. Jones, B. A., Tinsley, I. J., Wilson, G., and Lowry, R. R., Toxicology and brominated fatty acids: metabolite concentration and heart and liver changes, *Lipids*, 18, 327, 1983.

194. Laitinen, M., Lang, M., Hietanen, E., and Vainio, H., Enhancement of hepatic drug biotransformation rate by polychlorinated biphenyls in rats fed cholesterol-rich diet, *Toxicology*, 5, 79, 1975.

Chapter 8

REGULATION OF FAT DIGESTION AND ABSORPTION IN THE SMALL INTESTINE

A. G. D. Hoffman and A. Kuksis

TABLE OF CONTENTS

I. INTRODUCTION

The rate of digestion and absorption of fats in the small intestine depends on numerous factors. These include the growth and differentiation of the mucosa, the supply of energy, composition of the chyme, rate of gastric emptying, peristaltic activity in the small intestine, enzymatic degradation into absorbable products, and membrane transport. Evidence exists that all of these processes are subject to regulation by hormones as well as by regulatory peptides released either from endocrine cells or neurones located within the walls of the gastrointestinal tract. The ingestion of fat or other food substances triggers the release of these peptides from specialized cells distributed throughout the gastrointestinal mucosa.

The following chapter summarizes recent evidence for various physiological factors governing the digestion and absorption of food. In reviewing the extensive literature, emphasis has been placed upon those studies which have involved fats as substrates and measurements of lipases as secretion products by the various organs affected by contact with food or by the gastrointestinal peptides released in response to the food. Although it is not always clear as to whether the effects are specific for fats and whether the findings with exogenously administered peptides are physiological or pharmacological, the results to date promise a better understanding of the physiology of the fat absorption process in the future.

II. ENERGY METABOLISM

In order to permit both the replicative differentiation and absorptive functions of the enterocytes, the intestinal mucosa is efficiently supplied with metabolic fuels. While in general the energy metabolism of the intestine is similar to that of the liver[1] and other tissues, there are certain features which are peculiar to the intestine.[2,3]

A. Sources of Substrate

The small intestine is a unique organ in that it can potentially utilize metabolic substrates from both the lumen (luminal surface) and blood (serosal surface). The corresponding areas of enterocyte cell membrane are the microvillar membrane or brush border, and the basolateral membrane, respectively. The single layer of specialized epithelium, supported by the lamina propria, projects finger-like villi into the lumen, while the gland-like crypts of Lieberkuhn dip down from the surface. Both villus and crypt cells are well supplied by a portal capillary network, which would theoretically expose both the immature crypt epithelium and the mature villus cells to blood-borne substrates. However, the crypts are hidden from the lumen and are generally filled with mucous, which would make them less accessible to the luminal nutrients compared with the microvillus cells at the tips of the villi.[4]

The small intestine is a highly vascular organ, deriving its blood supply directly from the superior mesenteric artery. It has been estimated that the resting mucosal blood flow to canine intestinal mucosa[5] is 1 mℓ/(min × 100 g), making it one of the better perfused tissues of the body. There is a positive relationship between intestinal absorption and blood flow, with several investigators reporting increased flow in the superior mesenteric and celiac arteries in response to the ingested meal in a variety of species.[6-8] Hultman and Castenfors[9] have reported that vasodilation is specific for the absorptive process rather than for individual nutrients. Chou et al.[10,11] have demonstrated that glucose and micellar solutions of oleic acid and monooleoylglycerol are effective vasodilators of gut. As well, hyperosmotic solutions including glucose and mannitol are able to affect a regional hyperemia. This vasodilatory response appears to be an intramural cholinergic reflex. Several speculations have emerged regarding the ability of circulating compounds such as gastrin, CCK, and glucagon to enhance intestinal blood flow, but clear evidence for a physiological role of these mediators is lacking.

In vivo and in vitro studies have demonstrated that a wide variety of substrates is capable of being metabolized to CO_2. These include glucose, galactose, fructose, glutamine, glutamate, asparagine, aspartate, arginine, alanine, ketone bodies, Krebs cycle acids, free fatty acids, and lactate. The various animal models that have been used in studies on intestinal metabolism include intact whole living animals, exteriorized loops of intestine in vivo, vascularly and luminally perfused intestine in vivo and in vitro, everted sacs, sheets of mucosal tissue suspended between two bathing solutions, cross-sectional slices and rings of intestine, suspensions of isolated villi, isolated intestinal absorptive villus and crypt cells, and intestinal mucosal homogenates and subcellular fractions derived from them of various degrees of purification.[2,3,12]

B. Special Features of Intestinal Metabolism

Current understanding of intestinal energy metabolism stems from the pioneering work of Windmueller and Spaeth,[13] which demonstrated the importance of glutamine as a major substrate in the isolated, vascularly perfused rat intestine. Subsequently, they[14] quantitated the relative contributions of the various metabolic fuels to CO_2 production. Glutamine accounted for about one third, 3-hydroxybutyrate and acetoacetate about one quarter each, and glucose, lactate, and unesterified fatty acids the remainder. These metabolic fuels accounted for all the major circulating fuels of rat small intestinal metabolism, and no further quantitatively important source of CO_2 remained to be identified. Glutamine was the major source of incorporation of label into citrate and other organic acids, such as alanine, proline, citrulline, and ornithine, and into glucose. In contrast, glucose was primarily metabolized to CO_2 and lactate, although some label also appeared in amino acids and in lipids. Plasma-free fatty acids (palmitate and oleate) were both metabolized by β-oxidation to CO_2, but also contributed label to intestinal neutral and phospholipids. However, only about 1% of the fatty acid flux was extracted, suggesting that permeability across the basolateral membrane may be a limiting factor. Windmueller and Spaeth[15] also examined the fate of luminal glutamine with the auto-perfused rat jejunal segment preparation, and found it to be rapidly absorbed from the lumen. This was similar to that seen with arterial glutamine, with 34% of the original intraluminal molecules appearing in portal plasma unchanged. The presence of luminal glutamine or glutamate depresses the utilization of arterial glutamine,[3] thus providing a sparing effect.

Watford et al.[16] prepared isolated columnar absorptive cells from rat and chicken small intestine and found for both species that the preferred fuels of respiration (i.e., those that when added to the medium effected the highest rate of O_2 consumption) were glutamine, glutamate, and glucose. Acetoacetate had a small effect on O_2 consumption, but oleate, butyrate, propionate, acetate, β-hydroxybutyrate, alanine, asparagine, aspartate, succinate, and citrate had no major effect. Their work also suggested that alanine is formed in the small intestine by the oxidative degradation of glutamate to pyruvate, followed by transamination. Windmueller[3] has summarized much of this work on glutamine utilization by mammalian small intestine in a recent review, where he presents evidence for an important physiological role of this substrate as metabolic fuel. He also outlines the interrelationships between glutamine and other metabolic fuels in intermediary metabolism.

Utilization of ketone bodies by mammalian intestinal epithelium has been shown to be of importance in fasted rats.[14] Two of the enzymes required for ketone body metabolism, 3-hydroxybutyrate dehydrogenase and acetoacetyl-CoA acetyltransferase, are present in mitochondria in both villus and crypt cells. Iemhoff and Hulsmann[17] compared the activities of mitochondrial enzymes in rat small intestinal epithelial villus and crypt cells, and found equivalent 3-hydroxybutyrate dehydrogenase activity, but a 2.6:1 preponderance of the acetyltransferase enzyme, allowing them to conclude that the mitochondria in mature enterocytes were different from those in the immature crypt cells and may indeed be undergoing a differentiation paralleling that of the whole cell. Data have suggested that ketone bodies may account for up to 50% of the respiratory CO_2 in the small intestine of rats fasted overnight and, therefore, may be a major oxidative fuel during starvation.[14,18,19] There is also the possibility of a glucose sparing effect as both glucose[19] and pyruvate[20] oxidation is decreased, with the net result that almost all glucose utilized by the intestine in starvation is converted to lactate, which is then available via the portal circulation for gluconeogenesis by the liver. From available data and from a review of ketone body utilization in various mammalian tissues by Robinson and Williamson,[18] it can be concluded that the intestine now joins brain, heart, and kidney as a major user of these substrates as oxidative fuels, lipogenic precursors, and as regulators of metabolism.

Detailed reviews of intestinal energy metabolism have been provided previously by Porteus[12] and Hulsmann.[2] Both authors discuss the utilization of glutamine, ketone bodies, fatty acids, glucose, and lactate and review both the work with the perfused rat intestine and with isolated rat intestinal cells. The findings with both types of animal preparations are in good agreement with each other. Hulsmann[2] provides further evidence on the metabolic use of free fatty acids and includes data showing that increasing the molar ratio of oleate to albumin allows an increase in the rate of fatty acid oxidation by in vitro perfused rat intestine, to levels approaching that of perfused heart. Work by Bremer and Norum[21] provides evidence that fat feeding or fasting may induce peroxisome production by the intestine, thereby augmenting the β-oxidation capability of the gut. Whether the source of fatty acids is endogenous from stored or circulating pools or from diet, the small intestine has the ability to oxidize these substrates by both the mitochondrial and peroxisome mechanisms.

C. Hormonal Modulation

The effect of general hormones on the metabolism of the gut wall has been extensively investigated as has been that of the specific gut hormones. The discovery, chemistry, and structure-activity relationships of the gastrointestinal hormones and hormone-like peptides has been reviewed by Walsh[22] and Mutt,[23] while Nicholl et al.[24] have recently discussed the hormonal regulation of food intake and digestion and absorption in general. In this chapter we have reviewed this rapidly expanding area of research with specific emphasis on fat digestion and absorption.

1. Substrate Supply

Early work[25] had given largely negative or controversial effects with various potential hormonal modulators of intestinal mucosal metabolism. Levin and Syme[26] reported that triiodothyronine stimulated oxygen consumption in everted sacs of rat midjejunum in vitro. This effect was enhanced in the presence of glucose and sodium and was retarded in the absence of either. These workers suggested that thyroid hormone may directly stimulate [Na$^+$/K$^+$]-ATPase activities of enterocytes. Gaginella et al.[27] examined the state of respiration in mucosal mitochondria from hamster jejunum and found that certain long-chain fatty acids such as ricinoleate and oleate inhibited adenine nucleotide translocase and thereby adversely affected state 3 of mitochondrial respiration. In vitro studies with everted sacs showed a significant inhibition of water absorption by the same acids, presumably on the basis of a decrease in supply of mitochondrial ATP available for water transport. Apart from the above two studies, there have been no clear-cut observations on the hormonal or pharmacological regulation of substrate supply to the small intestine. Indeed, Hulsmann,[2] in his recent review, has concluded that the direct effects of insulin, glucagon, and catecholamines on small intestinal epithelium are either very modest or negligible.

2. Glycoprotein Biosynthesis

Forstner et al.[28] demonstrated stimulation of intestinal glycoprotein synthesis by adrenergic agents in intestinal segments using D[1-^{14}C]glucosamine as a tracer. This observation was confirmed by Lamont and Ventola[29] in rabbit colon. Using an improved preparation of enzyme-dispersed rat intestinal villus and crypt cells, Hoffman and Kuksis[30] demonstrated that isolated enterocytes were capable of responding to hormonal stimuli in vitro. The adrenergic agents, epinephrine and isoproterenol, as well as cyclic AMP (cAMP) and theophylline, all stimulated the synthesis and secretion of D-glucosamine-labeled glycoproteins. Previous work has shown that both colonic goblet[29] and jejunal epithelial cells from the villus and crypt regions[31] incorporated D-glucosamine into glycoproteins, which may be secreted from these cells. The marked response of the enzyme-dispersed cells to hormonal and pharmacological agents shows that the appropriate receptor sites survived the method of cell dispersion intact. These findings, therefore, support earlier reports of adrenergic receptors in intestinal mucosa, based on studies with everted sacs.[32] As the increases in the glycoprotein synthesis brought about by the hormones were manifested largely in the material found in the medium, both synthesis and secretion of labeled glycoproteins would appear to be mediated by the added hormonal agents. The possibility of gross cellular damage was ruled out by monitoring the leakage of appropriate cellular enzymes and by measuring increases in the permeability of the cell membrane to Nigrosin dye. It must, therefore, be concluded that the appearance of large amounts of radioactive glycoproteins in the medium represent an orderly secretion involving one or more of the processes previously demonstrated: the secretion of glucosamine-containing lipoproteins,[33] the release of absorptive cell glycocalyx,[34,35] and secretion of glycoproteins by goblet cells.[29] The present data do not distinguish among these possibilities. Forstner and Galland[36] have reported the influence of hydrocortisone on the synthesis and turnover of microvillus membrane glycoproteins in suckling rat intestine. More recently, Miura et al.[37] have shown that the degradation rates of both microvillus membrane glycoproteins and aminopeptidase were not changed by methyl prednisolone treatment. Methyl prednisolone, however, increased the maximum incorporation of [^{3}H]glucosamine into immunoprecipitable aminopeptidase in jejunum and ileum.

3. Lipogenesis

Previous work with citrate-dispersed intestinal cells has given no demonstrable modulation of lipogenesis following addition of several hormones and pharmacological agents to the incubation medium.[38] A hormonal response on the part of the intestinal cells was anticipated on the basis of their common embryological origin with the hepatocytes, which have given such a response in vitro.[24] Despite the improved cell preparation, Hoffman and Kuksis[30] also failed to obtain significant incorporation of either [^{14}C]acetate or [^{14}C]glucose into total lipids in response to hormonal stimulation. Likewise, a lack of response was noted in jejunal segments. In a preliminary experiment[39] there was, however, an apparent depression of the incorporation of [^{14}C]acetate into sterols, by isolated intestinal cells in the presence of glucagon. A comparable effect has been reported in organ culture of canine jejunal tissue after a 6-hr exposure to glucagon in vitro.[40] As indicated thus far by selected tracers and incubation conditions, mucosal lipogenesis in the mature animal is not susceptible to hormonal regulation compared to hepatic tissues. In this connection it is of interest to note that studies on the development of the cytoplasmic glucocorticoid receptors have shown[41] that they are present at high levels in the intestinal mucosa of the rat during the first two postnatal weeks, with much lower levels of activity thereafter. Possibly in the adult rat the intestinal lipogenesis is not regulated by cellular levels of cAMP as it is in the liver.[42]

In view of the present findings it may be necessary to revise the metabolic interpretation of certain in vivo effects of hormones upon the intestinal mucosa of the adult rat. Thus, the decreased fat absorption seen in adrenalectomy[43,44] may be due also to a suppression of glycoprotein synthesis resulting from a decrease in the concentration of catecholamines and concomitant mucosal cellular cAMP levels accompanying the loss of medullary hormones characteristic of this condition.[25] This interpretation of the present results concerning the regulation of glycoprotein synthesis necessary for transport of chylomicrons from the intestine is consistent with the well-known effects of cAMP levels on the transport phenomena in the intestine, as reviewed by Sernka.[45] The increased glycoprotein synthesis and secretion observed in response to hormonal stimulation indicates that the enzyme-dispersion of mucosal cells is compatible with the retention of the receptors for at least the adrenergic hormones and/or the capability of a physiological response with complex metabolic and secretory events.

III. INTESTINAL GROWTH AND DIFFERENTIATION

The mucosa of the gastrointestinal tract proliferates very rapidly and its cells have extremely short half-lives. The epithelial cell layer of the intestinal mucous membrane is constantly being renewed. The stem cells in the crypts of Lieberkuhn give rise to undifferentiated crypt cells, which migrate up the villi to the tips, where they are extruded.[46] It is generally accepted that the undifferentiated crypt epithelial cells give rise to columnar cells, Paneth cells, goblet cells, and enteroendocrine cells, as originally proposed by Cheng and Leblond.[47] Therefore, factors that alter its growth and differentiation are of great importance to fat absorption.

The growth of gastrointestinal mucosa is affected in general by such hormones as insulin and growth hormone, which regulate the growth of all body tissues, and specifically by the gastrointestinal hormones, which are secreted when food is ingested.[48] During the last decade it has become clear that these peptides are sequestered in secretory granules in nerve cells in mucosal endocrine cells and are released both by gut, the endocrine cells, and neurons. Indeed, some of these molecules may function as a hormone and a neurotransmitter. Many gut peptides are related to each other by similarities in amino acid sequence. The gastrin-CCK family of peptides shares a common C-terminal pentapeptide amide structure, while the secretin family, which includes glucagon, VIP, glucose-dependent insulin-releasing peptide, and growth hormone-releasing factor, has similarities in sequence throughout the peptide chain. An additional peptide belonging to this family has been given the name PHI (peptide with histidine at the N terminal and isoleucine amide at the C terminus).[23] This molecule is similar to VIP in its biological properties and is derived from the same precursor.[49] Many peptides exist in multiple molecular forms. Apparently, they are first made as large precursors, which are then cleaved within their cells or origin by proteolytic enzymes to give final active products.[50] The precise patterns of cleavage and, therefore, of the molecular forms can vary between cells. Table 1 summarizes the amino acid sequences of the gastrin/CCK, glucagon/secretin/VIP, and pancreatic polypeptide families of gut peptides.[51]

Table 1

AMINO ACID SEQUENCES OF GASTROINTESTINAL PEPTIDE FAMILIES STRUCTURE

Gastrin/Cholecystokinin Family[51]

Gastrin
    G 34 (human)   pQLGPQGPPHLVADPSKKQGPWLEEEEEAYGWMDF*
    G 17 (human)   QGPWLEEEEEAYGWMDF*
    G 14 (human)   WLEEEEEAYGWMDF*
    G 6   YGWMDF*
Cholecystokinin
    CCK 58   AQKVNSGEPRAHLGALLARYIQQARKAPSGRMSVIKNL*
    CCK 39   YIQQARKAPSGRVSMIKNLQSLDPSHRISDRDYMGMDF*
    CCK 33   RKAPSGRVSMIKNLQSLDPSHRISDRDYMGMDF*
    CCK 8   RDYMGMDF*

Glucagon/Secretin/VIP Family

PHI   HADGVFTSDFSRLLGQLSAKKYLESLI*
VIP   HSDAVFTDNYTRLRKQMAVKKYLNSILN*
Secretin   HSDGTFTSELSRLRDSARLQRLLQGLV*
Glucagon   HSQGTFTSDYSKYDSRRAQDFVQWLMNT*
GIP   YAEGTFISDYSIAMDKIRQQDFVNWLLAQKGKKSDWKHNITQ
GRF (human)   YADAIFTNSYRKVLGQLSARKLLQDIMSRQQGESNQERGA

Pancreatic Polypeptides Family

PYY   YPAKPEAPGEDASPEELSRYYASLRHYLNLVTRQRY*
PP   APLEPVYPGDDATPEQMAQYAAELRRYINMLTRPRY*

Note: The following abbreviations are used: A, alanine; D, aspartic acid; E, glutamic acid; F, phenylalanine; G, glycine; H, histidine; I, isoleucine; K, lysine; L, leucine; M, methionine; N, asparagine; Q, glutamine; R, arginine; S, serine; T, threonine; V, valine; W, tryptophan; Y, tyrosine; p, pyro; ..., no further sequence determined; *, amidated COOH terminal.

A. Growth

A summary of the trophic actions of the gastrointestinal peptides prepared by Johnson[52] has been reproduced in Table 2. These effects are reviewed in relation to signaling from fatty foods and to stimulation of growth of lipase-releasing tissues.

1. Epithelial Tissues

One of the first hormones found both in vivo[53] and in vitro[54] to have a direct trophic effect on gastric and duodenal mucosa was gastrin. Gastrin stimulates DNA, RNA, and protein synthesis in ileal mucosal tissue.[55] These effects are specific for the epithelial tissues of the gastrointestinal tract, and apart from the exocrine pancreas, no trophic effect has been demonstrated on extraintestinal tissues.[56] Different molecular forms of gastrin, including G17 I, G17 II, G34 II, and pentagastrin, all stimulate equally the incorporation of [^{3}H]thymidine into DNA of dog duodenal mucosa.[57] Growth of the mucosa in response to gastrin appears to be correlated with the development of the other systems involved in nutrient delivery regulated through the gastrin receptor.[58] Inhibitory factors normally at play include autoregulation of gastrin release at low intragastric pH, accounting for up to 50% total inhibition. Release of unknown inhibitors by fat is believed to be responsible for about 25% inhibition of responses to mixed meals.[59] Evidence to date suggests that secretin inhibits the trophic effects of gastrin,[60] but the mechanism of this effect is controversial.

Dembinski and Konturek[61] have provided evidence that naturally occurring prostaglandins of the E and F series and their methylated analogs can stimulate the growth of gastrointestinal mucosa and the pancreas and that this effect does not appear to be related to alterations in serum gastrin concentration.

Scheving et al.[62] have investigated the rhythmic variation which characterizes the circadian fluctuation in cell proliferation found in the intestinal tract of mice standard-

Table 2
TROPHIC ACTIONS OF GASTROINTESTINAL
PEPTIDES

Peptides	Tissue[52]	Action on growth
Gastrin	Oxyntic gland[a]	Strong stimulation
	Duodenum	Strong stimulation
	Colon	Strong stimulation
	Pancreas[b]	Stimulation
CCK	Pancreas	Strong stimulation
	Gall bladder	Strong stimulation
	Small intestine	Weak stimulation
Secretin	Oxyntic gland	Inhibition of gastrin
	Small intestine	Inhibition of gastrin
	Colon	Inhibition of gastrin
	Pancreas	Stimulation, potentiation of CCK
VIP	Oxyntic gland	Weak stimulation
	Colon	Inhibition of gastrin
Glucagon	Oxyntic gland	Weak stimulation
	Colon	Weak stimulation
EGF	Oxyntic gland	Strong stimulation
	Small intestine	Stimulation in suckling rats

[a]   Refers to mucosa only for all gastrointestinal tissues.
[b]   Pancreas refers to exocrine pancreas only.

ized to 12 hr of light alternating with 12 hr of darkness. The data show that every region of the digestive tract has a predictable daily trough and peak time when DNA synthesis is occurring, with some regions exhibiting as much as four- to fivefold variation over a 24-hr period. It was shown that this underlying circadian variation strongly influenced any effect seen when insulin, glucagon, somatostatin, gastrin, ACTH, and epidermal growth factor are administered at different circadian stages. Furthermore, the epidermal growth factor was shown under certain circumstances to advance the phase rhythm in DNA synthesis in both fasting and ad libitum-fed mice.

2. Pancreatic Growth

Pancreatic growth and maturation is essential for fat absorption because of the lipases and other enzymes that this organ releases into the intestine. Johnson[52] has reviewed the trophic effects of CCK on pancreatic, gastric, and duodenal tissue and has concluded that while growth of the exocrine pancreas is enhanced, there is no stimulating effect on the mucosa of the stomach or duodenum. Recent studies reported by Stock-Damge et al.[63] do not support the views that antral gastrin exerts a trophic action on the pancreas and that gastrin is implicated in postresectional hyperplasia of the gland. Secretin also increases pancreatic growth, although not as much as CCK.[64] Combinations of CCK and secretin potentiate the trophic response.

Guice et al.[65] have examined the early effects of various doses of cerulein (a CCK analog) administered continuously over a period of 4 days. Trophic changes were observed at day 1 and persisted through day 4. There appeared to be a phasic ultrastructural adaptation with continuous cerulein administration.

O'Loughlin et al.[66] have examined the effect of epidermal growth factor on the postnatal maturation of pancreas and small intestine in suckling rabbits. Daily i.p. injections of epidermal growth hormone were given from age 3 to 18 days. Morphometic measurements demonstrated that the increase in intestinal weight was associated with greater total thickness of the bowel wall due to a significant increase in thickness of the nonmucosal layers. Epidermal growth factor had no effect on the intestinal mucosa, mucosal weight, thickness or DNA content, and enzyme activities. These findings indicate that epidermal growth factor has a trophic effect on the developing stomach, pancreas, and nonmucosal elements of the small intestine and in addition appears to stimulate postnatal maturation of the pancreas, but not of the intestine.

Butzner et al.[67] have concluded that undernutrition prolongs recovery from experimentally induced viral enteritis in gnotobiotic piglets. Manifestations of acute disease in jejunum, significant decrease in villus/crypt ratio, lactase, and Na-K-ATPase activities, returned to noninfected values by 10 days in full diet controls, but remained significantly depressed for 15 days in undernourished pigs. This finding was believed to be due to enhanced initial injury, delayed mucosal repair, or both.

Miazza et al.[68] have presented results, which suggest that pancreatic-biliary diversion stimulates pancreatic growth in the rat. It was shown that CCK plays a central role in the development of the pancreatic adaptive response to pancreatic-biliary diversion. Foelsch and Creutzfeldt[69] have reported that feeding soybean trypsin inhibitor causes pronounced growth of pancreas, which is accompanied by increased enzyme content and increased CCK and gastrin concentration in plasma. McGuinness et al.[70] have observed that removal of proteases and bile salts from the upper small intestine results in pancreatic growth.

3. Mechanism

The cellular mechanism whereby trophic peptides influence normal and adaptive growth is not fully understood. Polyamine biosynthesis appears to have an essential role in cellular proliferation in many tissues. The rate-limiting step in this process is the decarboxylation of ornithine, which requires the enzyme ornithine decarboxylase. Luk and Baylin[71] have shown that increases in ornithine decarboxylase activity and in the polyamines, putrescine and spermidine, occur in rat ileal mucosa between days 1 and 4 after 50% intestinal resection. During this time there is initiation of mucosal cell hyperplasia, as measured morphologically and biochemically. Subsequently, Luk and Baylin[72] have presented data which indicate that the increases in ornithine decarboxylase activity and polyamine biosynthesis are critical for the process of adaptive postresectional crypt cell proliferation in vivo, and that the critical step mediated by polyamines in this adaptive process is the onset of new DNA synthesis. Dowling et al.[73] have reviewed the evidence for the effect of polyamines and have postulated a sequence of events as follows: luminal nutrients, particularly long-chain fats, reach the ileum and colon and stimulate increased enteroglucagon release. Enteroglucagon binds to cell receptors and triggers an intracellular cascade involving ornithine decarboxylase and the polyamines, which in turn stimulate RNA polymerase, DNA, RNA and protein synthesis, cell division, and adaptive growth.

B. Differentiation

Studies have also been undertaken to examine the role of hormones in intestinal development. The proliferation rate of crypt cells is thought to be under the control of several external factors, including growth hormone and insulin, which regulate the growth of all body tissues. In addition, it is affected by a host of events, which are brought about by the intake of food. These can be divided into direct effects of luminal nutrients and growth factors on the mucosa and indirect effects produced by hormones and secretions released or initiated by the presence of food in the gut.[56,74] Extensive literature has accumulated on the kinetics of crypt cell proliferation and villus crypt-morphometry.[75,76] Much of this deals with radiation-induced alterations in stem cell division or with specific pathological states affecting cell renewal and resulting in fat malabsorption. Holt et al.[77] have described a simple method for the measurement of the turnover of the epithelial cells of the small intestine, while Sjolund et al.[78] have reported an immunocytochemical study of the endocrine cell distribution in human intestine.

Some of the recent work on the hormonal effects on differentiation have been reviewed by Henning.[79] It appears that the intestinal mucosa of infant rats is sensitive to the presence of glucocorticoids both in vivo and in vitro.[80-84] Increases in activities of sucrase, maltase, trehalase, aminopeptidase, and alkaline phosphatase are all seen with the administration of glucocorticoids to suckling rats within the first two postnatal weeks. Similarly, addition of glucocorticoid to intestinal explants in culture also stimulates sucrase and maltase activities.[82,83] Apparently, the mechanism for the accelerated appearance of enzymes is crypt mediated in that new cells which arise in the crypts and subsequently migrate to the villi have the ability to develop brush border enzyme activities that were not present on the previous "mature" villus enterocytes. The effect of hydrocortisone is permissive rather than essential, since brush border enzyme development and appearance is delayed but not absent in adrenalectomized animals.[79,85] Interestingly, the adult intestinal mucosa does not appear to respond to glucocorticoids.[79,86] Recently, Majumdar and Nielsen[84] have reported that glucocorticoids induce growth and maturation of fetal gut and pancreas and development of certain endocrine cells of the gut. Hydrocortisone induced sucrase, stimulated lactase activity in prenatal rats, and in their first 3 weeks of life increased antral gastrin levels[87] and hastened the normal decline in pancreatic gastrin levels with age.[88]

C. Studies with Isolated Cells

The diffuse distribution of mucosal endocrine cells in the intestine has made in vitro studies on the regulation of gastrointestinal hormone secretion difficult. Barber et al.[89] have developed a technique for maintaining isolated cells in short-term culture to directly determine the regulation of the release of neurotensin-like immunoreactivity. For this purpose, mucosal cells were dispersed from canine ileum by sequential incubation in crude collagenase and EDTA. Centrifugal elutriation enriched NTLI-containing cells fivefold over unfractionated mucosal cells as determined by immunohistochemistry and C-terminal-specific radioimmunoassay. The cells were cultured in Dulbecco's medium and the NTLI-positive cells were found to adhere selectively to the collagen substrate and after 48 hr constituted 40% of the cells in culture. These cells secreted NTLI in response to epinephrine in a dose-dependent manner. In culture, the NTLI release was stimulated by β-adrenergic agonists and was inhibited by somatostatin and by the muscarinic agonist carbachol.

Extensive empirical work has been done to improve methods for in vitro propagation of human and animal gastrointestinal epithelial cells. Moyer and Aust[90] have described conditions for improved, longer-term culture of both human and animal gastrointestinal cells, primarily from the colon. These included rapid tissue processing, extensive rinsing, mechanical dissociation, and choice of culture medium and supplements. A combination of a low calcium medium (suspension culture modified minimal essential medium) with an equal volume of Leibowitz highly enriched medium L15 (which was developed to culture colon cancer cells) was the best tested. Successful growth of over 90% of the various gastrointestinal cultures initiated has been obtained in L15:SMEM supplemented with 2% fetal bovine serum, 1% pituitary extract, 5% yeast extract broth source, 5 μg/mℓ insulin, 5 μg/mℓ transferrin, 5 ng/mℓ selenium, 1 μg/mℓ hydrocortisone, 1 mM nonessential amino acids, 25 ng/mℓ pentagastrin, 10 ng/mℓ epidermal growth factor, and 5 mM glutamine. Gastrointestinal cell clusters grew well on or under rat tail collagen gels, but survived longer when maintained within the gels. The isolated cells permit new avenues for investigation of the physiological and pharmacological effects of the gastrointestinal hormones.

In recent years it has been demonstrated that the microvillus membrane of the isolated small intestinal cells undergoes marked postnatal developmental change with respect to both function and gross membrane composition.[91,92] The relation between these differences in membrane lipid structural determinants and the development of intestinal lipid-dependent enzyme and transport functions remain to be investigated.

IV. GASTROINTESTINAL MOTILITY AND SECRETION

The events that are brought about by the intake of food include a stimulation of the motility of the gastrointestinal tract and the triggering of various internal and external secretions. Although these effects are initiated by the presence of food in the gastrointestinal tract, their maintenance depends upon the continued secretion of the various gastrointestinal peptides.

The delivery of the fat to the absorbing mucosal surface of the small intestine results from a flow of the fluid contents caused by movements of the muscular walls of the gut at several different sites. The musuclar walls of the stomach move to regulate delivery of nutrient fluid to the small intestine, the muscle layers of the small intestine to regulate the flow of fluid along the length of the small intestine, and the mucosa of the small intestine to create microcirculation across the surfaces of the absorbing cells. Christensen[93] has summarized the basic facts that are known about the nature of these systems, their patterns and controls of motion, and the flows these systems produce.

A. Gastric Emptying

Recent animal studies have suggested that gastric emptying is dependent on the caloric[94] and osmotic[95] content of ingested food. In addition, the role of calcium as a determinant has been proposed.[96] Shafer et al.[97] have reinvestigated the role of calories, osmolality, and calcium on gastric emptying in ten normal human volunteers utilizing a standardized ^{99m}Tc-sulfur colloid scrambled egg meal washed down with 50 mℓ water plus added nutrients. Their findings suggest that neither calories nor osmolality alone determine gastric emptying, that both Ca^{2+} and calcium-chelation with EDTA prolong gastric emptying, and that a specific food does not usually produce the same effect on gastric emptying in different individuals.

Powerful inhibitory reflexes, both nervous and hormonal, operate from the duodenum to slow gastric emptying, and there is evidence that the neurally released VIP is most obviously involved. These peptides also affect the intestinal motility, but a unifying concept for their role is lacking. A number of peptides are capable of causing either contraction or relaxation of smooth muscle in the small bowel, and as a result the overall rate is probably governed by the sequential action of a number of peptides. The overall effects so far obtained with the gastrointestinal peptides on gastric emptying and intestinal motility are summarized in Table 3.[51]

The major physiological controls of net gastric emptying are believed to be neural. The rate at which the stomach empties varies greatly according to the composition of the gastric contents, probably as a consequence of activation of receptors in the duodenal mucosa that slow the rate. Until recently it was thought that fats, acids, and osmotic pressure constituted three independent kinds of stimuli. A recent theory,[96] however, proposes that all three share a common mechanism of action. It is not clear whether the receptor mechanism inhibits emptying by neural or hormonal mechanisms.

Table 3
EFFECT OF GASTROINTESTINAL PEPTIDES ON
GASTRIC EMPTYING AND INTESTINAL
MOTILITY[51]

Gastrointestinal peptide	Gastric emptying	Intestinal motility
Gastrin releasing	Not reported	Stimulates
Cholecystokinin	Inhibits	Stimulates
Enkephalin	Inhibits	Stimulates
Gastrin	Inhibits	Stimulates
GIP	Inhibits	Atypical
Glucagon	Not reported	Inhibits
Motilin	Stimulates	Stimulates
Neurotensin	Inhibits	Inhibits
Pancreatic polypeptide	Stimulates	Stimulates and inhibits
Secretin	Inhibits	Inhibits
Somatostatin	Inhibits	Inhibits
Substance P	Not reported	Stimulates
VIP	Stimulates	Inhibits

Miller[98] has characterized the CCK receptors of human gastric smooth-muscle tumors. The acceptor was found to bind both CCK and gastrin, while Gardner and Jensen[99] have discussed the action of various CCK receptor antagonists.

Liddle et al.[100] have presented data based on physiological postprandial blood levels of CCK reproduced by infusion, which show that CCK in physiological concentrations delays gastric emptying in humans and that the effect was dose dependent. Liddle et al.[101] have shown that the CCK response to food is species specific. In the rat, fat and amino acids did not appear to stimulate CCK release nor pancreatic secretion. Intact protein was the primary food stimulant. It had previously been shown in dogs and humans that intragastric and intraduodenal instillations of either protein, fat, or amino acids stimulate CCK release and pancreatic secretion. These studies were made possible by the development of a sensitive and specific bioassay to measure plasma cholecystokinin.[102]

Since conflicting findings have been reported, Mogard et al.[103] have reevaluated the effect of somatostatin on gastric emptying in duodenal ulcer patients and in healthy volunteers. It was found that somatostatin significantly increased the rate of gastric emptying of liquid protein meals in normal subjects and in duodenal ulcer patients, but that the differences between them were not significant. It was concluded that somatostatin could have a unique role in opposing the action of other peptides known to delay gastric emptying. A similar conclusion has been reached by Bech et al.,[104] who evaluated the effect of somatostatin on gastric acid secretion and antral motility in conscious dogs with gastric fistulae. Somatostatin inhibited dose dependently the stimulated acid secretion, whereas the effect on antral mobility was more complex, acting especially on the amplitude of the contractions. The effects of somatostatin were not altered by using α-adrenergic, β-adrenergic, dopaminergic, and secretonergic blocking drugs.

Finally, Sanders[105] has discussed the significance of endogenous prostaglandins in regulating the mechanical contraction of gastric muscles and some of the possible consequences of prostaglandin synthesis on gastric motility. In general, emptying of solids could be retarded by endogenous prostaglandins, whereas emptying of fluids may be facilitated. Overproduction of prostaglandins may produce abnormal motility patterns and affect gastric emptying. Likewise, Bueno et al.[106] have concluded that prostaglandins act centrally to control the pattern of intestinal motility in both rats and dogs, and that calcitonin and neurotensin, when injected intracerebroventricularly, affect the intestinal motor profile, probably by stimulating prostaglandin release within the brain.

The physiology and pathophysiology of gastric emptying in humans has been reviewed by Minami and McCallum.[107]

B. Pancreatic Secretion

The secretion of pancreatic lipases is essential for the digestion of fats. The response of pancreas to the ingestion of food and the administration of gastrointestinal hormones, however, has usually been estimated by measuring the secretion of total protein or trypsin, chymotrypsin, and amylase. Although the secretion of lipases has been assumed to parallel those of the proteolytic and glycolytic enzymes, there is evidence that the release of the various pancreatic secretory proteins may not parallel each other. Thus, an asynchrony in intracellular transport of newly synthesized rat pancreatic proteins has been observed by Rohr and Keim.[108] Using a combination of in vivo labeling of pancreatic proteins in conscious rats followed by cell fractionation, it was shown that the newly synthesized proteases appear earlier in zymogen granules and pancreatic juice compared to other proteins such as lipase or amylase. An even more pronounced result could be obtained after a 3-hr prestimulation of the pancreatic secretion with carbachol.

Kern et al.[109] have studied the regulation of the successive steps along the secretory pathway in the rat exocrine pancreas using the model of in vivo infusion of synthetic cerulein in conscious rats for periods up to 72 hr in combination with electron microscopy and in vitro analysis of protein synthesis, intracellular protein transport, and enzyme discharge using isolated pancreatic lobules. It was observed that one group of proteins (two forms of trypsinogen, chymotrypsinogen, and procarboxypeptidase, respectively) showed progressive increases in synthesis, while a second group (two amylases) revealed a decrease in synthesis to levels about tenfold lower than controls. A third group of proteins (one trypsinogen, lipase, proelastase) did not show changes in synthesis with hormone stimulation.

Since the direct evaluation of pancreatic exocrine function in infants and children is not easily accomplished due to technical difficulties and impracticality with triple lumen tube intubation and collection techniques, Rosenthal et al.[110] have a simple method which permits direct measurement of pancreatic enzyme activity in small aliquots (1 to 5 $\mu\ell$) of human or animal duodenal fluid obtained by the relatively noninvasive string test. In order to assay for chymotrypsin activity, duodenal fluid expressed from the string was reacted with *N*-benzoyl-L-tyrosyl-*p*-aminobenzoic acid and aliquots of the reaction mixture were mixed directly into glycerol on a copper probe tip and analyzed by liquid secondary ion mass spectrometry (MS) using a Kratos MS-50 mass spectrometer with a cesium ion gun. Intense ions identified as unreacted substrate and enzymatically released *N*-benzoyl-L-tyrosine and *p*-aminobenzoic acid were easily distinguished and their intensities related to the enzymatic activity present. A simultaneous panel of pancreatic enzyme activities in small amounts of duodenal fluid could be determined by quantitation of reaction mixtures arising from the following substrates: *N*-benzoyl-L-arginine-*p*-nitroanilide yielding *N*-benzoyl-L-arginine and *p*-nitroaniline as an indicator of trypsin activity, amylose yielding glucose and maltose for amylase activity, and trioleoylglycerol yielding oleic acid and glycerol for lipase activity.

1. Stimulation

The mechanism by which intraluminal stimulants cause pancreatic secretion is subject to continued debate. Controversy continues over the relative contributions made by hormonal and neural (cholinergic) mechanisms in the exocrine pancreatic response to ingested food. The discovery by Hahne et al.[111] of a specific CCK receptor antagonist has enabled the reevaluation of the role of CCK in this process. Stubbs and Stabile[112] have demonstrated a virtually complete inhibition to all doses of intraduodenal amino acids and fat with proglumide administration. These results support the hypothesis that CCK is the major mediator of the intestinal phase of exocrine pancreatic secretion. However, the antagonists are of low activity and the validity of the conclusion may be in doubt.

Konturek et al.[113] have demonstrated a potent effect on pancreatic exocrine secretion following administration of the frog skin peptide bombesin, which is structurally related to gastrin-releasing peptide. Khuhtsen et al.[114] have recently shown that gastrin-releasing peptide in conjunction with acetylcholine is likely to play a prominent part in parasympathetic regulation of pancreatic exocrine secretion. Howard et al.[115] found that gastrin-releasing peptide was 70% as efficacious as bombesin for the ability to cause residual stimulation of amylase release. Valenzuela et al.[116] have concluded that the effect of exogenous CCK on pancreatic secretion of enzymes is not affected by atropine, that intraintestinal oleate stimulates pancreatic enzyme secretion significantly by an atropine-sensitive mechanism, and that the atropine effect is probably a blockade of a cholinergic enteropancreatic reflex.

The foregut is considered the major source of hormones stimulating exocrine pancreas, but little is known about their release along the small intestine. Konturek et al.[117] have compared the responses of pancreatic secretory and plasma secretin, CCK, and pancreatic polypeptide to perfusion with HCl and oleate of the intact intestine and isolated Thiry loops of duodenojejunal and distal ileal regions of conscious dogs with pancreatic fistulae. The results showed a gradient of immunoreactive secretin and CCK (but not pancreatic polypeptide) release along the small intestine. Gullo et al.[118] have concluded that secretin stimulates pancreatic enzyme secretion in man by a direct action on the acinar cells, probably involving binding of secretin to the acinar cell receptor, activation of adenyl cyclase, increased cellular cAMP levels, activation of cAMP-dependent protein kinase, and after a series of undefined steps, stimulation of enzyme secretion.

Brugge et al.[119] have examined the role of CCK in the intestinal phase of pancreatic secretion using a newly developed sensitive and specific assay for CCK. Subsequently, Brugge et al.[120] have examined the role of CCK in controlling the pancreatic secretion that is associated with phase III of interdigestive motility. The gastrointestinal hormones that have been identified as being associated with Phase III do not appear to be responsible for the pancreatic secretion. There was no correlation between the change in CCK level and the increase in trypsin output during a jejunal infusion of amino acids in six healthy volunteers. Hence, the agent responsible for the stimulation of interdigestive pancreatic secretion remains to be identified. Brugge et al.[120] have also demonstrated that trypsin infusion increases the plasma levels of CCK without stimulating pancreatic secretion of chymotrypsin. Similar effects were obtained with heat-inactivated trypsin. It was concluded that the CCK increases seen in phases II and III are secondary to intraluminal trypsin, but independent of the tryptic activity.

Debas et al.[121] have shown that epidermal growth factor causes a significant stimulation of pancreatic protein secretion when administered over a background infusion of CCK-8. No such interaction was demonstrated between *it* and *secretin*. Graded doses of epidermal growth factor had no significant effect on pancreatic secretion when given alone in the anesthetized rat. Konturek et al.[122] have found that it is an effective inhibitor of H^+ secretion from the parietal cells and that it does not affect alkaline gastroduodenal or pancreatic secretion.

It has been shown previously that synthetic neurotensin stimulates exocrine pancreatic secretion in conscious dogs.[123] Baca et al.[124] have reported experiments which suggest an interaction of neurotensin with pancreatic receptors for secretin and CCK. Increasing doses infused intravenously caused a dose-dependent stimulation of exocrine pancreatic secretion.

Feurle et al.[125] have demonstrated in man that neurotensin stimulates exocrine pancreatic secretion in doses which may be released after a meal. The ileal peptide neurotensin was found to stimulate exocrine pancreatic secretion in dog and man. In man it is not known whether levels in the plasma after neurotensin infusion are in the same range as postprandial concentration or whether the molecular forms of circulating neurotensin after infusion and after oral food are similar. It was found that neurotensin had a significant stimulatory action on pancreatic secretion of volume, enzymes, and bicarbonate. The plasma levels of neurotensin after a fatty meal and after infusion of neurotensin were comparable. Furthermore, HPLC indicated the presence of the tridecapeptide known to be the active molecular form of neurotensin in postprandial plasma. These results suggest that neurotensin may also represent a pancreatic secretagogue in man. It is known that neurotensin releases pancreatic polypeptide and that both are simultaneously elevated after fat ingestion. Sakamoto et al.[126] have evaluated the role of neurotensin in the regulation of the intestinal phase of pancreatic polypeptide release induced by fat. They showed that plasma levels were significantly elevated after intraduodenal sodium oleate and Lipomul, but this rise was not accompanied by neurotensin release after Lipomul. Physiological doses of i.v. neurotensin did not significantly stimulate pancreatic polypeptide release. Therefore, neurotensin does not seem to be an intestinal phase secretagogue of pancreatic polypeptide induced by fat. Likewise, Saad et al.[127] have shown that neurotensin stimulates pancreatic secretion in man at supraphysiological plasma levels, suggesting that it does not play a physiological role in the control of pancreatic secretion. Furthermore, neurotensin might be stimulating pancreatic exocrine secretion of fluid-volume and bicarbonate, and to a much lesser extent, of enzymes.

Fujimura et al.[128] have reported that oleate perfusion of jejunum, but not ileum, increases portal plasma levels of neurotensin. In a follow-up study, these workers[129] have shown that oleate, or its mixture with bile and pancreatic lipase, ethanol, or HCl in direct contact with the lumen of the ileum, fails to release neurotensin from the ileum of dogs. Neurotensin in the ileum could be released only by stimulation of the proximal small bowel.

Pandol et al.[130] have reported that growth hormone-releasing factors stimulate in vitro pancreatic enzyme secretion. In dispersed acini from guinea pig pancreas, rat hypothalamic growth hormone-releasing factor caused a twofold linear increase in amylase release compared to control over a 40-min incubation. The results showed that growth hormone-releasing factor stimulates pancreatic enzyme secretion, that rat hypothalamic growth hormone-releasing factor is greater than 100 times more potent than the human pancreatic factor, and that cAMP mediates this action. The physiological significance of these observations depends on the demonstration that such factors occur in the gastrointestinal tract.

Noguchi et al.[131] have demonstrated that activation of protein kinase C can produce stimulation of pancreatic enzyme secretion and suggest that those secretagogues that mobilize cellular calcium stimulate pancreatic enzyme secretion by causing activation of protein kinase C.

Malagelada et al.[132] have shown that intraduodenal Ca^{2+} can stimulate pancreatic enzyme secretion. Inoue et al.[133] report experiments which suggest that the divalent cations (Ca^{2+}, Mg^{2+}, and Zn^{2+}) may have a nonspecific electrical action that results in an alteration of membrane permeability, which leads to release of gastrointestinal hormones.

2. Inhibition

Other work has shown[134,135] that trypsin and chymotrypsin in the duodenum exert a negative feedback regulation on pancreatic enzyme secretion in the rat. It was speculated that this is mediated through the release of CCK. Using a sensitive and specific bioassay based on amylase release from isolated pancreatic acini,[102] Louie et al.[136] have examined this role of CCK. Diversion of pancreatic juice or perfusion with trypsin inhibitor increased enzyme secretion and raised CCK plasma levels. The inhibitory effect of trypsin on CCK release was enzyme and site specific, since it was not observed with amylase in the duodenum or intraileal perfusion of trypsin. The release of CCK induced by diversion of pancreatico-biliary juice was inhibited by intraduodenal perfusion of lidocaine, infusion of tetradoxin into the superior mesenteric artery, and i.v. atropine. It was therefore concluded that the feedback regulation of pancreatic enzyme secretion is mediated by CCK and is neurally controlled involving a cholinergic pathway. Owyang et al.[135] have also investigated this mechanism by determining the effect of atropine and trypsin on phenylalanine-, volume-, and osmolality-stimulated pancreatic enzyme secretion. Duodenal chymotrypsin and lipase output were measured during the intraluminal perfusion with the appropriate agents. Only the phenylalanine-stimulated chymotrypsin and lipase were partially inhibited by trypsin. Administration of atropine completely inhibited the secretion stimulated by volume and osmolality, but only partially (60%) inhibited that by phenylalanine, while simultaneous administration of atropine and intraduodenal trypsin was completely inhibitory. It was concluded, therefore, that the intestinal phase of human pancreatic exocrine secretion is under both hormonal and neural control. Intraduodenal trypsin only inhibits hormonal- but not cholinergic-mediated pancreatic enzyme secretion. This study suggests that feedback regulation of pancreatic enzyme secretion is stimulus specific.

Somatostatin is known to inhibit pancreatic exocrine function.[137,138] Loud et al.[139] have shown that plasma concentrations corresponding to reported physiological variations support the concept that somatostatin participates in the hormonal control of the pancreatic endocrine and acid secretion. The effects of low doses of somatostatin in the present study, however, were less pronounced than those reported by Souquet et al.[138] Since the effect of somatostatin in man is not known, Ellison et al.[140] have studied the effect of somatostatin analog 201-995 on basal pancreatic exocrine secretion in a single blind trial in a patient with a mature traumatic pancreatic fistula. Fistula volume was measured every 6 hr, and blood samples were taken every 12 hr for determination of amylase, lipase, and glucose. Somatostatin analog reduced basal fistula output by 30 to 37% and abnormal serum amylase and lipase elevations by 50%. It was concluded that somatostatin reduces the volume of pancreatic secretion and may decrease the release of amylase and lipase. Hence, it may be a valuable agent for the treatment of diseases in which inhibition of pancreatic exocrine secretion is desirable. Konturek et al.[141] have shown in conscious dogs with pancreatic fistulae that the intestinal phase of pancreatic secretion is inhibited by atropine and abolished by somatostatin. Atropine inhibits plasma pancreatic polypeptide, but it does not affect plasma CCK and secretin responses to intestinal stimulation including oleate perfusion. Somatostatin suppresses plasma secretin, CCK, and pancreatic polypeptide secretion almost completely. De Graef and Woussen-Colle[142] have shown that sham feeding and bombesin release somatostatin. They suggest that somatostatin release is largely under vagal-cholinergic control. Esteve et al.[143] have concluded that somatostatin binds to specific plasma membrane receptors to form a slowly reversible complex that is highly reactive with Ca^{2+}. Cell calcium mobilizing agents decrease the affinity of acinar somatostatin receptors for somatostatin.

Sakamoto and Williams[144] have shown that somatostatin binding to pancreatic acinar cell membranes is directly regulated by CCK mediated via the CCK receptor, as it could be blocked with the CCK antagonist, dibutyryl cGMP, and was induced by various CCK analogs. The cholinergic agonist carbachol, which acts on intact acini like CCK and Ca^{2+}, the intracellular messenger for CCK, had no effect on somatostatin binding by plasma membranes. Thus, the effect of CCK in inhibiting somatostatin binding to its receptor is directly induced by occupancy of the CCK receptor and is one of the first effects of CCK in a broken cell preparation.

Totemoto[145] has isolated peptide YY from porcine intestine and has shown it to exhibit marked structural homology to pancreatic polypeptide and to inhibit gastric and pancreatic secretion. It was suggested that peptide YY may be the gastric and pancreatic inhibitor in ileocolonic mucosa that is released by perfusion of intestine with oleic acid.

C. Biliary Secretion

Biliary secretion makes available bile salts for the emulsification of lipids during digestion and for the formation of lipid micelles during fat absorption. It is well established that the introduction of fat into duodenum leads to the secretion of CCK, the contraction of the gall bladder, and the release of bile salts into the duodenum. The mechanism of this phenomenon, however, is not fully understood and the physiological significance of hormonal effects on bile secretion is unclear.

Gall bladder contraction after a regular meal of liquids and solids is biphasic, with an initial fast phase and a second slow phase. Lawson et al.[146] have concluded that gastric emptying of the solid portion of a meal containing fat maintains tonic gall bladder contraction through continued release of humoral mediators from small bowel and pancreas. When gastric emptying nears completion, humoral stimulation ceases and the gall bladder refills, but human pancreatic polypeptide does not appear to be responsible for it. Fried et al.[147] have provided strong evidence that vagotomy alters gall bladder and pancreatic protein secretion primarily via neural effects on the target organs, because it does not alter significantly the plasma CCK concentrations. However, pancreatic juice, which was excluded in the experiments of Fried et al.,[147] could have influenced the result. The effect would have been interesting to observe either of oleate mixed with pancreatic juice on CCK release or the effect of replacement of the pancreatic juice in the duodenum on oleate CCK-releasing efficiency.

You et al.[148] have evaluated the effect of intraduodenal Lipomul (a synthetic lipid emulsion) in human volunteers. Lipomul in three different doses produced significant increases in plasma CCK-like immunoreactivity and significant decreases in gall bladder volume during a 90-min period of observation. However, the responses were not dose dependent. The peak concentration of CCK-like immunoreactivity coincided with the maximum gall bladder contraction. Atropine did not significantly influence Lipomul-stimulated gall bladder contraction or plasma CCK-like immunoreactivity. Hence, no significant cholinergic effect was apparent.

There is evidence that a non-CCK stimulant of gall bladder contraction is present in human serum[149] and that its concentration is elevated in patients with gallstone disease.[150] Strah et al.[151] have used high-pressure liquid chromatography (HPLC) to further purify the substance and to compare the mechanism of its action in gall bladder contraction with that of CCK-8. It was possible to isolate and to partially purify a substance which is immunologically distinct from CCK and which causes gall bladder contraction. The action of neither CCK-8 nor of this non-CCK stimulant was significantly inhibited by atropine. The specific antiserum and proglumide, which are potent inhibitors of CCK, caused no significant inhibition of the action of the unknown substance.

Sakamoto et al.[152] have shown that neurotensin given intravenously stimulates gall bladder contraction in a dose-dependent manner. It was approximately 1:50 as potent as CCK-8, on a molar basis. Gall bladder stimulation of neurotensin was diminished significantly by atropine, whereas CCK-8-induced gall bladder contraction was not. Kortz et al.[153] have demonstrated that the stimulation of bile flow by bombesin during stable bile acid secretion was independent of bile formation. Bombesin significantly increased bile bicarbonate output, suggesting a choleretic effect at the ductular level. The mechanism of this choleresis remains to be established. Romanski and Bochenek[154] have reported that Boots secretin increases bile flow by 63% and bile acid, cholesterol, and phospholipid output by 75, 96, and 73%, respectively. The more highly purified Kabi secretin had practically no effect on bile secretion.

In contrast, somatostatin inhibits bile flow.[155] Kaminski and Deshpande[156] have compared the effects of somatostatin and bombesin on secretin-stimulated ductular bile flow in dogs. It was found that bombesin augmented bile flow stimulated by infusion of acid into the duodenum by increasing secretin release, while somatostatin decreased bile flow by directly inhibiting the release of secretin from the duodenal mucosa. Hanks et al.[157] have evaluated the effect of graded doses of i.v. somatostatin on taurocholate-stimulated bile flow in awake-fasting dogs. While bile salt output remained unchanged, the bile salt-independent cannalicular or ductural secretion was inhibited.

Miyasaka et al.[158] have demonstrated that intraileal infusion of sodium oleate significantly suppresses bile secretion. In contrast, intraduodenal infusion of sodium oleate significantly increased pancreatic juice flow, protein, and bicarbonate output, but had no effect on bile secretion in the rat. Ladas et al.[159] have investigated the effects of medium- and long-chain triacylglycerol test meals on gall bladder contraction, small intestinal bile acid concentrations, and ileal flow rates in normal individuals. It was shown that medium-chain triacylglycerols, when compared with long-chain triacylglycerols, produced a smaller input of bile acids into the small intestine and smaller volume of fluid delivered to the colon.

V. SPECIFIC EFFECTS ON FAT ABSORPTION

Gastrointestinal peptides play both a direct and indirect role in the regulation of food intake, digestion, and absorption. The rate at which food fat is digested and absorbed is dependent upon the rates of gastric emptying, intestinal transit, pancreatic and biliary secretion, lipolytic degradation, and membrane transport.

A. Food Intake

The hypothalamus is recognized as the major regulatory area for appetite. The lateral hypothalamus appears to be concerned with hunger and the initiation of feeding and the ventromedial hypothalamus, the satiety center, with the inhibition of feeding. Teitelbaum[160] and Grossman[161] have demonstrated that manipulation of the central nervous system produces disorders of food intake. However, the control of food intake is extraordinarily complex, with psychological factors, such as taste, smell, visual appeal, and conditioned reflexes, and hypothalamic centers all contributing. The hypothalamus receives input from vagal afferents responding to gastric or intestinal distension, from absorbed products of food digestion, and possibly from gastrointestinal hormones or messengers released from the gut. It receives rich peptidergic innervation and is also influenced by exogenous peptides.

The idea that gastrointestinal hormones released after a meal might reach the brain and stimulate the satiety center was derived from the observation that some form of message inhibiting feeding comes from the stomach and small intestine when they contain food. Food placed in an isolated transplanted stomach elicits satiety,[162] and the satiating effect of food in the small intestine is greater than that of food in the stomach.

Cholecystokinin has been shown to inhibit food intake when injected intraperitoneally in both animals and man.[163] The problem is whether these effects are physiological or merely pharmacological, as suggested by other investigators.[164]

The main evidence that CCK is a peripheral signal of satiety comes from the demonstration in the rat that i.p. administration of CCK-8 terminates feeding. Vagotomy abolishes the response. In dogs, i.v. doses of the peptide that inhibit food intake are pharmacological. Holubitsky et al.[165] have shown that CCK-8 inhibits food intake in the rat when given intraperitoneally, but not subcutaneously, implying that it either has access to receptors (e.g., vagal afferent) or that i.p. injection of the peptide has noxious influences (e.g., gut spasms), while s.c. injection does not. It was concluded that the basis of the claim that CCK-8 is a peripheral signal of satiety needs reassessment.

The doses of CCK administered into the brain that produce satiety have been small, while those used peripherally have been pharmacological. Pappas et al.[166] have compared the potencies of endogenous CCK and of exogenous CCK-8 for the satiety, cholecystokinetic, and pancreozymic actions. It was shown that duodenal perfusion of fat in doses producing maximal pancreatic stimulation and gall bladder contraction had no satiety effect. The D_{50} of exogenous CCK-8 for satiety was supramaximal for its effects on the pancreas and gall bladder, but when centrally administered it produced satiety effects in picomolar doses. It was concluded that CCK-8 alone is not a peripheral satiety signal in the dog. Garcia et al.[167] have compared induction of satiety by cerulein (a CCK-analog), bombesin, and pancreatic polypeptide on food intake in lean and obese mice. The peptides were administered intraperitoneally. It was shown that cerulein is a more potent satiety-inducing factor than either bombesin or pancreatic polypeptide.

Recent studies have shown that CCK affects appetite by a peripheral rather than a central action, and that the effect of injections of CCK is reduced by vagotomy.[168] In addition, there is no CCK in the stomach to transmit a hormonal agent which decreases appetite.[160] Therefore, satiety hormones other than CCK have been postulated, including glucagon,[169] somatostatin,[170] and bombesin.[171] Garcia et al.[167] have compared the effects of cerulein, bombesin, and pancreatic polypeptide on food intake in lean and obese mice at 3, 9, and 27 nmol/kg. It was found that cerulein significantly inhibited food intake at all levels of the hormone tested. Although the highest dose of bombesin significantly decreased food intake in both obese and lean mice, the lower doses were only effective in obese mice. In contrast, no dose of pancreatic polypeptide had a significant effect on food intake in either. However, others have dissented against bombesin as a satiety factor.[172] The presence of CCK in the brain and in the gut and its well-documented primary action of triggering gall bladder contraction continues to support interest in its potential role as a satiety hormone. Taylor and Garcia[173] have shown that much larger doses of bovine pancreatic polypeptide than those used by Malaisse-Lagae et al.[174] are required to inhibit food intake acutely in obese mice and indicate that a reevaluation is necessary of the role of pancreatic polypeptide in the control of body weight. There appears to be a marked discrepancy between the reduction of body weight and the decrease in food intake observed by the two pancreatic polypeptide treatment groups described by the previous workers.[174] These observations, coupled with the present findings, suggest that pancreatic polypeptide may have effects on body weight that are independent of any it might have on appetite and satiety. Reidelberger et al.[175] have demonstrated that food intake significantly increases plasma levels of pancreatic digestive enzymes. Because the plasma trypsinogen responses to identical meals at the beginning and end of a 24-hr experimental period were very different, the authors believe that cephalic mechanisms may play an important role in mediating this effect. It was suggested that pancreatic enzymes enter the blood from the pancreatic interstitium rather than from the gut.

Furthermore, others[176,177] point out that CCK exists in the body in multiple molecular forms, which differ both quantitatively and qualitatively in their actions, and that the physiological equivalence of various commercial test preparations is not clear. Thus, the role of CCK as a satiety hormone acting by way of the bloodstream is not proved.[162,178]

Collins and Weingarten[179] have demonstrated that the glutamic acid derivative proglumide inhibits CCK-induced satiety in the rat. Since CCK influences also gastric motility in the rat, an antagonism of this action by proglumide may account for its effect on feeding. To determine the extent to which the action of proglumide on feeding is due to its effect on gastric emptying, Collins and Weingarten[179] have examined the ability of proglumide to influence food intake in sham-feeding rats. In this preparation, the presence of a gastric fistula prevents the accumulation of food in the stomach, thus, eliminating the influence of changes in gastric emptying on food intake. The results indicate that the ability of proglumide to modulate food intake is independent of its effect on gastric emptying, is specific for CCK, and must be due to the inhibition of the action of CCK at the hitherto unidentified receptor responsible for its satiety effect. Shillabeer and Davison[180] have shown that proglumide increases food intake only when injected after a food preload. They also have investigated whether the proglumide effect is dependent on the vagus nerve. The effects were compared in sham-operated vagotomized and normal rats and the results were rechecked for delayed gastric emptying. After proglumide, normal and sham-operated rats increased meal size significantly, while vagotomized rats were unaffected. Hence, proglumide acts on a factor, possibly CCK, produced by a food preload to increase food intake, and this action is dependent on the vagus nerve.

Calcitonin and prostaglandin E_2 are known to inhibit food intake centrally,[106] and it has been proposed that the site of action of prostaglandin E_2 is located in the medial and the lateral hypothalamic areas that contain several times greater concentrations than other brain regions. Bueno et al.[106] undertook to evaluate the effects of central administration of prostaglandin E_2 on the postprandial motility in rats and dogs, and to compare the response to neurotensin and calcitonin to that of the prostaglandin. The similarities in the digestive motor responses to the central administration of the two neuropeptides and of the prostaglandin suggest that prostaglandin E_2 mimics the effects of calcitonin in rats and dogs and serves as a common mediator of their central action.

B. Digestion

The rate of digestion of nutrients depends upon the composition of the chyme and the delivery of the various digestive enzymes and biliary emulsifiers to the intestine at appropriate times and in appropriate quantities. The synchronization of these processes is subject to regulation by peptides released either from endocrine cells or neurons located within the walls of the gastrointestinal tract. Thus, the control of pepsin secretion is intimately linked to that of acid secretion and stimulatory peptides, such as gastrin and bombesin.[181] Inhibitors of acid secretion such as gastric inhibitory peptide, VIP, secretin, and glucagon also control pepsin secretion.[182] Very little is known about the control of gastric lipase secretion. It appears that most of lipase activity in the stomach is of lingual origin and that the gastric component is an esterase that acts on short-chain triacylglycerols and is secreted in parallel to pepsin.[183] Possibly, the peptidergic control of its secretion may be coupled to that of pepsin. Fink et al.[184] have demonstrated that a lipase that hydrolyzes long-chain triacylglycerols is secreted by rabbit stomach mucosa and that the secretion of gastric lipase is stimulated by two different receptor mechanisms and is released in response to stimulation by various agents such as CCK and carbachol in a dose-dependent manner. Pancreatic amylase release by isolated acini is shown to be dose related in response to these secretogogues, also[184] suggesting that a common mechanism of enzyme (lipase, pepsinogen, and amylase) secretion may exist in both gastric and pancreatic tissues.

Evidence that the endocrine pancreas is capable of regulating the level of intestinal digestive enzymes has come from studies of diabetic man and experimentally induced diabetes in animals. In vivo and in vitro studies have shown that experimental diabetes is associated with increased transport of sugars and amino acids in the small intestine. Ghishan et al.[185] have demonstrated increased ^{32}P uptake into jejunal brush border membrane vesicles in control compared to diabetic rats. It was concluded that the reason for this was the increase in the number of Na^+-P cotransporters.

Rutgeerts et al.[186] have evaluated the ^{13}C-mixed triacylglycerol breath test by comparing 6-hr cumulative $^{13}CO_2$ breath excretion with duodenal lipase output, measured with a marker-corrected duodenal perfusion technique, after maximal stimulation with pancreozymin-secretin. It was found that the quantitative ^{13}C-mixed triacylglycerol breath test correlated very well with the lipase output in the duodenum after maximal pancreatic stimulation and is, therefore, a good test of pancreatic function. The mixed triacylglycerol 1,3-distearoyl-2-^{13}C-octanoyl-glycerol was combined with butter and taken with 100 g of bread as breakfast for the purpose of the test.

It is well established that the enzyme component of pancreatic secretion is controlled by the peptide CCK and the aqueous component secretin. Nutrient absorption can be affected by the endocrine pancreas and by somatostatin. Control of luminal enzyme secretion is increased by CCK, secretin, gastric inhibitory polypeptide, VIP, glucagon, and gastrin. Peptides influence the rate and direction of electrolyte and attendant water movement. The secretory actions of VIP are well documented. Furthermore, peptides such as VIP probably influence digestion and absorption via blood flow changes. Likewise, evidence has accumulated that gut hormones stimulate insulin release from the pancreas.

Gallavan et al.[187] have observed that oleic acid increases both blood flow and intestinal hormone production only when present in the lumen in micellar form, and suggest that VIP could play a role in jejunal vascular response to fat. The study was done to determine whether intestinal hormones could play a role in the regulation of the postprandial intestinal hyperemia by acting as paracrine rather than endocrine substances.

C. Cellular Uptake and Resynthesis of Dietary Fats

Evidence has been presented for a role of somatostatin in the control of nutrient absorption. Intravenous somatostatin inhibited the absorption of triacylglycerols.[188] When delivered intraportally, concentrations as low as 50 ng/min inhibited postprandial rises in triacylglycerol levels.[189] Antisomatostatin antiserum infusion augmented the plasma triacylglycerol levels obtained following a meal.[190] Although it is unclear as to which digestive or absorptive event is influenced, a local effect on the enterocyte is possible, since even luminally administered somatostatin can inhibit triacylglycerol uptake into circulation.[189]

Boivin et al.[191] have observed that chronic fatty diets surprisingly increase human interdigestive (fasting) secretion of bile acids and exocrine pancreatic enzymes, including lipase. Subjects who ate a high-fat as compared to a low-fat diet for 2 weeks not only increased postprandial, but also increased interdigestive biliary and pancreatic secretion. Varying the proportion of carbohydrate or protein chronically or acutely within a high-fat diet did not alter the secretory response to fat.

Dennison et al.[192] have examined the effect of transplantation on motility and absorption of various nutrients by the small intestine of the dog. It was found that intestinal absorption of water, glucose, electrolytes, medium-chain triacylglycerols, amino acids (alanine), and short-chain fatty acids (lauric) is unimpaired by intestinal transplantation. Absorption of the long-chain fatty acid oleic acid required lymphatic regeneration and did not occur until day 28. Intestinal regeneration and propulsive function were markedly depressed initially, but became active during week 2.

Thompson and Rajotte[193] have determined the effect of once-daily injections of NPH insulin and islet cell transplantation on the enhanced uptake of lipids into the intestine of diabetic rats. The enhanced uptake of a homologous series of saturated medium-chain length fatty acids in untreated diabetic rats was reduced with NPH insulin and islet cell transplantation. The uptake of cholesterol was lower in diabetic rats treated with NPH insulin and islet cell transplantation than in the untreated, over a wide range of concentrations of cholesterol, bile acid, and bile acid/cholesterol. In contrast to the small intestine, there was no effect on colonic uptake of these probes. Thus, injection of exogenous or endogenous insulin supplied by islet cell transplantation in diabetic rats reduced the enhanced uptake of lipids, normalized the effective resistance of the unstirred water layer, and reduced the enhanced passive permeability of the jejunum.

Zimmerman et al.[194] have tested the effects of hormones and drugs on human intestinal lipid absorption, synthesis, and secretion. Methylprednisolone, insulin, pentagastrin, and nicotinic acid did not affect fatty acid uptake or triacylglycerol biosynthesis by small intestinal biopsies. Preincubation with ethanol or cimetidine decreased triacylglycerol biosynthesis, resulting in concomitant increases in tissue fatty acid levels. The cimetidine-induced inhibition of intestinal triacylglycerol synthesis was thought to be related to its interaction with enzymatic systems.

D. Transport

There is evidence that villous motility is also involved in intraluminal transport and that villous movement is controlled by a humoral factor, which has been called villikin.[195] It has been suggested that villi agitate the chyme and that absorption of fat into lacteals is facilitated by alternate contraction and relaxation of the villi.[195] Villikin has not yet been purified.

Armstrong et al.[196] have tested whether neurotensin can affect the transport of oleic acid from the intestinal lumen into the mesenteric lymph. It was shown that the infusion of neurotensin into the superior mesenteric artery increases the translocation of [^{3}H]oleic acid from a micellar solution instilled in the small intestinal lumen into lymph triacylglycerols. On this basis it was suggested that neurotensin plays a role in the intestinal processing of absorbed dietary fat. Tso et al.[197] have recently underscored the necessity of using steady-state lymphatic lipid output to assess factors affecting the cellular packaging and release of chylomicrons in the small intestine. The chylomicron appearance time lengthened as lymph flow fell. Chen et al.[198] have shown that gut regulatory peptides are present in thoracic duct lymph and their concentrations and outputs increase after feeding canned dog food. These immunoassay results support previous bioassay studies suggesting that gut hormones are present in the lymph. Clearly, further biochemical and physiological investigation is warranted.

Previous work has demonstrated[199] that neurotensin, which is present in a subpopulation of cells distributed throughout the epithelium of the small intestine, is released into the circulation during intraluminal administration of oleic acid, whereas perfusions of glucose and amino acids have no effect on circulating levels of neurotensin-like immunoreactivity. Ferris et al.[200] have determined the necessity for a minimal chain length or the presence of a carboxylic acid for the stimulation of neurotensin release during perfusion of the small intestine. It was shown that perfusions of fatty acids with four or more carbons resulted in a four- to fivefold increase in neurotensin levels in plasma as compared to neurotensin values from saline and taurodeoxycholate perfused groups. The four- and eight-carbon alcohols were as effective in increasing plasma neurotensin levels as fatty acids of the same chain length. It was interesting that perfusion with ethanol also caused an elevation in plasma neurotensin since in other studies perfusion with acetic acid did not increase plasma neurotensin-like immunoreactivity.[200] Rosell et al.[201] have investigated possible involvement of nervous pathways and circulating hormones in the neurotensin response to fat in anesthetized rats. Fat in the form of fatty acids was the nutrient most important in increasing the release of neurotensin from storage sites in the small intestine and indicated that the effect is potentiated by nervous pathways of adrenergic receptors. In addition, corticosteroids (dexamethasone) seem to have a pronounced influence on the release of neurotensin. Sakamoto et al.[202] have investigated the effect of atropine, vagotomy, and somatostatin on the endogenous release of neurotensin by fat. It was found that neurotensin increases in a dose-dependent manner in response to intraduodenal Lipomul (a commercial fat emulsion). Atropine suppresses the rapid release of neurotensin, but atropine does not reduce the integrated neurotensin response to Lipomul. The mechanism of neurotensin release by fat appears to be independent of vagal cholinergic innervation. Somatostatin suppresses neurotensin release in response to fat. Barber et al.[203] have developed a method that allows direct assessment of the regulation of neurotensin release from canine ileal cells indicating modulation by bombesin.

It is likely that the overall absorptive process is also modified by peptide-induced changes in the gastrointestinal circulation. Intraduodenal corn oil increased both pancreatic and jejunal blood flow[204] and cholecystokinin or secretin have been implicated in this response.[205] There appears to be both a functional hyperemia and redistribution of blood flow from the muscularis to the mucosa during feeding. Gallavan et al.[206] have investigated whether the increase in the intestinal blood flow, which follows the instillation of micellar lipids into the jejunum lumen, is associated with an increase in neurotensin release. Neither oleic acid nor bile alone altered either blood flow or neurotensin release when placed in the jejunal lumen of dogs. The data indicate that the presence of micellar lipid in the jejunum results in a rapid release of neurotensin, which decreases over time and which generally parallels the change in jejunal blood flow. The study suggests that neurotensin could play a role in the initiation of the lipid-induced jejunal hyperemia. Harper et al.[207] have presented data which implicate neurotensin as a possible mediator of both the increase in intestinal capillary permeability and the postprandial increase in intestinal blood flow seen during fat absorption.

Gallavan et al.[187] have demonstrated that the luminal placement of micellar fat significantly increased both jejunal blood flow (+21%) and VIP production (+100%) in the anesthetized dog. In a subsequent study, Gallavan et al.[208] have examined the effect of exogenously administered VIP on the jejunal vascular resistance. Given the limitations of exogenous drug administration, it was not possible to rule out VIP as a mediator of the fat-induced hyperemia. However, it did not appear to contribute directly to the vascular response and must have acted indirectly as part of a more complex regulatory pathway.

VI. SUMMARY AND CONCLUSIONS

Fat assimilation is a multistep process involving the coordinated action of several organs. Their normal development and maturation is essential for fat absorption as is the regeneration of the absorptive sites following resection of the intestine. The small intestine also has special energy requirements which must be met for efficient normal function. All of these processes are under hormonal and nervous control. The physiological and biochemical effects are initiated with the ingestion of food and are continued during the digestion, absorption, and transport phases of the process.

The fat absorption depends on food intake. Intake appears to be regulated through the hypothalamus, which receives signals from the peripheral signaling system that is located in the upper gastrointestinal tract. This system involves physical, nervous, and hormonal factors that become activated after the ingestion of food. Gastrointestinal hormones released after a meal are believed to reach the brain and stimulate the satiety center. CCK, thyrotropin-releasing hormone, enkephalins, and other peptides have been postulated as being involved in appetite regulation. Most evidence favors a possible role for CCK, which is released when food enters the small intestine. Numerous studies have shown that injection or infusion of this hormone decreases food intake in both animals and man. Whether these effects of CCK are physiological or pharmacological have been extensively investigated in recent years, but the results have remained inconclusive. The stomach also transmits a humoral agent, which decreases appetite, but there is no CCK in the stomach. For this and other reasons, other satiety hormones have been postulated, including glucagon, somatostatin, and bombesin. Work is in progress to determine whether the effects of these gastrointestinal peptides are physiological or pharmacological. The problem of initiated effects would not be expected to be complicated by the presence of exorphins, which are biologically active peptides either present in protein-containing foods or generated from them during proteolyisis in the intestine. These peptides appear to be identical to or closely similar to the peptides found in the gut and brain.

The absorption of food and fat depend upon proper gastric emptying and on normal intestinal motility, which also are controlled by a complex interplay of both neural and endocrine factors. The majority of the gastrointestinal peptides so far isolated inhibit gastric emptying, with the VIP possibly possessing a physiological role. Other peptides are capable of causing either contraction or relaxation of smooth muscle of the small bowel. In addition, the intraluminal transport is affected by a humoral factor controlling the motility of the villi. It has been proposed that absorption of fat into lacteals is facilitated by alternate contraction and relaxation of the villi.

The composition of the chyme, which is the complex contents of the stomach, reaching the absorptive sites in the small intestine depends upon the digestion of the food by the enzymes secreted by the stomach and pancreas and upon normal contraction of the gall bladder and discharge of the emulsifying agents. The secretion of pepsinogen and activation to pepsin are under direct and indirect endocrine control via the control of acid secretion. Gastrin and bombesin stimulate this process; gastric inhibitory polypeptide, VIP, secretin, and glucagon inhibit it. Recent work has established the presence of a gastric lipase, which, along with lingual lipase, is responsible for the initial hydrolysis of dietary fatty esters. The peptidergic control of its secretion presumably is similar to that demonstrated for pepsin.

The control of pancreatic enzyme secretion is under the control of CCK, while secretin is the major bicarbonate secretagogue. There is evidence that the endocrine pancreas is capable of regulating the level of intestinal digestive enzymes. The proteolytic enzymes are released in the form of zymogens from which active enzymes are generated by proteolysis. The form of release of pancreatic lipase is less well defined. It requires a co-lipase for activity. The control of the digestion may be exercised through the release and activation of the digestive enzymes and any co-factors. Since measurements of immunoreactivity estimate the total and those of biochemical activity only the amount of active enzyme, it is theoretically possible to distinguish between the active and inactive forms of the enzyme released. Such experiments, however, have not been carried out.

The overall absorptive process is also modified by peptide-induced changes in cell membrane permeability and in the gastrointestinal circulation. Dietary fat as well as amino acids and HCl increase pancreatic and jejunal blood flow, and both cholecystokinin and secretin have been implicated in this phenomenon. Although some controversy exists, there appears to be both a functional hyperemia and a redistribution of the blood flow from the muscles to the mucosa during feeding.

In the past, it has been generally accepted that no differences exist among different foods as triggers of gastrointestinal peptide secretion, but recent evidence suggests that each food constituent may bring about a characteristic set of signals resulting in specific physiological and biochemical responses. It has been demonstrated that a difference exists between large and small molecules and between the original food molecules and their digestion products. In other instances differences have been seen between the hydrolysis and the absorption products of a substrate. The fatty acids and simple carbohydrates clearly exhibit different responses, and there are differences between amino acids and fatty acids. There is also an indication that differences exist between short- and long-chain fatty acids as triggers of gastrointestinal function. Finally, evidence has been obtained that at least in the case of the simple carbohydrates, differences may exist in the anomeric carbohydrate-induced stimulation of the enteroglucagon release.

Much of the work remains controversial because of the uncertainty of the relationship of the administered dose to physiological or pharmacological levels. Furthermore, the extreme complexity of the system makes it nearly impossible to provide appropriate control experiments. Hence, the interpretation of the results becomes more a matter of opinion rather than a direct extrapolation from experimental facts. The introduction of the immunochemical and receptor methodology to the identification and characterization of the action of the gastrointestinal peptides has strengthened the experimental evidence and sharpened the arguments. The gradual extension of these studies to include methods of molecular biology promises to resolve the remaining controversies and to improve the understanding of the role of the gastrointestinal peptides in the absorption of all foodstuffs including fats.

Only CCK, cerulein, and somatostatin preparations have thus far found therapeutic use, mainly in the control of pancreatic function, including the levels of lipase. The roles of gastrin, VIP, glucagon, and somatostatin in clinical disease states are clearly established and other gastrointestinal hormones may become available in the near future.

ACKNOWLEDGMENTS

The studies by the authors and collaborators referred to in this review were supported by funds from the Ontario Heart Foundation, Toronto, Ontario and the Medical Research Council of Canada, Ottawa, Ontario.

REFERENCES

1. Zakim, D., Integration of energy metabolism by the liver, in *The Role of the Gastrointestinal Tract in Nutrient Delivery*, Bristol-Myers Nutrition Symp., Vol. 3, Green, M. and Green, H. L., Eds., Academic Press, New York, 1984, 157.
2. Hulsmann, W. C., Energy metabolism in small intestine, in *Intestinal Toxicology*, Schiller, C. M., Ed., Raven Press, New York, 1984, 57.
3. Windmueller, H. G., Glutamine utilization by the small intestine, in *Advances in Enzymology*, Vol. 53, Meister, A., Ed., John Wiley & Sons, New York, 1982, 201.
4. Trier, J. S. and Madara, J. L., Functional morphology of the small intestine, in *Physiology of the Gastrointestinal Tract*, Vol. 1, Johnson, L. R., Ed., Raven Press, New York, 1981, 925.
5. Tepperman, B. L. and Jacobson, E. D., Mesenteric circulation, in *Physiology of the Gastrointestinal Tract*, Vol. 2, Johnson, L. R., Ed., Raven Press, New York, 1981, 1317.

6. Brant, J. L., Castleman, L., Ruskin, H. D., Greenwald, J. J., and Kelly, J., The effect of oral protein and glucose feeding on splanchnic blood flow and oxygen utilization in normal and cirrhotic subjects, *J. Clin. Invest.*, 34, 1017, 1955.

7. Reininger, E. J. and Sapirstein, L. A., Effect of digestion on distribution of blood flow in the rat, *Science*, 126, 1176, 1957.

8. Vatner, S. F., Franklin, D., and Van Citter, R. L., Mesenteric vasoactivity associated with eating and digestion in the conscious dog, *Am. J. Physiol.*, 219, 170, 1970.

9. Hultman, E. and Castenfors, H., Effect of injection of hypertonic glucose on splanchnic blood flow and oxygen consumption, *Scand. J. Clin. Lab. Invest.*, 13, 503, 1961.

10. Chou, C. C., Kvietys, P., Gallevan, R., and Nyhof, R., Blood flow, oxygen consumption and absorption of glucose and oleic acid in the canine jejunum, *Gastroenterology*, 76, 1114, 1979.

11. Chou, C. C., Kvietys, P., and Sit, S. P., Constituents of chyme responsible for postprandial intestinal hyperemia, *Am. J. Physiol.*, 235, H677, 1978.

12. Porteus, J. W., Intestinal metabolism, *Environ. Health Perspect.*, 33, 25, 1979.

13. Windmueller, H. G. and Spaeth, A. E., Uptake and metabolism of plasma glutamine by the small intestine, *J. Biol. Chem.*, 249, 5070, 1974.

14. Windmueller, H. G. and Spaeth, A. E., Identification of ketone bodies and glutamine as the major respiratory fuels *in vivo* for postabsorptive rat small intestine, *J. Biol. Chem.*, 253, 69, 1978.

15. Windmueller, H. G. and Spaeth, A. E., Intestinal metabolism of glutamine and glutamate from the lumen as compared to glutamine from the blood, *Arch. Biochem. Biophys.*, 171, 662, 1975.

16. Watford, M., Lund, P., and Krebs, H. A., Isolation and metabolic characteristics of rat and chicken enterocytes, *Biochem. J.*, 178, 589, 1979.

17. Iemhoff, W. G. J. and Hulsmann, W. C., Development of mitochondrial enzyme activities in rat small intestinal epithelium, *Eur. J. Biochem.*, 23, 429, 1971.

18. Robinson, A. M. and Williamson, D. H., Physiological roles of ketone bodies as substrates and signals in mammalian tissues, *Physiol. Rev.*, 60, 143, 1980.

19. Hanson, P. J. and Parsons, D. S., Factors affecting the utilization of ketone bodies and other substrates by rat jejunum: effects of fasting and diabetes, *J. Physiol. (London)*, 278, 55, 1978.

20. Lamers, J. M. J. and Hulsmann, W. C., The effects of fatty acids on oxidative decarboxylation of pyruvate in rat small intestine, *Biochim. Biophys. Acta*, 343, 215, 1974.

21. Bremer, J. and Norum, K. R., Metabolism of very long chain monounsaturated fatty acids (22:1) and the adaptation to their presence in the diet, *J. Lipid Res.*, 23, 243, 1982.

22. Walsh, J. H., Gastrointestinal hormones and peptides, in *Physiology of the Gastrointestinal Tract*, Vol. 1, Johnson, L. R., Ed., Raven Press, New York, 1981, 59.

23. Mutt, V., Chemistry of the gastrointestinal hormones and hormone-like peptides and a sketch of their physiology and pharmacology, *Vitam. Horm.*, 39, 231, 1982.

24. Nicholl, C. G., Polak, J. M., and Bloom, S. R., The hormonal regulation of food intake, digestion and absorption, *Ann. Rev. Nutr.*, 5, 213, 1985.

25. Levin, R. J., The effects of hormones on the absorptive, metabolic and digestive functions of the small intestine, *J. Endocrinol.*, 45, 315, 1969.

26. Levin, R. J. and Syme, G., Thyroid control of small intestinal oxygen consumption and the influence of sodium ions, oxygen tension, glucose and anaesthesia, *J. Physiol. (London)*, 245, 271, 1975.

27. Gaginella, T., Bass, P., Olsen, W., and Shug, A., Fatty acid inhibition of water absorption and energy production in the hamster jejunum, *FEBS Lett.*, 53, 347, 1975.

28. Forstner, G. G., Shih, M., and Lukie, B., Cyclic AMP and intestinal glycoprotein synthesis: the effect of beta-adrenergic agents, theophylline and dibutyryl cyclic AMP, *Can. J. Physiol. Pharmacol.*, 51, 122, 1973.

29. Lamont, J. T. and Ventola, A., Stimulation of colonic glycoporotein biosynthesis by dibutyryl cyclic AMP and theophylline, *Gastroenterology*, 72, 82, 1977.

30. Hoffman, A. G. D. and Kuksis, A., *In vitro* response of enzyme-dispersed rat intestinal villus and crypt cells to hormones, *Can. J. Physiol. Pharmacol.*, 57, 1393, 1979.

31. Weiser, M. M., Intestinal epithelial cell surface membrane glycoprotein biosynthesis. I. An indicator of cellular differentiation, *J. Biol. Chem.*, 248, 2536, 1973.

32. Aulsebrook, K. A., Intestinal absorption of glucose and sodium: effects of epinephrine and norepinephrine, *Biochem. Biophys. Res. Commun.*, 18, 165, 1965.

33. Yousef, I. M. and Kuksis, A., Release of chylomicrons by isolated cells of rat intestinal mucosa, *Lipids*, 7, 380, 1972.

34. Forstner, G. G., [1-¹⁴C]Glucosamine incorporation by subcellular fractions of small intestinal mucosa, *J. Biol. Chem.*, 245, 3584, 1970.

35. Weiser, M. M., Neumeier, M. M., Quaroni, A., and Kirch, K., Synthesis of plasmalemmal glycoproteins in intestinal epithelial cells, *J. Cell Biol.*, 77, 722, 1978.

36. Galland, G. and Forstner, G. G., Isolation of microvillus plasma membranes from suckling-rat intestine. The influence of premature induction of digestive enzymes by injection of cortisol acetate, *Biochem. J.*, 144, 293, 1974.

37. Miura, S., Morita, A., Erickson, R. H., and Kim, Y. S., Content and turnover of rat intestinal microvillus membrane aminopeptidase, *Gastroenterology*, 85, 1340, 1983.

38. Shakir, K. M. M., Sundaram, S. G., and Margolis, S., Lipid synthesis in isolated intestinal epithelial cells, *J. Lipid Res.*, 19, 433, 1978.

39. Hoffman, A. G. D., Comparative Study of Fat Absorption and Lipid Metabolism in Isolated Villus and Crypt Cells of Rat Jejunum, Ph.D. thesis, University of Toronto, 1981.

40. Goodman, M. W., Prigge, W. F., and Geghard, R. L., Glucagon inhibits intestinal mucosal cholesterol biosynthesis, *Gastroenterology*, 76, 1141, 1979.

41. Henning, S. J., Ballard, P. L., and Kretchmer, N. A., A study of the cytoplasmic receptors for glucocorticoids in intestine of pre- and post-weanling rats, *J. Biol. Chem.*, 250, 4131, 1975.

42. Geelen, M. J. H., Groener, J. E. M., De Haas, C. G. M., Wisserhof, T. A., and Van Golde, L. M. G., Influence of insulin and glucagon on the synthesis of glycerolipids in rat hepatocytes, *FEBS Lett.*, 90, 57, 1978.

43. Watson, W. C. and Murray, E., Fat digestion and absorption in the adrenalectomized rat, *J. Lipid Res.*, 7, 236, 1966.

44. Rogers, J. B., Riley, E. M., Drummey, G. D., and Isselbacher, K. J., Lipid absorption in adrenalectomized rats: the role of altered enzyme activity in the intestinal mucosa, *Gastroenterology*, 53, 547, 1967.

45. Sernka, T. K., Gastrointestinal mucosal metabolism, in *Gastrointestinal Physiology*, Jacobsen, E. D. and Shanbour, L. L., Eds., University Park Press, Baltimore, 1974, chap. 2.

46. Leblond, C. P. and Stevens, C. E., The constant renewal of the intestinal epithelium of the albino rat, *Anat. Rec.*, 150, 357, 1948.

47. Cheng, H. and Leblond, C. P., Origin, differentiation and renewal of the four main epithelial cell types in the mouse small intestine. V. Unitarian theory of the origin of the four cell types, *Am. J. Anat.*, 141, 537, 1974.

48. Dowling, R. K., Intestinal adaptation and its mechanisms, in *Topics in Gastroenterology, 10*, Jewell, D. P. and Selby, W. S., Eds., Blackwell Scientific, Oxford, 1982, 135.

49. Itoh, N., Obata, K.-I., Yanaihara, N., and Okamoto, H., Human preprovasoactive intestinal polypeptide contains a novel PHI-27-like peptide, PHM-27, *Nature (London)*, 304, 547, 1983.

50. Docherty, K. and Steiner, D. F., Posttranslational proteolysis in polypeptide hormone biosynthesis, *Ann. Rev. Physiol.*, 44, 625, 1982.

51. Tache, Y., Nature and biological actions of gastrointestinal peptides: current status, *Clin. Biochem.*, 17, 77, 1984.

52. Johnson, L. R., Regulation of gastrointestinal growth, in *The Role of the Gastrointestinal Tract in Nutrient Delivery*, Bristol-Myers Nutrition Symp., Vol. 3, Green, M. and Greene, H. L., Eds., Academic Press, New York, 1984, 1.

53. Johnson, L. R., Aures, D., and Hokanson, R., Effect of gastrin on the *in vivo* incorporation of [^{14}C]leucine into protein of the digestive tract, *Proc. Soc. Exp. Biol. Med.*, 132, 996, 1969.

54. Johnson, L. R., Aures, D., and Yuen, L., Pentagastrin-induced stimulation of protein synthesis in the gastrointestinal tract, *Am. J. Physiol.*, 217, 251, 1969.

55. Johnson, L. R., New aspects of the trophic effects of gastrointestinal hormones, *Gastroenterology*, 72, 788, 1977.

56. Johnson, L. R., Regulation of gastrointestinal growth, in *Physiology of the Gastrointestinal Tract*, Vol. 1, Johnson, L. R., Ed., Raven Press, New York, 1981, 179.

57. Johnson, L. R. and Guthrie, P. D., Stimulation of DNA synthesis by big and little gastrin (G-34 and G-17), *Gastroenterology*, 71, 599, 1976.

58. Takeuchi, K., Speir, G. R., and Johnson, L. R., Mucosal gastrin receptor. I. Assay standardization and fulfillment of receptor criteria, *Am. J. Physiol.*, 237, E284, 1979.

59. Johnson, L. R. and Guthrie, P. D., Proglumide inhibition of trophic action of pentagastrin, *Am. J. Physiol.*, 246, G62, 1984.

60. Johnson, L. R. and Guthrie, P. D., Secretin inhibition of gastrin-stimulated deoxyribonucleic acid synthesis, *Gastroenterology*, 67, 601, 1974.

61. Dembinski, A. and Konturek, S. J., Effects of E, F, and I series prostaglandins and analogues on growth of gastroduodenal mucosa and pancreas, *Am. J. Physiol.*, 248, G170, 1985.

62. Scheving, L. E., Vener, K. J., Tsai, T. H., and Scheving, L. A., Differential responses in the amount of DNA synthesis in the mouse intestinal tract to several hormones and epidermal growth factor administered at different times of the day or night (circadian stages), *Gastroenterology*, 88(Abstr.), 1574, 1985.

63. Stock-Damge, E., Aprahamian, M., and Pousse, A., Gastrin modulation of pancreatic growth, *Scand. J. Gastroenterol.*, 20(Suppl. 112), 68, 1985.

64. Dembinski, A. B. and Johnson, L. R., Stimulation of pancreatic growth by secretin, caerulein and pentagastrin, *Endocrinology*, 106, 323, 1980.

65. Guice, K. S., Oldham, K. T., Cox, R., Townsend, C. M., Jr., and Thompson, J. C., Early effects of continuous caerulein on the hamster pancreas: a biochemical and morphological study, *Gastroenterology*, 88(Abstr.), 1407, 1985.

66. O'Loughlin, E., Zahavi, I., Chung, M., Hayden, J., Hollenberg, M., and Gall, G., Effect of epidermal growth factor on the developing small intestine and pancreas, *Gastroenterology*, 86(Abstr.), 1201, 1984.

67. Butzner, D., Butler, D. G., Miniats, O. P., and Hamilton, J. R., Impact of chronic protein-calorie malnutrition on small intestinal repair after acute viral enteritis: a study in gnotobiotic piglets, *Pediatr. Res.*, 19, 476, 1985.

68. Miazza, B. M., Turberg, Y., Guillaume, P., Hahne, W., Chayvialle, J. A., and Loizeau, E., Mechanism of pancreatic growth induced by pancreatico-biliary diversion in the rat. Inhibition by proglumide, benzotript and ranitidine, *Scand. J. Gastroenterol.*, 20(Suppl. 112), 75, 1985.

69. Foelsch, U. R. and Creutzfeldt, W., Adaptation of the pancreas during treatment with enzyme inhibitors in rats and man, *Scand. J. Gastroenterol.*, 20(Suppl. 112), 54, 1985.

70. McGuinness, E. E., Morgan, R. G. H., and Wormsley, K. G., Trophic effects on the pancreas of trypsin and bile salt deficiency in the small-intestinal lumen, *Scand. J. Gastroenterol.*, 20(Suppl. 112), 64, 1985.

71. Luk, G. D. and Baylin, S. B., Polyamines and intestinal growth — increased polyamine biosynthesis after jejunectomy, *Am. J. Physiol.*, 245, G656, 1983.

72. Luk, G. D. and Baylin, S. B., Inhibition of intestinal epithelial DNA synthesis and adaptive hyperplasia after jejunectomy by suppression of polyamine biosynthesis, *J. Clin. Invest.*, 74, 698, 1984.

73. Dowling, R. H., Hosomi, M., Stace, N. H., Lirussi, F., Miazza, B., Levan, H., and Murphy, G. M., Hormones and polyamines in intestinal and pancreatic adaptation, *Scand. J. Gastroenterol.*, 20(Suppl. 112), 84, 1985.

74. Sagor, G. R., Ghatei, M. A., O'Shaughnessy, D. J., Al-Mukhtar, M. Y. T., Wright, N. A., and Bloom, S. R., Influence of somatostatin and bombesin on plasma enteroglucagon and cell proliferation after intestinal resection in the rat, *Gut*, 26, 89, 1985.

75. Willems, G., Cell population kinetics in the mucosa of the gastrointestinal tract, in *Scientific Basis of Gastroenterology*, Duthie, H. L. and Wormsley, K. G., Eds., Churchill Livingstone, Edinburgh, 1979, 2.

76. Wright, N. A., Cell proliferation in the normal gastrointestinal tract. Implications for proliferative responses, in *Cell Proliferation in Gastrointestinal Tract*, Appleton, D. R., Sunter, J. P., and Watson, A. J., Eds., Pitman Medical, Tunbridge Wells, Kent, England, 1980, 1.

77. Holt, P. R., Kotler, D. P., and Pascal, R. R., A simple method for determining epithelial cell turnover in small intestine. Studies in young and aging rat gut, *Gastroenterology*, 84, 69, 1983.

78. Sjolund, K., Sanden, G., Hakanson, R., and Sundler, F., Endocrine cells in human intestine: an immunocytochemical study, *Gastroenterology*, 85, 1120, 1983.

79. Henning, S. J., Hormonal and dietary regulation of intestinal enzyme development, in *Intestinal Toxicology*, Schiller, C. M., Ed., Raven Press, New York, 1984, 17.

80. Kedinger, M., Simon, P. M., Raul, F., Grenier, J. F., and Haffen, K., The effect of dexamethasone on the development of rat intestinal brush border enzymes in organ culture, *Dev. Biol.*, 74, 9, 1980.

81. Majumdar, A. P. N., Bilateral adrenalectomy: effect of hydrocortisone and pentagastrin on structural and functional properties of the stomach of weanling rats, *Scand. J. Gastroenterol.*, 16, 151, 1981.

82. Simon, P. M., Kedinger, M., Raul, F., Grenier, J. F., and Haffen, K., Organ culture of suckling rat intestine: comparative study of various hormones on brush border enzymes, *In Vitro*, 18, 339, 1982.

83. Simon-Assman, P. M., Kedinger, M., Grenier, J. F., and Haffen, K., Control of brush border enzymes by dexamethasone in the fetal rat intestine cultured *in vitro*, *J. Pediatr. Gastroenterol. Nutr.*, 1, 257, 1982.

84. Majumdar, A. P. N. and Nielsen, H., Influence of glucocorticoids on prenatal development of the gut and pancreas in rats, *Scand. J. Gastroenterol.*, 20, 65, 1985.

85. Daniels, V. G., Hardy, R. N., and Malinowska, K. W., The effect of adrenalectomy or pharmacological inhibition of adrenocortical function on macromolecule uptake by the newborn rat, *J. Physiol. (London)*, 229, 697, 1973.

86. Henning, S. J. and Leeper, L. L., Coordinate loss of glucocorticoid responsiveness by intestinal enzymes during postnatal development, *Am. J. Physiol.*, 242, G89, 1982.

87. Majumdar, A. P. N. and Rehfeld, J. F., Effect of hydrocortisone on gastrin cell function in various tissues of suckling rats, *Digestion*, 27, 165, 1983.

88. Majumdar, A. P. N., Gastric ontogeny and gastric mucosal growth during development, *Scand. J. Gastroenterol.*, 19(Suppl. 101), 13, 1984.

89. Barber, D. L., Buchan, A. M. J., Walsh, J. H., and Soll, A. H., Regulation of neurotensin release from isolated mucosal cells maintained in short-term culture, *Gastroenterology*, 88(Abstr.), 1711, 1985.

90. Moyer, M. P. and Aust, J. B., Development of methods for *in vitro* culture of gastrointestinal epithelial cells, *Gastroenterology*, 88, 1509, 1985.

91. Schwarz, S. M., Lind, S., Hostetler, B., Draper, J. P., and Watkins, J. B., Lipid composition and membrane fluidity in the small intestine of the developing rabbit, *Gastroenterology*, 86, 1544, 1984.

92. Schwarz, S. M., Hostetler, B., Ling, S., Mone, M., and Watkins, J. B., Intestinal membrane lipid composition and fluidity during development in the rat, *Am. J. Physiol.*, 248, G200, 1985.

93. Christensen, J., Gastrointestinal motility: the regulation of nutrient delivery, in *The Role of the Gastrointestinal Tract in Nutrient Delivery*, Bristol-Myers Nutrition Symp., Vol. 3, Green, M. and Greene, H. L., Eds., Academic Press, New York, 1984, 83.

94. McHugh, P. R. and Moran, T. H., Calories and gastric emptying: a regulatory capacity with implications for feeding, *Am. J. Physiol.*, 236, R254, 1979.

95. Brener, W., Hendrix, T. R., and McHugh, P. R., Regulation of the gastric emptying of glucose, *Gastroenterology*, 85, 76, 1983.

96. Hunt, J. N., Does calcium mediate slowing of gastric emptying by fat in humans?, *Am. J. Physiol.*, 248, G89, 1983.

97. Shafer, R. B., Levine, A. S., Marlette, J. M., and Morley, J. E., Do calories, osmolality, or calcium determine gastric emptying?, *Am. J. Physiol.*, 248, R479, 1985.

98. Miller, L. J., Characterization of cholecystokinin receptors on human gastric smooth muscle tumors, *Am. J. Physiol.*, 247, G402, 1984.

99. Gardner, J. D. and Jensen, R. T., Cholecystokinin receptor antagonists, *Am. J. Physiol.*, 246, G476, 1984.

100. Liddle, R. A., Morita, E., Conrad, C. K., and Williams, J. A., CCK in physiologic concentrations delays gastric emptying in humans, *Gastroenterology*, 88(Abstr.), 1476, 1985.

101. Liddle, R. A., Green, G. M., Conrad, C. K., and Williams, J. A., CCK response to food is species specific: fat and amino acids do not stimulate CCK release in the rat, *Gastroenterology*, 88(Abstr.), 1476, 1985.

102. Liddle, R. A., Goldfine, I. D., and Williams, J. A., Bioassay of plasma cholecystokinin in rats: effects of food, trypsin inhibitor and alcohol, *Gastroenterology*, 87, 542, 1984.

103. Mogard, M., Maxwell, V., Van Deventer, G., Elashoff, J., Yamada, T., and Walsh, J. H., Somatostatin enhances emptying of liquid meals in man, *Gastroenterology*, 86(Abstr.), 1186, 1984.

104. Bech, K., Ladegaard, L., and Andersen, D., Effect of somatostatin on bethanechol-stimulated gastric acid secretion and gastric antral motility in digs with gastric fistula, *Scand. J. Gastroenterol.*, 20, 470, 1985.

105. Sanders, K. M., Role of prostaglandins in regulating gastric motility, *Am. J. Physiol.*, 247, G117, 1984.

106. Bueno, L., Fargeas, M. J., Fioramonti, J., and Primi, M. P., Central control of intestinal motility by prostaglandins: a mediator of the actions of several peptides in rats and dogs, *Gastroenterology*, 88, 1888, 1985.

107. Minami, H. and McCallum, R. W., The physiology and pathophysiology of gastric emptying in humans, *Gastroenterology*, 86, 1592, 1984.

108. Rohr, G. and Keim, V., Asynchrony in intracellular transport of newly synthesized rat pancreatic proteins, *Gastroenterology*, 86(Abstr.), 1222, 1984.

109. Kern, H. F., Adler, G., and Scheele, G. A., Structural and biochemical characterization of maximal and supramaximal hormonal stimulation of rat exocrine pancreas, *Scand. J. Gastroenterol.*, 20(Suppl. 112), 20, 1985.

110. Rosenthal, P., Whitney, J., Ling, V., and Thomas, D., Detection of exocrine pancreatic insufficiency by liquid secondary ion mass spectrometry (LSIMS), *Gastroenterology*, 88(Abstr.), 1560, 1985.

111. Hahne, W. F., Jenson, P. T., Lemp, G. F., and Gardner, I. D., Proglumide and benzotript: members of a different class of cholecystokinin receptor antagonists, *Proc. Natl. Acad. Sci. U.S.A.*, 78, 6304, 1981.

112. Stubbs, R. S. and Stabile, B. E., Role of cholecystokinin in pancreatic exocrine response to intraluminal amino acids and fat, *Am. J. Physiol.*, 248, G347, 1985.

113. Konturek, S. J., Krol, R., and Tasler, J., Effect of bombesin and related peptides on the release and action of intestinal hormones on pancreatic secretion, *J. Physiol. (London)*, 257, 663, 1976.

114. Khuhtsen, S., Holst, J. J., Jensen, S. L., Knigge, U., and Nielsen, O. V., Gastrin-releasing peptide: effect on exocrine secretion and release from isolated perfused porcine pancreas, *Am. J. Physiol.*, 248, G281, 1985.

115. Howard, J. M., Jensen, R. T., and Gardner, J. D., Bombesin-induced residual stimulation of amylase release from mouse pancreatic acini, *Am. J. Physiol.*, 248, G196, 1985.

116. Valenzuela, J. E., Lamers, C. B., Modlin, I. M., and Walsh, J. H., Cholinergic component in the human pancreatic secretory response to intraintestinal oleate, *Gut*, 24, 807, 1983.

117. Konturek, S. J., Bielanski, J., Tasler, J., Bilski, J., Jansen, J., de Jong, A., and Lamers, C., Release of secretin, CCK and PP by regional perfusion of intestine with HCl and oleate, *Gastroenterology*, 86, 1140, 1984.

118. Gullo, L., Priori, P., Costa, P. L., Mattioli, G., and Labo, G., Action of secretin on pancreatic enzyme secretion in man. Studies on pure pancreatic juice, *Gut*, 25, 867, 1984.

119. Brugge, W. R., Burke, C. A., Izzo, R., and Praissman, M., The role of cholecystokinin (CCK) in the control of pancreatic secretion during the inter-digestive state, *Gastroenterology*, 86(Abstr.), 1036, 1984.

120. Brugge, W. R., Burke, C. A., Praissman, M., and Izzo, R., The role of cholecystokinin (CCK) in the control of inter-digestive pancreatic secretion, *Gastroenterology*, 88(Abstr.), 1337, 1985.

121. Debas, H. T., Garcia, R., and Soon-Shiong, P., Epidermal growth factor (EGF): interactions with CCK and secretin on exocrine pancreatic secretion, *Gastroenterology*, 86(Abstr.), 1058, 1984.

122. Konturek, S. J., Cieszkowski, M., Jaworek, J., Konturek, J., Brzozowski, T., and Gregory, H., Effects of epidermal growth factor on gastrointestinal secretions, *Am. J. Physiol.*, 246, G580, 1984.

123. Baca, I., Feurle, G. E., Knauf, W., and Carraway, R., Relationships between neurotensin and pancreatic secretion in the dog, *Ann. N.Y. Acad. Sci.*, 400, 400, 1982.

124. Baca, I., Feurle, G. E., Haas, M., and Mernitz, T., Interaction of neurotensin, cholecystokinin, and secretin in the stimulation of the exocrine pancreas in the dog, *Gastroenterology*, 84, 556, 1983.

125. Feurle, G. E., Hoffmann, G., Baca, I., and Carraway, R., Neurotensin stimulates exocrine pancreatic secretion in man in doses which may be released after a meal, *Gastroenterology*, 88(Abstr.), 1382, 1985.

126. Sakamoto, T., Newman, J., Townsend, C. M., Greeley, G. H., Jr., and Thompson, J. C., Role of neurotensin in the intestinal phase of pancreatic polypeptide release induced by fat, *Gastroenterology*, 86(Abstr.), 1229, 1984.

127. Saad, C., Mogard, M. H., Dooley, C., Elashoff, J. D., Valenzuela, J. E., and Walsh, J. H., Neurotensin stimulates pancreatic secretion in man, *Gastroenterology*, 88(Abstr.), 1565, 1985.

128. Fujimura, M., Sakamoto, T., Khalil, T., Greeley, G. H., Jr., Townsend, C. M., Jr., and Thompson, J. C., Role of the jejunum and ileum in the release of secretin and cholecystokinin by acid or fat in conscious dogs, *Gastroenterology*, 86(Abstr.), 1083, 1984.

129. Fujimura, M., Lluis, F., Lonovics, J., Greeley, G. H., Jr., Townsend, C. M., Jr., and Thompson, J. C., Inability of neurotensin (NT) secretogogues to release NT by direct luminal stimulation of the ileum in dogs, *Gastroenterology*, 88(Abstr.), 1390, 1985.

130. Pandol, S., Thomas, M., Rivier, J., and Vale, M., Growth hormone-releasing factors stimulate *in vitro* pancreatic enzyme secretion, *Gastroenterology*, 86, 1205, 1984.

131. Noguchi, M., Adachi, H., Gardner, J. D., and Jensen, R. T., Activation of protein kinase C causes stimulation of pancreatic enzyme secretion, *Gastroenterology*, 86(Abstr.), 1196, 1984.

132. Malagelada, J.-R., Holtermuller, K. H., McCall, J. T., and Go, V. L. W., Pancreatic, gallbladder, and intestinal responses to intraluminal magnesium salts in man, *Dig. Dis. Sci.*, 23, 481, 1978.

133. Inoue, K., Fried, G. M., Wiener, I., Sakamoto, T., Lilja, P., Greeley, H., Jr., Watson, L. C., and Thompson, J. C., Effect of divalent cations on gastrointestinal hormone release and exocrine pancreatic secretion in dogs, *Am. J. Physiol.*, 248, G28, 1985.

134. Owyang, S., Leksell, K., May, D., Kothary, P., and Vinik, A. I., Intraduodenal trypsin inhibits cholecystokinin release and pancreatic enzyme secretion in man, *Gastroenterology*, 84, 1268, 1983.

135. Owyang, S., Dismond, D., and May, D., Feedback regulation of pancreatic enzyme secretion in man is stimulus specific, *Gastroenterology*, 86(Abstr.), 1205, 1984.

136. Louie, D., May, D., Miller, P., and Owyang, C., Cholecystokinin mediates feedback regulation of pancreatic enzyme secretion in rats, *Am. J. Physiol.*, 250, G252, 1986.

137. Bottcher, W., Yamada, T., and Kauffman, C. L., Somatostatin is an enterogastrone, *Gastroenterology*, 84, 1112, 1983.

138. Souquet, J. C., Rambliere, R., Riou, J. P., et al., Hormonal and metabolic effects of near physiological increase of plasma immunoreactive somatostatin 14, *J. Clin. Endocrinol. Metab.*, 56, 1076, 1983.

139. Loud, F. B., Holst, J. J., Egense, E., Petersen, B., and Christiansen, J., Is somatostatin a humoral regulator of the endocrine pancreas and gastric secretion in man?, *Gut*, 26, 445, 1985.

140. Ellison, E. C., O'Dorisio, T. M., Lott, J., Beaver, B., Cloutier, C. T., and Carey, L. C., The effect of long-acting somatostatin analog (SMS 201995) on basal pancreatic exocrine secretion in man, *Gastroenterology*, 88(Abstr.), 1373, 1985.

141. Konturek, S. J., Bielanski, W., Tasler, J., Bilski, J., Jansen, J., de Jong, A., and Lamers, C., Effect of atropine and somatostatin on the hormonal regulation of the intestinal phase of pancreatic secretion, *Gastroenterology*, 86(Abstr.), 1140, 1984.

142. De Graef, J. and Woussen-Colle, M. C., Effects of sham feeding, bethanechol, and bombesin on somatostatin release in dogs, *Am. J. Physiol.,* 248, G1, 1985.

143. Esteve, J. P., Susini, C., Vaysse, N., Antoniotti, H., Wunsch, E., Berthon, G., and Ribet, A., Binding of somatostatin to pancreatic acinar cells, *Am. J. Physiol.,* 247, G62, 1984.

144. Sakamoto, C. and Williams, J. A., CCK directly regulates somatostatin binding to pancreatic acinar cell membranes, *Gastroenterology,* 86, 1228, 1984.

145. Totemoto, K., Isolation and characterization of peptide YY (PPY) a candidate gut hormone that inhibits pancreatic exocrine secretion, *Proc. Natl. Acad. Sci. U.S.A.,* 79, 2514, 1982.

146. Lawson, M., Everson, G. T., Klingensmith, W., and Kern, F., Jr., Coordination of gastric and gallbladder emptying after ingestion of a regular meal, *Gastroenterology,* 85, 866, 1983.

147. Fried, G. M., Ogden, W. D., Greeley, G., and Thompson, J. C., Correlation of release and actions of cholecystokinin in dogs before and after vagatomy, *Surgery,* 93, 786, 1983.

148. You, C. H., Lee, S. I., and Chey, W. Y., Gallbladder contraction and plasma cholecystokinin in response to fat in the duodenum, *Gastroenterology,* 88, 1639, 1985.

149. Cox, K. L., Rosenquist, G. L., and Iwahashi-Hosoda, C. K., Noncholecystokinin peptides in human serum which cause gallbladder contraction, *Life Sci.,* 31, 3023, 1982.

150. Soon-Shiong, P., Cox, K. L., Iwahashi-Hosoda, C. K., Rosenquist, G., and Leung, R., Increased gallbladder bioactivity in cholelithiasis due to noncholecystokinin substances, *Gastroenterology,* 84(Abstr.), 1317, 1983.

151. Strah, K., Soon-Shiong, P., Cox, K., Rosenquist, G., Reeve, J., Walsh, J. H., and Debas, H. T., Non-CCK stimulant of gallbladder contraction in human serum: chromatographic, immunologic and biological comparisons to CCK, *Gastroenterology,* 86(Abstr.), 1267, 1984.

152. Sakamoto, T., Mate, L., Greeley, G. H., Jr., Townsend, C. M., Jr., and Thompson, J. C., Effect of neurotensin on gallbladder contraction in dogs, *Gastroenterology,* 86(Abstr.), 1229, 1984.

153. Kortz, W. J., James, R. B., Nashold, J. R. B., Rotolo, F. S., and Myers, W. C., Effects of bombesin on bile secretion, *Gastroenterology,* 86(Abstr.), 1142, 1984.

154. Romanski, K. W. and Bochenek, W. J., Effect of secretin, pancreozymin OP-CCK, and glucagon on bile flow and bile lipid secretion in rats, *Gut,* 24, 803, 1983.

155. Holm, I., Thulin, L., Samnegard, H., et al., Anticholeretic effect of somatostatin in anaesthetized dogs, *Acta Physiol. Scand.,* 104, 241, 1978.

156. Kaminski, D. L. and Deshpande, M., Effect of somatostatin and bombesin on secretin-stimulated ductular bile flow in dogs, *Gastroenterology,* 85, 1239, 1983.

157. Hanks, J. B., Kortz, W. J., Andersen, D. K., and Jones, R. S., Somatostatin suppression of canine fasting bile secretion, *Gastroenterology,* 84, 130, 1983.

158. Miyasaka, K., Kitani, K., Watanabe, S., and Takeuchi, T., The different effect of oleate between infused into the duodenum and the terminal ileum on pancreas and bile secretion in the rat, *Gastroenterology,* 88(Abstr.), 1504, 1985.

159. Ladas, S. D., Isaacs, P. E. T., Murphy, G. M., and Sladen, G. E., Comparison of the effects of medium and long chain triglyceride containing liquid meals on gall bladder and small intestinal function in normal man, *Gut,* 25, 405, 1984.

160. Teitelbaum, P., Disturbances of feeding and drinking behaviour after hypothalamic lesions, in *Nebraska Symposium on Motivation,* Jones, M. R., Ed., University of Nebraska Press, Lincoln, 1961, 39.

161. Grossman, S. P., Neurophysiological aspects: extra-hypothalamic factors in the regulation of food intake, *Adv. Psychosom. Med.,* 7, 49, 1972.

162. Koopmans, H. S., The role of the gastrointestinal tract in the satiation of hunger, in *The Body Weight Regulatory System: Normal and Disturbed Mechanisms,* Cioffi, L. A., James, W. P. T., and Van Itallie, T. B., Eds., Raven Press, New York, 1981, 45.

163. Smith, G. P., The peripheral control of appetite, *Lancet,* 2, 88, 1983.

164. Sturdevant, R. A. L. and Goetz, H., Cholecystokinin both stimulates and inhibits human food intake, *Nature (London),* 261, 713, 1976.

165. Holubitsky, D., Garcia, R., and Dabas, H. T., Intraperitoneal but not subcutaneous CCK-8 inhibits food intake in the rat, *Gastroenterology,* 88(Abstr.), 1421, 1985.

166. Pappas, T. N., Melendez, R. L., Strah, K. M., and Debas, H. T., CCK is not a peripheral satiety signal in the dog, *Am. J. Physiol.,* 249, G733, 1985.

167. Garcia, R., Elashoff, J., and Taylor, I. L., Induction of satiety in congenitally obese mice by gastrointestinal hormones, *Gastroenterology,* 86, 1085, 1984.

168. Lorenz, D. N. and Goldman, S. A., Vagal mediation of the cholecystokinin satiety effect in rats, *Physiol. Behav.,* 29, 599, 1982.

169. Martin, J. R. and Novin, D., Decreased feeding in rats following hepatic-portal infusion of glucagon, *Physiol. Behav.,* 19, 461, 1977.

170. Lotter, E. C., Krinsky, R., McKay, J. M., Treener, C. M., Porter, D., Jr., and Woods, S. C., Somatostatin decreases food intake of rats and baboons, *J. Comp. Physiol. Psychol.,* 95, 278, 1981.

171. Gibbs, J., Fauser, D. J., Rowe, E. A., Rolls, B. J., Rolls, E. T., and Madison, S. P., Bombesin suppresses feeding in rats, *Nature (London)*, 282, 208, 1979.

172. Deutsch, J. A., Bombesin — satiety or malaise?, *Nature (London)*, 285, 592, 1980.

173. Taylor, I. L. and Garcia, R., Effects of pancreatic polypeptide, cerulein, and bombesin on satiety in obese mice, *Am. J. Physiol.*, 248, G277, 1985.

174. Malaisse-Lagae, F., Carpenter, J. L., Patel, Y. C., Malaisse, W. J., and Orci, L., Pancreatic polypeptide: a possible role in the regulation of food intake in the mouse. Hypothesis, *Experientia*, 33, 915, 1977.

175. Reidelberger, R. D., O'Rourke, M., Durie, P. R., Geokas, M. C., and Largman, C., Effects of food intake and cholecystokinin on plasma trypsinogen levels in dogs, *Am. J. Physiol.*, 246, G543, 1984.

176. Harvey, R. F. and Oliver, J. M., Hormonal content and effect on gall bladder of commercial preparations of cholecystokinin-pancreozymin, *Horm. Metab. Res.*, 13, 446, 1981.

177. Stacher, G., Steinringer, H., Schmierer, G., Schneider, C., and Winklehner, S., Cholecystokinin octapeptide decreases intake of solid food in man, *Peptides*, 3, 133, 1982.

178. Harvey, R. F., Gut peptides and the control of food intake, *Br. Med. J.*, 287, 1572, 1983.

179. Collins, S. M. and Weingarten, H. P., The effect of cholecystokinin antagonist on satiety is independent of gastric emptying, *Gastroenterology*, 86, 1052, 1984.

180. Shillabeer, G. and Davison, J. S., Vagatomy abolishes the effect of the cholecystokinin-antagonist, proglumide, on food intake, *Gastroenterology*, 86(Abstr.), 1249, 1984.

181. Thompson, J. C. and Marx, M., Gastrointestinal hormones, *Curr. Problems Surg.*, 21, 1, 1984.

182. Brown, J. C., McIntosh, C. H. S., and Pederson, R. A., The gastrointestinal peptides and nutrition, *Can. J. Physiol. Pharmacol.*, 61, 282, 1983.

183. Patton, J. S., Gastrointestinal lipid digestion, in *Physiology of the Gastrointestinal Tract*, Vol. 1, Johnson, L. R., Ed., Raven Press, New York, 1981, 1123.

184. Fink, C. S., Hamosh, M., Hamosh, P., DeNigris, J., and Kasbekar, D. K., Lipase secretion from dispersed rabbit gastric glands, *Am. J. Physiol.*, 248, G68, 1985.

185. Ghishan, F. K., Mulberg, A., Borowitz, S., and Robinson, A., Transport of phosphate by jejunal brush border membrane vesicles of the diabetic rat, *Gastroenterology*, 86, 1087, 1984.

186. Rutgeerts, P., Vantrappen, G., and Ghoos, Y., The quantitative ^{13}C-mixed-triglyceride breath test is a good pancreatic function test, *Gastroenterology*, 86, 1227, 1984.

187. Gallavan, R. H., Jr., Chen, M. H., Joffe, S. N., and Jacobson, E. D., Vasoactive intestinal polypeptide, cholecystokinin, glucagon, and bile-oleate-induced jejunal hyperemia, *Am. J. Physiol.*, 248, G208, 1985.

188. Pointner, H., Hengl, G., Bayer, P. M., and Flagel, U., Inhibition of postprandial increase in serum triglycerides by somatostatin in man, *Wien Klin. Wochenschr.*, 89, 224, 1977.

189. Schusdziarra, V., Rouiller, D., and Unger, R. H., Oral administration of somatostatin reduces postprandial plasma triglycerides, gastrin and gut glucagon-like immunoreactivity, *Life Sci.*, 24, 1595, 1979.

190. Schusdziarra, V., Zyznar, E., Rouiller, D., Boden, G., Brown, J. C., Arimura, A., and Unger, R. H., Splanchnic somatostatin: a hormonal regulator of nutrient homeostasis, *Science*, 207, 530, 1980.

191. Boivin, M., Lanspa, S. J., Zinsmeister, A. R., Go, V. L. W., and DiMagno, E. P., Chronic fatty diets surprisingly increase human inter-digestive secretion of bile acids and exocrine pancreatic enzymes, *Gastroenterology*, 88(Abstr.), 1651, 1985.

192. Dennison, A. R., Watkins, R. M., Collin, J., and Morris, P. J., Motility and absorption in autotransplants of canine small intestine, *Gastroenterology*, 86(Abstr.), 1061, 1984.

193. Thompson, A. B. R. and Rajotte, R., Insulin and islet cell transplantation — effect on intestinal uptake of lipids, *Am. J. Physiol.*, 246, G627, 1984.

194. Zimmerman, J., Eisenberg, S., and Rachmilewitz, D., Ethanol and cimetidine inhibit triglyceride synthesis and secretion by human small intestinal mucosa, *Gastroenterology*, 86, 1308, 1984.

195. Kokas, E., Pisano, J. J., and Crepps, B., Villikin: characterization and function, in *Gastrointestinal Hormones*, Glass, G. B. J., Ed., Raven Press, New York, 1980, 899.

196. Armstrong, M. J., Ferris, C. F., and Leeman, S. E., Neurotensin increases the translocation of ^{3}H-oleic acid from the intestinal lumen into lymph in rats, *Dig. Dis. Sci.*, 29(Abstr.), 6S, 1984.

197. Tso, P., Pitts, V., and Granger, D. N., Role of lymph flow in intestinal chylomicron transport, *Gastroenterology*, 88(Abstr.), 1618, 1985.

198. Chen, Y. K., Richter, H. M., III, Kelly, K. A., Jay, J. M., and Go, V. L. W., Postprandial gastrin (G), neurotensin (NT), vasoactive intestinal peptide (VIP), substance P (SP) and bombesin (BLI) output in canine thoracic duct lymph, *Gastroenterology*, 86(Abstr.), 1046, 1984.

199. Ferris, C. F., Carraway, R. E., and Leeman, S. E., Lipid stimulation of neurotensin release from rat small intestine, *Ann. N.Y. Acad. Sci.*, 400, 433, 1982.

200. Ferris, C. F., Armstrong, M. J., George, J. K., Stevens, C. A., Carraway, R. E., and Leeman, S. E., Alcohol and fatty acid stimulation of neurotensin release from rat small intestine, *Dig. Dis. Sci.*, 29(Abstr.), 26S, 1984.

201. Rosell, S., Al-Saffar, S., and Theodorsson-Norheim, E., Neural and hormonal modifications of the neurotensin response to intraduodenally administered fat, *Dig. Dis. Sci.,* 29(Abstr.), 71S, 1984.
202. Sakamoto, T., Fujimara, M., Walker, J. P., Greeley, G. H., Jr., Townsend, C. M., Jr., and Thompson, J. C., Regulation of endogenous release of neurotensin by fat: the effect of atropine, vagatomy and somatostatin, *Dig. Dis. Sci.,* 29(Abstr.), 72S, 1984.
203. Barber, D., Buchan, A., and Soll, A. H., Bombesin stimulation of neurotensin release from isolated canine ileal cells, *Dig. Dis. Sci.,* 29(Abstr.), 7S, 1984.
204. Fara, J. W., Rubinstein, E. H., and Sonnenschein, R. R., Intestinal hormones in mesenteric vasodilation after intraduodenal agents, *Am. J. Physiol.,* 223, 1058, 1972.
205. Fara, J. W., Effects of gastrointestinal hormones on vascular smooth muscle, *Am. J. Dig. Dis.,* 20, 346, 1975.
206. Gallavan, R. H., Jr., Shaw, C., Murphy, R. F., Joffe, S. N., and Jacobson, E. D., The lipid-induced jejunal hyperemia and neurotensin release, *Dig. Dis. Sci.,* 29, 29S, 1984.
207. Harper, S. L., Barrowman, J. A., Kvietys, P. R., and Granger, D. N., Effect of neurotensin on intestinal capillary permeability and blood flow, *Am. J. Physiol.,* 247, G161, 1984.
208. Gallavan, R. H., Jr., Chen, M. H., Tabata, M., Joffe, S. N., and Jacobson, E. D., Vasoactive intestinal polypeptide (VIP) and the fat-induced jejunal hyperemia in dogs, *Gastroenterology,* 86(Abstr.), 1084, 1984.

INDEX

Printed and bound by CPI Group (UK) Ltd, Croydon, CR0 4YY

22/10/2024

01777630-0011